Tools to Aid Environmental Decision Making

Springer
*New York
Berlin
Heidelberg
Barcelona
Hong Kong
London
Milan
Paris
Singapore
Tokyo*

Virginia H. Dale
Mary R. English
Editors

Tools to Aid Environmental Decision Making

With 30 Illustrations

 Springer

Virginia H. Dale
Senior Scientist
Environmental Sciences Division
Oak Ridge National Laboratory
Oak Ridge, TN 37831-6036

Mary R. English
Research Leader
Energy, Environment and Resources Center
University of Tennessee
Knoxville, TN 37996-4134

Library of Congress Cataloging-in-Publication Data
Tools to aid environmental decision making / Virginia H. Dale and Mary
　R. English, editors.
　　　p.　cm.
　　Includes bibliographical references and index.
　　ISBN 0-387-98555-7 (alk. paper).—ISBN 0-387-98556-5 (pbk.:
alk. paper)
　　　1. Environmental policy—Decision making—Mathodology.　2. Problem
solving—Methodology.　I. Dale, Virginia H.　II. English, Mary R.
GE170.T66　1998
363.7'05—dc21　　　　　　　　　　　　　　　　　　　　　　　　　　　　　　98-7730

Printed on acid-free paper.

©1999 Springer-Verlag New York Inc.
All rights reserved. This work may not be translated or copied in whole or in part without the written permission of the publisher (Springer-Verlag New York, Inc., 175 Fifth Avenue, New York, NY 10010, USA), except for brief excerpts in connection with reviews or scholarly analysis. Use in connection with any form of information storage and retrieval, electronic adaptation, computer software, or by similar or dissimilar methodology now known or hereafter developed is forbidden.
The use of general descriptive names, trade names, trademarks, etc., in this publication, even if the former are not especially identified, is not to be taken as a sign that such names, as understood by the Trade Marks and Merchandise Marks Act, may accordingly be used freely by anyone.

Production coordinated by Chernow Editorial Services, Inc., and managed by Tim Taylor; manufacturing supervised by Jeffrey Taub.
Typeset by Best-set Typesetter Ltd., Hong Kong.
Printed and bound by R.R. Donnelley and Sons, Harrisonburg, VA.
Printed in the United States of America.

9 8 7 6 5 4 3 2 1

ISBN 0-387-98555-7 Springer-Verlag New York Berlin Heidelberg　SPIN 10681010 (hardcover)
ISBN 0-387-98556-5 Springer-Verlag New York Berlin Heidelberg　SPIN 10657867 (softcover)

Foreword

Environmental decision making is, like politics, mostly local. In fact, making decisions about the environment at the subnational level—in state, regional, and local jurisdictions—is a lot like politics. For resolving environmental issues demands, but often resists, a balance between deeply held feelings and stark confrontations among opposing views.

This volume describes tools that should make the decision maker's lot a bit more tolerable. The authors would be the last to suggest that these decision-aiding tools will somehow bring a benign order to issues that reach to people's fundamental values. What they *can* help do is to keep the debate focused on the important issues, to serve up useful options, and to narrow the range of disagreement. Even this is a challenging assignment.

Still, why bother? The chief reason is that the locus of environmental decision making has, in the past decade or so, shifted from the national to the subnational level (a convenient, if colorless, term to denote the hurly-burly of environmental controversy outside the Washington Beltway). For example, New England has taken a regional stand on tropospheric ozone control, and California requires automotive pollution controls that some other jurisdictions have partially adopted. This shift is a profound but not unexpected result of the way environmental policy has evolved since the modern environmental movement began around the late 1960s. Back then, the pendulum was swinging the other way. Regulatory standards varied significantly from state to state, and some advocates feared that the competition for economic development would drive future standards to the level of the lowest common denominator. Enforcement of the standards was weak in most places, often dependent upon complex settlements among parties of sharply different negotiating power. Even though a few states had reasonably strong environmental programs, some problems simply resisted a subnational solution. For example, California had an aggressive program of smog control, but even it could not adequately regulate automobile emissions.

The remedy for this situation seemed obviously to lie in federal preemption of state and local environmental standards. The policy reasons for this strategy seemed solid enough, given the unsettling experience with regulation before 1970. The political reasons were perhaps even stronger. Environmental issues had become embroiled in presidential politics, creating a competition for environmental credentials among some very powerful figures: Richard Nixon, Edmund Muskie, and Henry "Scoop" Jackson, to name three. Congress and the newly formed Environmental Protection Agency thus became convenient one-stop shops for tightening pollution laws, and the large interest groups quite naturally focused their attention there. It was a cost-effective strategy because one stop is easier to tend than several, although one doubts that the advocates would have characterized it thus.

On balance, striking the pollution mule with the national two-by-four was probably the only workable way to launch the new age of environmental regulation (which was, and is, mostly pollution-control regulation). If nothing else, brute force had the salutary effect of getting many polluters up to a minimum standard of compliance and correcting a number of clearly bad environmental situations. The problem, of course, was that this preemptive strategy did not easily accommodate the real differences in spatial and temporal settings, physical conditions, and constituency preferences that shape the environmental debate in various regions of the country. So, the closer we came to achieving uniform national standards, the more important subnational differences became. That the action is now returning to environmental decision makers at the subnational level is, in a real sense, the inevitable result of a successful national program. No longer is the issue whether a state wants to do less than its brethren, but rather whether it cares to do more.

Unfortunately, success at the national level does not make the subnational decision maker's life any easier. An important reason is that for a couple of decades most analytic tools to support environmental decision making have naturally focused on large national problems, where the constraints of funding, data availability, and trained support personnel are less pressing than in subnational decision situations. And even where these constraints are not serious, tools that are useful to national decision makers are not uniformly applicable in the more varied, less deliberative world of local environmental policy.

The National Center for Environmental Decision-Making Research (NCEDR) was established in the fall of 1995 to respond to the emerging needs of subnational decision makers. Its aim is to provide those decision makers with the information, techniques, and processes needed to address environmental issues. In doing so, NCEDR engages the national community of scholars in environment-related disciplines, along with practitioners, in its activities. The Decision-Aiding Tools Research Program has been one of the main thrusts of NCEDR's work.

The mission of the Tools Research Program is not only to promote the use of decision-aiding tools, but also to identify and help develop tools that are user-friendly and appropriate for a variety of decision-making situations. The program therefore has given priority to:

- Developing an overview of existing environmental decision-aiding tools.
- Conducting research on needs for decision-aiding tools within various subnational environmental decision-making settings and on psychological, social, economic, and technological barriers to the adoption of various tools.
- Assembling a tool kit that can meet a variety of decision-making needs, and is adaptable to different user capabilities, by refining existing tools or developing new ones.
- Promoting the use of decision-aiding tools.

Ideally, this program will help build a bridge between tools and practitioners. The initial focus is on analytic tools, their availability, use, and adaptability to different decision-making settings. Other studies will need to take up communication tools and tools that facilitate citizen involvement in environmental decision-making processes. This volume is hence a product of NCEDR's early research, but it is also an important part of the research process itself because the reaction of the user will test whether the bridge in fact reaches all the way to the immediate needs of subnational decision makers.

Still, this volume will not interest everyone. It is neither sufficiently rigorous for the pure scholar nor easily accessible to busy senior decision makers. Its intended audience is rather the considerable body of participants in the decision-making process lying between these extremes. Applied researchers, especially those whose work cuts across the usual scientific disciplines, should find this work helpful; so should their students. Policy and analytic staff to decision makers, whether regular employees or outside consultants, should find value here, as well. Persons playing similar roles in advocacy groups and other associations are also an intended audience.

Interestingly, NCEDR research has found that many in these positions do not think of themselves as part of the decision-making process. Yet they are the only real link between analytic (and other) tools and the ultimate decision maker; the tools simply will not get used to improve the process unless these people use them. Perhaps the confusion in role arises because the reason for using better tools is not to supplant the decision maker, but rather to help focus his or her attention on the important issues and the useful options. That is the underlying purpose of this volume and of the work that NCEDR is delivering to decision makers.

ROBERT FRI
National Museum of Natural History
Smithsonian Institution

Acknowledgments

This volume was created under the auspices of the National Center for Environmental Decision-Making Research (NCEDR). NCEDR was created with funds from the National Science Foundation and is administered by the University of Tennessee, Oak Ridge National Laboratory, and the Tennessee Valley Authority as an activity of their Joint Institute for Energy and Environment. The mission of NCEDR is to improve environmental decision making at the subnational level. The center conducts research and sponsors outreach programs intended to put better tools and information in the hands of public- and private-sector decision makers at regional, state, and local levels. We appreciate the support and guidance of Milton Russell, the first director of NCEDR, and Robert Turner, his successor.

An earlier draft of the entire book was reviewed by Howard Kunreuther and Charles Van Sickle, who provided written comments on all of the chapters and discussed these comments with the editors and the contributing authors. The book is much improved by their efforts. We appreciate Granger Morgan's suggestion that key resources be included in each chapter. In addition, members of NCEDR's external advisory board and a number of other people too numerous to list here (both internal and external to NCEDR) provided comments on various chapters and on the structure of the book as a whole. We also extend our thanks to them.

Frederick M. O'Hara Jr. edited all of the chapters and suggested ways to organize the material to better convey the information. Wendy Hudson Ramsey and Claire Van Riper-Geibig provided support for the workshop that brought together the contributing authors with other researchers and participants in subnational environmental decision making; they also helped determine the format of the book and identified authors. Nancy Watlington and Jeannette Cox helped at the workshop and assisted with much of the correspondence. In addition, Jeannette Cox drafted some of the figures. All of these efforts were essential to getting the book into print.

Finally, we greatly appreciate the patience and creative hard work of the contributing authors, who gave unstintingly of their time in preparing drafts, revising drafts, and participating in the workshop and in conference calls. Their efforts are, obviously, at the heart of this book.

This work was supported by the National Science Foundation under Grant No. SBR-9513010 with additional support from Oak Ridge National Laboratory, Tennessee Valley Authority, and the University of Tennessee. However, any opinions, findings, and conclusions or recommendations expressed in this material are those of the authors and do not necessarily reflect the views of the National Science Foundation.

Contents

Foreword by Robert Fri v

Acknowledgments ix

Contributors xiii

1. Overview 1
   Mary R. English, Virginia H. Dale, Claire Van Riper-Geibig, and Wendy Hudson Ramsey

2. Identifying Environmental Values 32
   Robin Gregory
   Decision-Maker Response: *Joseph W. Lewis*

3. Tools to Characterize the Environmental Setting 62
   Virginia H. Dale and Robert V. O'Neill
   Decision-Maker Response: *William R. Miller III*

4. Tools for Understanding the Socioeconomic and Political Settings for Environmental Decision Making 94
   William R. Freudenburg
   Decision-Maker Response: *Roy Silver, Elizabeth Ungar Natter, and Chetan Talwalkar*

5. Characterizing the Regulatory and Judicial Setting 130
   Mary L. Lyndon
   Decision-Maker Response: *Dean Hill Rivkin*

6. Integration of Geographic Information 161
   Jeffrey P. Osleeb and Sami Kahn
   Decision-Maker Response: *Surya S. Prasad*

xi

7. Forecasting for Environmental Decision Making 192
   J. Scott Armstrong
   Decision-Maker Response: *Julia A. Trevarthen*

8. Assessment, Refinement, and Narrowing of Options 231
   Miley W. Merkhofer
   Decision-Maker Response: *Lynn C. Maxwell*

9. Post-Decision Assessment 285
   Gilbert Bergquist and Constance Bergquist
   Decision-Maker Response: *Katharine Jacobs*

10. Next Steps for Tools to Aid Environmental
    Decision Making 317
    Mary R. English and Virginia H. Dale

Biographies of Contributing Authors 329

Index 335

Contributors

J. Scott Armstrong
Associate Professor of Marketing, The Wharton School, University of Pennsylvania, Philadelphia, PA 19102, USA

Constance Bergquist
Vice President, Programs and Education, Commercial Services Corporation, Tallahassee, FL 32308, USA

Gilbert Bergquist
Associate Director, Environmental Management, Florida Center for Public Management, Florida State University, Tallahassee, FL 32306-4025, USA

Virginia H. Dale
Senior Scientist, Environmental Sciences Division, Oak Ridge National Laboratory, Oak Ridge, TN 37831-6036, USA

Mary R. English
Research Leader, Energy, Environment and Resources Center, University of Tennessee, Knoxville, TN 37996-4134, USA

William R. Freudenburg
Professor of Rural Sociology, University of Wisconsin-Madison, Madison, WI 53706, USA

Robert Fri
Director, National Museum of Natural History, Smithsonian Institution, Washington, DC 20560, USA

Robin Gregory
Senior Researcher, Decision Research, Inc., and Value Scope Research, North Vancouver, B.C. V7P 1Z9 Canada

Katharine Jacobs
Director, Tucson Active Management Area, Arizona Department of Water Resources, Tucson, AZ 85701, USA

Sami Kahn
Science Education Specialist, Rutgers University, Westfield, NJ 07090, USA

Joseph W. Lewis
Staff Economist, Forest Health Protection, USDA, Washington, D.C. 20090-6090, USA

Mary L. Lyndon
Professor of Law, School of Law, St. John's University, Jamaica, NY 11439, USA

Lynn C. Maxwell
Manager, Systems Integration, Tennessee Valley Authority, Chattanooga, TN 37402-2801, USA

Miley W. Merkhofer
Principal, Applied Decision Analysis, Inc., Menlo Park, CA 94025, USA

William R. Miller III
Manager, Environmental Affairs, Saturn Corporation, Spring Hill, TN 37174, USA

Elizabeth Ungar Natter
Executive Director, Democracy Resource Center, Inc., Lexington, KY 40503, USA

Robert V. O'Neill
Corporate Fellow, Environmental Sciences Division, Oak Ridge National Laboratory, Oak Ridge, TN 37831-6036, USA

Jeffery P. Osleeb
Professor of Geography, Hunter College of the City University of New York, New York, NY 10021, USA

Surya S. Prasad
Environmental Engineer, DESCIM Program Office, Alexandria, VA 22332-2300, USA

Wendy Hudson Ramsey
Environmental Research Consultant, 601 Wethersfield Lane, Knoxville, TN 37922, USA

Dean Hill Rivkin
Professor of Law, School of Law, University of Tennessee, Knoxville, TN 37996, USA

Roy Silver
Associate Professor of Sociology, Southeast Community College, Cumberland, KY 40823, USA

Chetan Talwalkar
Democracy Resource Center, Inc., Lexington, KY 40503, USA

Julia A. Trevarthen
Assistant Director, Southeast Florida Regional Planning Commission, Hollywood, FL 33021, USA

Claire Van Riper-Geibig
Research Associate, Energy, Environment and Resources Center, University of Tennessee, Knoxville, TN 37996-4134, USA

1
Overview

MARY R. ENGLISH, VIRGINIA H. DALE, CLAIRE VAN RIPER-GEIBIG, and WENDY HUDSON RAMSEY

The Need for Decision-Aiding Tools

At any single moment, environmental decisions are being made: a homeowner decides where to dump old house paint; a city government decides whether to issue a permit for a new subdivision; a state agency decides where to reroute a state highway; a business decides whether to expand its operations; the federal government decides how to revise an air-quality standard. Each of these actions has environmental effects that reach far beyond the person or group making the decision, yet, at present, there are few widely used tools are available to help make these decisions.

Environmental decision makers of today are faced with greater difficulties than ever before. Some of these difficulties are social and political in nature; they arise partly because of controversial but deeply held views on how decision-making processes should be conducted and what their outcomes should be. Other difficulties are caused by uncertainties regarding how environmental and social systems will change, and also about future goals and budgets. Still other difficulties are caused by a lack of resources; they arise partly because decision makers do not have the time or means to systematically analyze the problems they face. This last type of difficulty is especially likely at the subnational level, the level of environmental decision making that is the primary focus of this book.

People and groups involved in subnational environmental decision making come from all walks of life and include citizens as well as government officials and business representatives. Most environmental decisions at the local or regional level involve one or more of these types of participants. Their experience and expertise in the use of decision-aiding tools ranges greatly. Some individuals regularly log onto the Internet; others rarely turn on a computer. Some businesses and government agencies use expensive, highly sophisticated analytic systems; others operate with "back of the envelope" analyses. Some people are "tool-savvy"; others are unfamiliar with many contemporary methods of gathering, organizing, and analyzing information. Because of this diversity, it is difficult but important to define "tool" as the term is used in this book.

Definition of Tools for Environmental Decision Making

When building a house, one has tools for different tasks (e.g., saws, hammers, and drills) and tools for variations on the same task, such as Phillips-head or flat-head screwdrivers. Similarly, different tools are appropriate for different environmental decision-making situations. A survey may be needed to elicit the values of a large group of people, but focus groups or other small-group meetings may be needed to reach a detailed understanding of people's values. An elaborate multi-attribute utility analysis may be used to assess options when the issue is complicated, the budget is large, and the decision is not urgent, but a simplified decision-aiding model may suffice in other situations.

The underlying concept of a tool is that it is a means to an end, it is not the end itself. Thus, the term "tool" can be defined as anything regarded as necessary to carrying out one's tasks or mission. In its everyday usage, the term "tool" is rarely defined but usually well understood. Within the realm of environmental decision making, however, the meaning of "tool" is much less clear. The term has described everything from a computer to a printed procedure to an entire policy approach (Office of Technology Assessment, 1995). It can also include anything from a formal, systematically applied technique to an ad hoc method appropriate to certain situations.

In this book, we consider three types of tools for aiding environmental decision making: bits of information, or data; tools to gather data; and tools to organize and analyze data, including models to describe relationships among units of information. Data, like other meaningful representations of our physical surroundings or our thoughts and feelings, are themselves tools. They may be quantitative or qualitative. They may include, for example, measurements or observations of environmental conditions, such as soils, weather, and vegetation; of social and economic conditions, such as population size, income, education; and of legal and regulatory conditions, such as compliance histories and court cases.

Physical scientists collect environmental data using such tools as pH meters, vegetation surveys, and atmospheric tests. Social scientists collect economic, political, and social data using such tools as surveys, interviews, and systematic investigations of public records. As is noted frequently throughout this book, data vary in their precision and reliability. In addition, pieces of data do not become information (in the sense that they do not inform) until they are organized and analyzed. Formerly, data organization and analysis was accomplished by employing a few relatively straightforward methods: conceptual tools, such as taxonomy (categorization by similarities and dissimilarities); and mathematical tools, such as statistical analysis (analyzing information according to its distribution and bivariate or multivariate relationships). Today, while the same basic principles apply, the conversion of data into usable information may be accomplished through a wide variety of tools, depending upon the types of data and the uses to which they will be put. The selection of these tools and organizing constructs is critical because they can influence the results.

As large amounts of data become available, organization becomes all the more essential. Increasingly, the organization of data for environmental decisions is being done through spatial means with, for example, geographic information systems. Statistical, graphic, and other methods for information analysis are becoming more and more sophisticated with increased use of high-powered computers. In addition, conceptual models are increasingly used to understand and predict natural, technological, and social systems as these systems become more complex and tightly coupled and as their interactive effects become recognized. Models can range from simple concepts of how a "puzzle" fits together to detailed simulation models. The overall goal of the modeling exercise is to combine the necessary information into a framework that can help guide decision making.

The Purpose of This Book

The development and routine use of tools for environmental decision making is still in an early stage because of the diversity of both environmental issues and potential tool users. To date, the most sophisticated tools to gather, organize, and analyze information relevant to environmental decisions have, for the most part, been borrowed from other policy arenas or developed to address national problems where constraints (such as a lack of time, money, and trained staff) are not typically as pressing as in most subnational situations. In addition, the settings for environmental decisions are extremely varied, as are the participants in these decisions. A wide variety of tools is likely to be needed by people with different backgrounds, skills, access to information and equipment, and degrees of involvement in decision-making processes.

This book is addressed to all participants in environmental decision making. It reviews some of the most significant tools, categorizes them by the kinds of functions they serve, and provides assessments of how useful and appropriate they are likely to be. Our goal is to examine data-gathering and analytic tools that can aid environmental decision making and to assess their strengths and limitations. In doing so, we hope to make tools more accessible to both "savvy" and novice tool users and to clarify which aspects of environmental decision making can be improved by using which kinds of tools.

We do not see decision-aiding tools as a panacea. No tool, however sophisticated, can remedy situations where goodwill and common sense are scarce and where values differ greatly. Furthermore, we recognize that decision making cannot be completely "decomposed" into a discrete set of functions with a handy tool for each function. Instead, environmental decision making is inevitably political, in both the worst and the best senses of that term. It cannot escape the inequalities and struggles for domination of values that plague many aspects of our society. And it benefits from attempts to make basic changes in our collective understanding of who gets to set the decision-making agenda, who gets to participate in the decision-making process, and how decisions are finally reached. All of these issues fall outside the scope of this book, yet they are, quite likely, of much more fundamental importance than any tool to aid decision making.

But tools to aid environmental decision making are tools for knowledge and thus for power. If only for this reason, it behooves us to understand how these tools operate, how they can be improved, and how they can be made available to the many, rather than to the privileged few.

Environmental Decision-Making Settings: Four Dimensions

Tools are only useful if they suit the needs of those involved in environmental decision making. The settings in which decision-aiding tools might be used are infinitely varied, but they can be better understood by considering four dimensions: (1) the types of environmental issues on which decisions might be made; (2) the physical setting of the prospective environmental decision, including its spatial scale; (3) the types of individuals and groups who might interact in a process leading up to an environmental decision; and (4) the time frame within which the decision must be made.

When faced with an environmental problem or issue, preliminarily characterizing the issue along these four dimensions can help to launch the decision process in the right direction with the right tools. In effect, these dimensions can serve as a conceptual tool for ferreting out "gaps and blinders" in one's thinking about the issue. As discussed in more detail in Chapter 4, gaps and blinders (which might also be called oversights and tunnel vision) can be one of the most serious impediments to good environmental decision making.

Types of Environmental Issues

Virtually all human actions have environmental consequences. A shopper's decision to buy chicken rather than beef affects the relative strength of the poultry and cattle industries, which in turn affects grains grown, land devoted to grazing and feedlots, and runoff to rivers and streams. A corporation's decision to locate an auto-parts manufacturing plant in Tennessee rather than Michigan affects housing starts in each state, as well as patterns of transportation and vehicular air pollution. A city's decision to build a new baseball stadium affects land use and transportation patterns.

Decisions like these are all, in a sense, "environmental decisions." But they are all indirect environmental decisions—although they have environmental consequences, they currently are determined primarily by such factors as jobs, profit, and personal taste. If their environmental consequences become more widely recognized, they may be reconstrued as environmental issues. For example, the severe air pollution that three decades ago was seen as an inevitable consequence of industry in Chattanooga, Tennessee, has become a focus of the city's attention in recent years. To a large extent, environmental issues, like other issues, are socially constructed; that is, they are a product of society's collective consciousness, which can differ with locality and change over time (Hannigan, 1995).

Despite the dynamic nature of environmental problems, a categorization of key contemporary issues can help frame what is meant by an "environmental decision" within the context of this book. The following ten clusters of issues are listed in Sidebar 1.1.

- Natural-resource management.
- Critical natural areas.
- Growth management and infrastructure.
- Air-quality control.
- Water-quality control.
- Water allocation.
- Waste management.
- "Green" technologies.
- Energy production and distribution.
- Historic, cultural, and aesthetic resources.

This list is meant to be suggestive rather than exhaustive. Its structure is similar to other recent compilations of environmental issues (The Conservation Foundation 1987; Miller 1992; McKinney, and Schoch, 1996), but it may omit some types of issues. In addition, the categories in this list are not mutually exclusive. Instead, many of the issues are interrelated. Finally, while the categories are presented here as discrete, they should not be treated as discrete in decision-making situations.

Types of Physical Settings

The physical setting is the "subject matter" of the environmental decision. As the clusters of environmental issues listed above would indicate, these settings can differ greatly. Some environmental decisions may address vast tracts of wilderness, such as the canyonlands of Utah; others may deal with a small urban park or an historic building. The physical settings of environmental decisions can be characterized by (1) the extent to which they involve the *natural* or the *built* environment and (2) their *spatial scale*. Understanding these attributes can help to characterize the issue at hand,

Sidebar 1.1
Ten Clusters of Environmental Issues

Natural-resource management. Issues concerning the use of trees and other plants, minerals, soils, fish, and wildlife for purposes such as materials and food, as well as for consumptive and nonconsumptive recreation. As human population size has increased, people have become more aware that natural resources are finite and that economic growth is tied to resource availability. The need to move beyond exploitation to holistic management strategies is increasingly recognized.

Critical natural areas. Issues concerning the identification and protection of coastal areas, flood plains, wetlands, ecological "bioreserves," parks, the habitats of endangered species, and other specialized locales. Certain ecosystems are especially vital to human and ecological well-being. Some of these areas are highly susceptible to disruption from activities not only within, but surrounding them.

Growth management and infrastructure. Issues concerning the type, intensity, and distribution of land uses (e.g., agricultural, forest, residential, commercial, office, and industrial) and of infrastructure (e.g., utilities and transportation systems). Population growth patterns, including patterns of migration, and the search for environmentally sustainable, economically viable forms of development underlie many of these issues.

Air-quality control. Issues concerning criteria pollutants, chlorofluorocarbons (CFCs), and greenhouse gases (especially carbon dioxide and methane). Indoor as well as outdoor air pollution and accidental as well as routine releases into the atmosphere must be considered, as must the contribution from nonpoint as well as point sources. Air quality is a function, however, not only of the number and frequency of the use of pollution sources, but also of the sophistication of pollution-control technologies. For example, vehicular emissions in the United States have decreased during the past 25 years even though the vehicular miles driven now far exceed those in 1970 (McKinney and Schoch, 1996).

Water-quality control. Issues concerning contaminants in groundwater and surface water, including sewage treatment, sludge management, controlled and uncontrolled releases of contaminants, and thermal discharge. Accidental and routine releases from point and nonpoint sources must be considered. As with air quality, water quality can deteriorate with an increase in the burden of pollution sources, especially nonpoint sources such as road and agricultural runoff, unless pollution-control measures are taken.

Water allocation. Issues concerning the provision of water, including aquifers, aquifer recharge areas, rivers, and dams. Central to this cluster of issues are debates concerning water-use rights and responsibilities, and land-use rights and responsibilities affecting the quantity and quality of water. Western regions of the United States have been encountering either the reality or the prospect of severe water shortages, especially as population grows, agricultural and industrial uses of water increase, and aquifers are drawn down faster than they can be replenished. Water-allocation issues are not limited to the western United States, however, and although dams can control flooding, generate power, and provide measured supplies of water, they also affect water tables and aquifer-recharge areas.

Waste management. Issues concerning solid waste (e.g., garbage, yard wastes, construction and demolition material), chemically hazardous waste, and low- and high-level radioactive waste, including spent nuclear fuel. Waste-management issues encompass both how to treat, store, and dispose of the current waste stream flowing from commercial and industrial enterprises, utilities, households, governmental institutions, etc., and how to clean up land and water that has become contaminated by past inadequate waste management. Related issues include how to reduce the quantity and toxicity of wastes now being produced.

Green technologies. Issues concerning technologies and practices used in manufacturing, construction, agriculture, etc., that are less environmentally burdensome than conventional practices. Whereas pollution-control technologies capture contaminants as they come out of the pipe or up the stack, green technologies avoid the use of materials and methods that result in contaminants. In addition to pollution prevention, green technologies may have the goal of not depleting scarce materials. The total life-cycle cost of the product or process, especially its internalized and externalized environmental costs, becomes a central focus with green technologies.

Energy production and distribution. Issues concerning conventional energy sources (e.g., coal-fired, gas-fired, nuclear, and hydro plants), alternative energy sources (e.g., solar, wind, geothermal, and biomass), and energy conservation. The production and distribution of electricity, heating fuel, fuel for vehicles, etc., have important environmental effects in terms of both pollution and the depletion of exhaustible natural resources. The electricity industry, like the natural gas industry before it, is now in a period of change with the prospect of deregulation; the environmental implications of electric-utility deregulation remain to be seen.

Historic, cultural, and aesthetic resources. Issues concerning the identification and protection of historic buildings and districts, archeological artifacts, sacred places, "viewsheds," and other sensitive

> areas. Of the ten issue clusters, this one is especially dependent upon the eye (and mind and heart) of the beholder. Individual and group values are central to determining which resources should be protected and in what manner. One particularly pervasive and contentious issue is the proliferation of signs and billboards across the landscape; another is the protection of large natural areas identified as sacred within the spiritual traditions of some Native Americans.

which in turn will help to identify both whom should participate in the decision-making process and what tools will be most useful. Failing to understand these attributes will lead to a flawed and possibly obstructed decision-making process. For example, a seemingly local-scale, built-environment issue, such as a new shopping center, may have implications for the natural environment and for people in the surrounding region, yet this may become clear only when objections are raised.

The Natural and Built Environments

Virtually no part of this planet is completely "natural" and untouched by human actions; our effects are present even in Antarctica. Furthermore, as one among many species, humans are in many ways part of the natural environment. Nevertheless, humans have a well-developed capacity to create and use technology that alters the natural environment. In some places, the effects are imperceptible without highly sensitive instruments; in other places, they are dramatic. The Manhattan skyline and the Rocky Mountain skyline are both built on bedrock, but they may appear to have little in common except height.

The "natural environment" and the "built environment" are thus two hypothetical poles on a continuum. Most environmental decisions involve elements of both; the question is the mix. Toward the "natural" end of the continuum are decisions concerning acquiring wilderness areas, protecting endangered species, managing forests, regulating to prevent over-fishing, and so forth. Toward the "built" end of the continuum are decisions concerning, for example, constructing city plazas, regulating commercial land uses, and detoxifying contaminated buildings.

But even issues that seemingly concern the natural environment can be embedded in the built environment. For instance, the cleanup of the Great Lakes involves identifying and eliminating or reducing anthropogenic sources of pollution, such as polychlorinated biphenols (PCBs), dioxins, mercury, phosphates, and coliform bacteria (www.epa.gov). And even an issue that seemingly concerns mainly the built environment—for example, the emergency cleanup of a chlorine spill at an industrial plant—can also involve the surrounding natural environment (if contaminants from the spill spread to soil and groundwater) as well as the natural environment elsewhere (at the site where the chemical wastes are disposed).

Spatial Scale

On the one hand, environmental decisions can focus on specific places, like New York State's decision to create the Adirondack Park or Miami Beach's decision to establish its historic Art Deco district. On the other hand, decisions can lead to broadly applicable policies, like the international decision to curtail the use of chlorofluorocarbons (CFCs). Regardless of whether an environmental decision concerns a place or a policy, its spatial scale can vary.

In the case of an industrial plant chemical spill, the spatial scale is primarily limited to the area where the accident and the emergency response has its greatest impact. In contrast, the decision to clean up the Great Lakes covers an area that contains approximately 10 percent of the population of the United States and 40 percent of its industry, as well as a significant portion of Canada's people and industry. But, as with the natural-versus built-environment distinction, characterizing an environmental decision as small or large in spatial scale can be difficult. Cleanup of the industrial plant's chemical spill is regulated by state and federal laws, and the Great Lakes cleanup inevitably involves hundreds of small, local actions.

Types of Participants in Environmental Decision Making

Environmental decision making can be even more complicated than decision making on other public issues. First, environmental impacts do not respect property or jurisdictional lines; they often cross boundaries. Second, environmental decisions involve government agencies in two capacities, as managers and as regulators. And third, environmental issues can provoke especially heated value conflicts that require value trade-offs. For these reasons, it is essential to understand who participates in environmental decision-making processes.

Environmental decisions are made by people both as individuals and as members of organizations. For the purposes of this book, the latter is the main focus. With decisions made by organizations, it is not always clear exactly who the decision maker is. While the final decision may rest with the head of an agency or the chief executive officer of a corporation, it is often supported by advice from assistant directors or vice presidents and by analysis from support staff.

In addition, decisions made by organizations are likely to involve not only internal but also external dialogue and debate. The extent of such interaction can be seen in a decision made in the late 1980s about water quality in the Pigeon River, which flows from North Carolina into Tennessee. This decision, which has been much revisited, has involved the U.S. Environmental Protection Agency (particularly its Region 4 field office); governors and environmental protection agency personnel in both states; the U.S. District Court, the Tennessee Supreme Court, and the U.S. Supreme Court; Champion Paper Company, the main source of the pollutants of concern; and environmental and "pro-jobs" groups on either side of the state line (Bartlett, 1995).

The lines between organizations are not always clear, especially with formal and informal alliances. In the case of the Pigeon River, local citizen groups protesting pollution from Champion Paper Company have included the Dead Pigeon River Council and the Hartford Environmental League Project (HELP), most of whose members live down river from Champion; the Pigeon River Action Group, led by a western North Carolina resident; and the Knoxville-based Foundation for Global Sustainability; as well as national groups like Greenpeace, the Environmental Defense Fund, and the Izaak Walton League. These organizations have not all been involved at the same time and to the same extent, but they have all played a role, as have others who have entered the fray, even though the final decision has rested with the U.S. Environmental Protection Agency (EPA) and the judicial system.

One or more of the following types of organizations are likely to participate in environmental-decision-making situations:

- Local citizen groups.
- State, regional, and national citizen groups.
- Small businesses.
- Business associations and large businesses.
- Local government.
- State government.
- Regional government.
- Federal government.

These eight types of organizations are briefly described in Sidebar 1.2. The descriptions are general; there undoubtedly are exceptions. In addition, the types of groups that participate in environmental decision making and their constituents change over time. Today, groups representing the environmental-justice concerns of low-income and racial and ethnic minorities are far more central to environmental decision making than they were 20 years ago (Dunlap and Mertig, 1992; Hofrichter, 1993).

Decision-Making Time Frame

"Not to decide is to decide." Pop philosophy reminds us that not doing something is, in effect, doing something: We are choosing to let the situation stay the same or evolve without our intervention. Going with the "no action alternative" can be a conscious choice (as in an EPA decision to rely on natural attenuation of ground water contamination at a Superfund site), or it can result from procrastination or ignorance about the problem.

Sidebar 1.2
Organizations That Typically Participate in Environmental Decision Making

Local citizen groups. These groups may be neighborhood-based or they may include people from different areas within a municipality or county. A group may have several different concerns, such as education, health care, and environmental quality, or it may have a single focus, such as greenways, housing, or economic development. It may have a handful of members or a roster in the hundreds. While local citizen groups may have dues and may apply for and receive grants, they typically have limited funding and few, if any, paid workers, relying instead on volunteer time and skills.

State, regional, or national citizen groups. These groups typically have a paid staff and a large number of members who pay individual or institutional membership fees. They may also rely on other sources of funding, such as grants from foundations or corporations. They may comprise a single organization, sometimes with local, state, or regional chapters or offices (e.g., the Sierra Club and The Nature Conservancy), or a coalition of groups linked in common cause by an umbrella group, such as the Center for Health, Environment and Justice. They usually have an overarching goal and one or more strategies, including lobbying; negotiating on legislative bills and agency regulations; releasing informational materials and otherwise publicizing their cause; and bringing lawsuits. Their goal may or may not be environmental protection; it may, for example, be property-rights protection, the multiple use of public lands, or economic development.

Small businesses. Some small businesses have environmental services and products as their enterprise. Most, however, are businesses of a different sort (dry cleaners, auto shops, etc.) that get involved in environmental decision making only because they are themselves making a decision, perhaps about their waste management practices, or stand to be affected by one such as a zoning decision or a decision requiring an environmental cleanup.

Large businesses and business associations. As with small businesses, some large corporations, such as waste-management companies, are in the "environmental business," but most are in other enterprises like automobile manufacturing, paper products, chemical products, and agriculture. Large businesses (and associations of small or large businesses) typically get involved in environmental decision making because they themselves make environmental decisions or because their businesses stand to be affected by the decisions of others. In the latter regard, they often seek to influence broad public policies (e.g., state or federal laws and regulations) as well as immediate, local decisions.

Local government. The legislative and executive arms of municipal or county governments routinely make environmental decisions of various sorts. They do so either as managers of public property (roads, parks, water treatment plants, solid waste disposal facilities, etc.) or as regulators of private property through zoning and subdivision controls; local health-department regulations; special regulations concerning signs, billboards, historic districts; etc. In addition, municipal courts decide cases involving local-ordinance violations. Like businesses and citizen groups, local government agencies and officials team up through state, regional, and national associations such as the U.S. Conference of Mayors and the National Association of County Health Officials.

State government. As with local governments, state governments make diverse environmental decisions. Elected officials and administrative agencies serve as public-property managers (e.g., by acquiring and managing a state park or wildlife preserve) and as regulators of private activities (e.g., by enacting legislation and regulations concerning air and water quality). State governments also may provide grants to local governments for their environmental activities. In many ways, state government is the intermediary between broad federal policies on the one hand and local activities on the other; in addition, state government may take policy initiatives of their own, for example, setting more stringent standards than federal standards. The state judicial system decides cases of an intrastate character, interpreting them in light of state laws and state and federal constitutional provisions. The U.S. Supreme Court receives appeals from state supreme court decisions if they present federal questions.

Regional government. Regional governmental entities can be *intra*state (e.g., the Massachusetts Metropolitan District Commission was established by state legislation in 1893 to provide parks, roadways, police protection, sewage disposal, and clean drinking water to cities and towns in the Boston area). Or regional governmental entities can be *inter*state (e.g., the Appalachian Regional Commission was established by federal legislation in 1965 to assist in the region's economic development). Unification of municipalities or counties is usually for general-purpose government, but typically, intrastate and interstate regional authorities, commissions, etc. do not replace the existing governmental structure. The authority of the regional entity usually is limited to one or a few functions, such as water supply, transportation, economic development, or environmental protection, although within its functional area, it may have a great deal of authority to act.

An intrastate regional governmental body may act as either a public-property manager or a private-property regulator. For example, while the Massachusetts Metropolitan District Commission acquires and manages parks and reservoirs, the San Francisco Bay Conservation and Development Commission regulates proposed development in and along the Bay. In contrast, an interstate regional governmental body usually serves as a property manager rather than as a regulator; the power system of the Tennessee Valley Authority includes 11 coal-fired plants, 29 hydroelectric dams, 48 combustion-turbine units, one pumped-storage facility, and several nuclear-power plants, as well as 16,000 miles of transmission lines (Tennessee Valley Authority, 1995). Typically, special-purpose intrastate or interstate regional entities are led by appointed boards or commissions; their members usually are not elected.

Federal government. As with the state and local government, all three branches of government play important but different roles in environmental decision making. Congress is responsible for enacting laws concerning the environment. During the past 30 years, Congress has passed more than 20 acts that have expanded the federal government's role as an environmental regulator. The EPA has major responsibility for promulgating regulations and enforcing federal environmental laws; in addition, federal agencies, such as the Department of Health and Human Services, the Department of Agriculture, and the Occupational Safety and Health Review Commission, assume regulatory roles on issues related to the environment. The U.S. Department of Justice enforces the regulations of such agencies as EPA; the federal court system considers cases of an interstate character or concerning federal law, with the U.S. Supreme Court serving as the final arbiter of disputes about interpretations of federal statutes or the U.S. Constitution.

More than one-quarter of the land in the continental U.S. is part of the federal public domain and is managed by such agencies as the U.S. Bureau of Land Management (BLM), the U.S. Forest Service (USFS), and the National Park Service (NPS) following the general directives of Congress. Other federal agencies, particularly the Department of Energy, the Department of Defense, and the Department of Transportation, have responsibility for managing federal property that, while smaller than properties managed by the BLM, USFS, and NPS, have major environmental and economic impacts. For example, it is estimated that the cleanup of radioactive and hazardous waste contamination of the Department of Energy sites will total about $150 billion (U.S. Department of Energy, 1998).

If a decision (even a decision to take no action) is made consciously, then individuals and groups usually are aware that they should reach their decision within a certain period of time. That period can be very short if action is urgently needed, as when a spill of volatile chemicals has occurred or a dam is threatening to break. In other situations, the period can be much longer and the decision-making process much more deliberative. The plans and policies put into place in the Great Lakes pollution-control agreements took years to craft.

As with the other dimensions discussed above, the "urgent" versus "deliberative" distinction is not absolute. Deliberative issues may become urgent as the need to reach closure draws near, and people may disagree on how urgent an issue actually is. Furthermore, environmental decisions are not singular; they usually are part of larger sequences. Joint pollution-control agreements concerning the Great Lakes region had been in place between the United States and Canada since the 1970s; in 1991, the U.S. government reinforced those agreements by requiring accelerated cleanups. And even a decision at the time of a chemical spill or a dam failure is part of a stream of past and future decisions concerning emergency preparedness and after-the-fact repairs.

Who Needs Tools and When?

When categorizing environmental decision-making situations, being arbitrary is virtually impossible. Not only do cases vary along the four dimensions discussed, they also vary by the attributes of the decision-making process itself. Tonn, English, and Travis (in preparation) have identified six common modes of environmental decision making: routine, analytic, "élite corps," conflict management, collaborative learning, and emergency response. Furthermore, the ultimate goal of the decision-making process may vary. Ideally, the goal should be to reach a decision that is durable, fair, technically credible, widely supported, efficient, and effective (Feldman, 1997; Feldman and Nicholas, 1997). In fact, however, the decision-making process may be in thrall to factors like back-room politics and deal making.

A deepened understanding of how environmental decision making actually operates will improve our collective sense of when and how tools can help the most. Meanwhile, for the purposes of thinking about who is likely to need decision-aiding tools, we have developed the following thumbnail sketches of typical environmental decision-making situations in the United States. These sketches, summarized in Table 1.1, are organized by spatial scale and provide brief, generalized descriptions using the four dimensions discussed above. These sketches are based on observations of *current* environmental decision-making practices, which could easily

TABLE 1.1. Typical environmental decision-making situations.*

Type of situation		Type of participant							
		Local citizen groups	Regional/national citizen groups	Small businesses**	Large businesses and business associations**	Local government	Regional authorities, commissions, etc.	State government	Federal government
Environmental setting	Natural	✓	✓		✓	✓	✓	✓	✓
	Built	✓	✓	✓	✓	✓	✓	✓	✓
Spatial scale	Local	✓				✓			
	Small regional	✓	✓	✓	✓		✓		
	Large regional		✓		✓			✓	✓
	National		✓		✓			✓	✓
Decision-making timeframe	Deliberative	✓	✓		✓	✓	✓	✓	✓
	Urgent	✓		✓	✓	✓			

*Relates to such issues as natural-resource management; identification and protection of critical natural areas; growth management and infrastructure; air-quality control; water-quality control; waste management; green technologies; water allocation; energy production and distribution; and historic, cultural, and aesthetic preservation.

**Does not typify businesses that are environmental service providers.

change as new decision-making modes are adopted or as new issues come to the fore and old issues are reconstructed.

Local Decisions

Decisions involving a few acres or a few square miles can be urgent or deliberative and can focus on either the built or the natural environment. At the local scale, typical issues include water-quality control; the protection of critical natural areas; waste management; growth management and infrastructure concerns; and historic, cultural, and aesthetic preservation. They may also involve natural-resource management; air quality; water-allocation issues; green technologies; and energy production and distribution.

State and federal laws, as well as local ordinances, provide the legal context within which local-scale decisions take place, and state or federal governments may be direct participants in those decisions, especially if state or federal property is involved. Nevertheless, local-scale decisions often are limited to local government, local citizen groups, and major local businesses. Some small businesses are intensively involved in local environmental decisions, especially if their owners are active in community affairs. Most, however, get involved only on a limited basis, and then only if they are directly and immediately affected.

Regional Decisions

Decisions involving a shared ecosystem (such as a bay) or a shared investment (such as a waste facility) can be urgent, but more typically are deliberative, and can focus on the built environment, but more typically focus on the natural environment. At this scale, typical issues include natural-resource management, the protection of critical natural areas, and water and air quality. Issues at the small regional scale may also include waste management; growth management and infrastructure concerns; water allocation; energy production and distribution; and historic, cultural, or aesthetic preservation. They are less likely to include green technologies.

Again, local, state, and federal laws and regulations help to provide the context within which regional decisions take place, and to the extent that state or federal property is involved, these levels of government are direct participants in environmental decision making. Often, however, local governments, citizens groups, and businesses are the most active participants in environmental decisions at this scale. The City of San Jose, area businesses, and a coalition of environmental groups called CLEAN South Bay became involved in a water-quality issue concerning the southern end of the San Francisco Bay. CLEAN South Bay sampled water in the South Bay and found high concentrations of nickel and copper coming from the local wastewater treatment facility. An independent study found that businesses in the area were the source of the pollution, even though they were complying with local water standards. This prompted the city to lower its limits for nickel and copper, thus affecting the pollution abatement methods practiced by local businesses.

Other issues at a somewhat larger regional scale may involve state or federal government, as well as local government and regional or national citizens groups. Such expanded commitment was marshaled for the Chesapeake Bay, the nation's largest estuary. Since the 1960s, the bay has been the target of collective action to reverse a decline in its estuarine grasses and fishery resources and to promote integrated, baywide ecosystem management. This effort has included citizen and environmental groups (e.g., the Citizens' Program for the Chesapeake Bay, the Chesapeake Bay Foundation, the Sierra Club, Clean Water Action, Save Our Streams, the Maryland Waste Coalition, the Maryland Conservation Council, the Environmental Policy Institute, and the League of Women Voters); state governors (of Maryland, Pennsylvania, and Virginia); the mayor of Washington, D.C.; the administrator and staff of EPA; members of Congress and of the three states' legislatures; members of the scientific community; and the news media. This effort culminated in the Chesapeake Bay Agreement of 1983, which, as expanded through a 1987 agreement, included regional commitments concerning a phosphate ban; more strict pollution controls; a moratorium on striped bass fishing; critical-areas protection; non-tidal wetlands and forestlands preservation; and growth management (Fraites and Flanigan, 1993).

Environmental decisions concerning airsheds or river systems in a multistate region tend to be deliberative rather than urgent, and they tend to focus on the natural environment, although they often involve the built environment by implication. Issues typically include air and water quality; water allocation; energy production and distribution; and natural-resource management. While growth management is not likely to be addressed at the large regional scale, infrastructure issues sometimes arise, such as the location of interstate highway systems, as do some waste-management issues, such as the disposal of low-level radioactive wastes through interstate-compact systems. Issues concerning the protection of critical natural areas, tall-grass prairies for example, increasingly are being considered on a large regional scale as well as at the local scale.

At the large regional scale, participants in environmental decision-making processes typically are state, regional, or national citizens groups; large businesses and business associations; and regional, state, and federal governments. All of these types of groups have been involved with a complicated water-allocation issue involving Arizona, California, and Nevada. Under water rights obtained through a 1928 Congressional act apportioning the lower Colorado River basin's waters among Arizona, California, and Nevada and subsequently adjudicated in the 1963 case, *Arizona v.*

California, decided by the U.S. Supreme Court (National Research Council, 1992), the Central Arizona Project pumps water from the Colorado River to Arizona. As of the mid-1990s, however, the water provided through the newly completed Central Arizona Project was being underutilized within Arizona because of its relatively high cost, among other reasons, and California and Nevada were seeking to lease unused supplies. Arizona officials were concerned that leasing the waters might jeopardize Arizona's entitlement to it. Out-of-state pressure mounted to get access to the water, while within Arizona, the issue involved not only the governor, state officials, and water district officials, but also others such as Native American tribes (Gelt, 1993).

National Decisions

Environmental decisions at the national scale, such as those to protect endangered species or to manage toxic wastes, are general in their intent and diffuse in their effects. They tend to be deliberative rather than urgent, and they tend to focus on the natural environment but have implications for the built environment. National-scale issues typically include most of the ten clusters identified above, but as policy rather than on-the-ground decisions. With the current trend toward devolution of federal governmental responsibilities to state and local governments, however, some of these issues may soon no longer be regarded as national-scale decisions.

Currently, participants in decisions at the national scale are likely to include federal and state governments; large businesses or business and professional associations; and regional or national citizens groups; as well as people at universities, nonprofit institutes, and consulting firms conducting research within the physical and social sciences. As one example, recent discussions about the reauthorization of the Superfund have included diverse groups at various points: the EPA, legislative committees and their staff in the U.S. House and Senate; industrial associations such as the American Petroleum Institute and the Chemical Manufacturers Association; environmental management and consulting companies and their associations, such as the Hazardous Waste Treatment Council; small-business coalitions such as the Small Business Survival Committee; associations of municipalities such as Local Governments for Superfund Reform; insurer groups such as the Alliance of American Insurers; research organizations such as Resources for the Future; and environmental groups as well as environmental-justice groups. The list is extensive.

Global Decisions

Although beyond the scope of this book, environmental decisions do not stop at the national scale. Broad policy decisions are made about smaller-scale actions that, taken in the aggregate, have supranational or global effects. Examples include climate change, depletion of stratospheric ozone, and deforestation. These broad policy decisions typically are preceded by extensive research and deliberation, and they most often concern major aspects of the natural environment, as well as many non-environmental factors. Participants in these decisions are likely to include representatives of nations and international organizations; representatives of citizens groups that have formed international networks; multinational corporations and business associations; as well as physical-and social-science researchers.

Clearly, the sketches presented above will need to be revisited as events play out and our collective understanding of environmental decision making deepens. Furthermore, these sketches are not prescriptive; they simply summarize the current situation. Nor do they address the extent to which different participants are (or should be) involved in environmental decision-making processes. Since some may be involved only tangentially or at certain points, while others may be involved centrally or for the duration, their needs for the results of decision-aiding tools and the tools themselves may differ.

A Functional Analysis of Decision-Aiding Tools

In all decisions at the local, regional, national, and global level, information-gathering and analytic tools offer the potential to improve the input to the decision-making process and thus its outcome. As noted at the outset to this chapter, however, not all tools serve the same function; rather, they fall into eight functional categories (see Figure 1.1).

Each of these categories is discussed in the eight following chapters, and an example is given in Sidebar 1.3 of how these categories of tools might help participants in an environmental decision-making process reach a well-informed decision.

While presented as linear and sequential in the example, as well as in the book, tools within these categories frequently will be used iteratively so that tentative decisions can be modified as more information is obtained. An understanding of values may lead to a redefinition of the information needed, or an awareness of limited options may lead to more modest goals. In addition, by planning ahead on how a post-decision assessment will be conducted, prudent modifications to the process can be made early on. Thus, although these categories are treated individually here, they should be thought of as iterative and interdependent.

20 M.R. English et al.

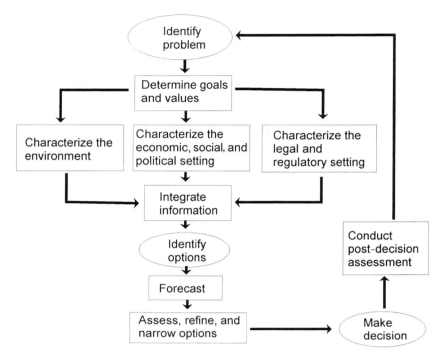

FIGURE 1.1. Eight functional categories of tools for environmental decision making.

Sidebar 1.3
An Example

Consider a new industrial plant being established to manufacture plastics in a rural community. Environmental hazards are associated with the manufacture of these plastics, and local citizens are concerned about potential impacts on environmental, economic, and social conditions in their area. Decisions revolve around the size and location of the manufacturing plant; environmental safeguards to meet local, state, and federal regulations; and efforts to communicate with local residents about environmental hazards and other impacts. Financial commitments are part of these decisions, and the company wants to be as cost-effective as possible while maintaining good community relations. The discussion below focuses on the company's use of decision-aiding tools, but other participants in the decision-making process (e.g., the local government, local citizens groups) might also use tools from one or more of these categories.

First, the company may seek to determine local citizens' goals and values on environmental and other issues related to the proposed plant using tools such as preference surveys and establishing a local advisory group (Category 1). Environmental data can either be collected from field studies or obtained from information sources such as U.S. Geological Survey maps and lists of local rare and endangered species (Category 2). To configure and locate the facility appropriately and plan its operation, given both business and community considerations, the local economic, political, and social setting must be considered (Category 3). Some of this information can be obtained from available sources (e.g., from census data or from the local government); however, the company's analysts may wish to incorporate this data into their own data systems to enable assessments of specific sites. The company may conduct an evaluation of legal considerations to identify potentially applicable local, state, and national laws and regulations and to assess how cases concerning the hazards associated with this type of industry have been decided in the past (Category 4).

At this point, the company may decide to use tools such as a geographic information system to integrate information into a single, spatially explicit database (Category 5). This integration may be especially revealing and important if pollution from the plant can affect different environmental pathways or different jurisdictions having different regulations. Next, the company may want to use forecasting tools to analyze the future effects that the manufacturing plant might have on the surrounding area and vice versa (Category 6). For instance, are there going to be population influxes associated with the plant, creating pressures on local public services such as schools and roads? Will the local work force be able to meet the plant's growing labor demands? Over time, will effluents from the plant be dispersed through groundwater, and if so, with what effects? Are future recreational activities in the area likely to be adversely affected by the plant's presence? What types of people will be living in the vicinity of the plant 20 years from now, and will air pollutants have adverse cumulative effects on them? Addressing all of these questions may require the use of such tools as scenario analysis, computer simulation, time series forecasting, and uncertainty analysis.

In conjunction with forecasting, the company may decide to employ decision-aiding tools to assess, refine, and narrow the options that are available (Category 7) by evaluating their relative costs and benefits and other implications. Finally, if a decision is reached to locate the manufacturing plant in the area in question, plant managers, as well as local officials and citizens, may decide to conduct a post-decision assessment to determine whether the pre-decision estimates were valid and the plant lives up to their mutual expectations. This analysis might employ tools such as the collection of environmental and socio-economic data on certain performance indicators (Category 8).

Why Any Categories?

These eight categories are used to classify different decision-making *aids*, but they do not attempt to describe the decision-making process itself. Although the categories are displayed in Figure 1.1 as components in a flow chart, each of these functions could be undertaken in different ways by different participants in a decision-making process at different points in the process. Furthermore, these categories cover only a limited set of aids. As explained at the outset of this chapter, the tools highlighted in this book are primarily for information-gathering and analysis. Not all tools useful to environmental decision making are covered in depth here. For instance, only passing attention is given to tools designed to involve people in the decision-making process or to communicate the results of information gathering and analysis.

These categories are grounded on an analytical approach, one compatible with (although not necessarily reliant upon) formal decision analysis. Decision analysis began to flourish in the 1950s, building upon its World War II origins, and is multidisciplinary, drawing upon fields such as economics, psychology, statistics, and operations research. It is not without its critics and skeptics, but it is by now fairly well established. To a large extent, the categories used here draw on the core ideas of decision analysis.

One core idea is that, for decisions to be rational, they must have *order* rather than randomness or arbitrariness, and that every rational decision builds some kind of order: "Every decision is a creative act; it brings together observations, judgments, values, and norms into a particular concept" (Diesing, 1962, p. 240).

Another core idea is that, while the total process of decision making includes finding both decision-making occasions and possible courses of action, the analytic phase focuses on choosing among possible courses of action by *decomposition*: that is, by breaking down the decision "problem"[1] into component parts (Bunn, 1984). This process of "decomposing" is complicated when uncertainty, multiple objectives, multiple options, and sequential effects are present, as they often are. But one of the advantages of a decision-analytic approach is its ability to handle (or at least clarify) complexities like these. In addition, a decision-analytic approach can help *define objectives, identify and reformulate options*, and *provide a common language for communication* about the decision, including pinpointing areas of agreement and disagreement (von Winterfeldt and Edwards, 1986). All

---

[1] The conventional use of the term "problem" as the catalyst for a decision has been criticized as being unduly negative and overlooking the fact that opportunities can also necessitate decisions (Keeney, 1992). Although we have used the conventional terminology here, this point is well taken.

analyses are incomplete (the concept of "bounded rationality" [Simon, 1957] suggests that we necessarily limit our scopes of thought and inquiry), and all analyses ultimately include subjective judgments. Nevertheless, the process of making decisions can be improved and possibly expedited if we "think, decompose, simplify, specify, and rethink" (Behn and Vaupel, 1982), while also remembering that decomposition must be followed by reintegration.

Contrasted with those who advocate a decision-analytic approach are those who argue that, especially in the realm of public policy-making, this approach is inappropriate and destructive. These arguments attack the notion of "decisionism": "the vision of a limited number of political actors engaged in making calculated choices among clearly conceived alternatives" (Shklar, 1964, p. 13). The decision-analytic approach is criticized as assuming a unitary decision maker; downplaying conflicts between different groups; translating all relevant factors into present-day terms; being preoccupied with outcomes and ignoring the processes by which outcomes are produced; being mechanical; and not allowing for the role of argument (Majone, 1989).

In addition to these criticisms, numerous debates have persisted about essential attributes of decision-making processes. One debate concerns the nature of individual rationality. Is the rational individual one who maximizes current and expected utility, including both subjective and disinterested preferences (Harsanyi, 1977), or is rationality something that goes beyond utility maximization to admit the rationality of other grounds for behavior, such as commitment (Sen, 1977)? Underlying this debate are polarized world views that could be called cynical on the one hand ("strategic thinking is the art of outdoing an adversary, knowing that the adversary is trying to outdo you" [Dixit and Nalebuff, 1990, p. ix]) and idealistic on the other ("the injunction to love one's neighbor involves a widening of agendas ... It implies, that is to say, a kind of Copernican Revolution and an abandonment of perspective" [Boulding, 1966, p. B-168]).

Related to the issue of rationality, then, is the issue of perspective. We are cautioned that decisions take place within "frames"—conceptions of the acts, outcomes, and contingencies associated with a particular choice—that may not be shared and that, when changed, can significantly alter preferences concerning the decision at hand, although these preference reversals are not necessarily irrational (Tversky and Kahneman, 1981). Other debates concern whether individually rational behavior is also rational for the group (Barry and Hardin, 1982, Bacharach and Hurley, 1991) and whether institutions need to be intentionally *crafted* so that the answer to this question is "yes" (Bacharach and Hurley, 1991). Also relevant is the old question of collective behavior: Are groups and institutions simply aggregates of individuals, or is society the source of much that defines individuals?

These debates and qualifications all concern the *context* within which decisions take place; other qualifications concern the appropriate or pragmatically feasible *role* of decision-analytic approaches. Numerous refinements have been made to decision-analytic approaches, such as the "value-focused thinking" espoused by Keeney (1992), which is adopted as a guiding precept here. Cautions also have been raised that decision-analytic approaches can be peripheral to the way that actual decisions are made, especially in large public agencies (Feldman, 1989), or that they *should* be peripheral if the decisions are essentially political in nature (House and Shull, 1988).

All these objections and qualifications are important. But even the most severe critics of decision-analytic approaches see some merit to systematic inquiry.

The Importance of Systematic Inquiry

Systematic inquiry is not unique to formal decision analysis; it also is characteristic of related approaches, such as cost-benefit analysis and risk assessment. Both of these approaches, together with decision analysis, are discussed in detail in Chapter 8. Because of their central importance to recent thinking about environmental decision making they also merit mention here, along with another approach, adaptive management, which is receiving increasing attention.

Cost-benefit analysis provides a means of comparing the pros and cons of an environmental decision's prospective impacts. This approach has been widely espoused because it uses a common metric—money—to simplify comparisons and provide a clear financial picture. A key issue with this approach, however, is the challenge of taking values that are not easily monetized, such as environmental and social well-being, and converting them into dollar figures. (See Chapter 2 for further discussion of methods intended to achieve this conversion.) In addition, this approach requires that the temporal and spatial dimensions of the environmental problem at hand (as well as discount rates related to these dimensions) be precisely specified.

Risk assessment provides a means of predicting and evaluating the consequences of future events. Within the context of environmental decision making, risk assessment has been developed during the past two decades as a formal approach to assessing the likelihood and magnitude of effects (especially adverse effects) of toxic substances on human health, and has been conceptualized as having four components: hazard identification, dose-response assessment, exposure assessment, and risk characterization (National Research Council, 1983). A companion approach, ecological risk assessment, is sometimes conceptualized as having three components—problem formulation, analysis, and risk characterization—with the task of characterizing exposure and ecological effects running through these three components (U.S. Environmental Protection Agency, 1992). Ecological risk assessment has also been conceptualized as having four components comparable to those for human health risk assessment (National Research Council, 1993). In any case, ecological risk assessment has been broadened to consider effects not only on single organisms but also on populations, ecosystems, and large regions (Suter, 1993). Regardless of whether the risk assessment is concerned mainly with humans or with the environment, its culminating point, risk characterization, serves as both an input to and an integral part of the larger risk decision process (National Research Council, 1996).

Adaptive management provides another, somewhat different paradigm for decision making. Its emphasis is on decision making as a continuing process, not a discrete endpoint (Heifetz, 1994). Within the realm of environmental decision making, adaptive management has been applied especially to the management of natural resources (Christensen et al., 1996; Stanford and Poole, 1996). The critical elements for adaptive ecosystem management include: (1) reviewing and synthesizing existing information; (2) defining the ecosystem based on available science; (3) identifying goals based on scientific synthesis and public values; (4) developing a peer-review management system; (5) implementing management actions that meet stated goals within the parameters of acceptable risks and consequences; and (6) conducting research (basic science and monitoring) to reduce uncertainties and to evaluate management actions. Because of its ongoing, iterative nature, the process enables adaptation to new information, to changing societal goals, and to long-term environmental change.

Each of these approaches to environmental decision making takes a different slant. Each shares, however, the fundamental characteristic of systematic inquiry, of understanding *components* of both the issue at hand and the process by which the issue will be addressed. The differences in these approaches are mainly matters of emphasis.

We believe that the taxonomy used in this book captures the essential components common to most forms of systematic environmental decision making and environmental planning. Yet the taxonomy used here is a modest one. It does not claim to provide the structure for a complete decision-making process. As indicated in the above discussion about variation in environmental decision-making situations, many different processes are likely to be needed and to be tailored to the situations at hand. This taxonomy is also flexible. The various categories of tools and the different tools within each category can be combined in many different ways. But the taxonomy does rest upon the notion that systematic inquiry can improve decision making, particularly on complex issues such as those that concern the environment.

Thus, these functional categories identify important *inputs* to environmental decision making. They are not the only inputs, and they should not be thought of as the exclusive province of a single decision maker or small set of decision makers. But the issues addressed ("What do we need and want?" "What do we know about present conditions, and what do we predict for the future?" "What are the likely outcomes of alternative actions?" "How did we do?") are or should be important with virtually any decision.

Topics Addressed

In Chapters 2 through 9, the categories of tools noted in Figure 1.1 are addressed in depth. Each chapter takes a somewhat different slant on environmental decision making, depending on the perspective of the author.

Chapter 2, "Identifying Environmental Values," focuses on environmental values while noting that other values are also likely to be important. Chapter 3, "Tools to Characterize the Environmental Setting," emphasizes natural-resource management, including the actions that public agencies take as resource managers and regulators. Chapter 4, "Tools for Understanding the Socioeconomic and Political Settings for Environmental Decision Making," stresses tools commonly (or not so commonly) used by researchers to get at social rather than purely economic complexities. Chapter 5, "Characterizing the Regulatory and Judicial Setting," focuses mainly on laws governing environmental pollution, rather than laws concerning how natural resources are managed by the public and private sectors. Chapter 6, "Integration of Information," emphasizes the integration of geographic information, but does not delve into other, nonspatially explicit means of combining large and diverse data sets. Chapter 7, "Forecasting for Environmental Decision Making," concentrates on methods for forecasting economic and other information relevant to environmental decision making, while leaving models for predicting the behavior of environmental systems to Chapter 3. Chapter 8, "Assessment, Refinement, and Narrowing of Options," discusses three "megatools" that have been used for options analysis on complex national problems, thereby providing a conceptual underpinning for (but not an extensive discussion of) simplified tools for analyzing options. Chapter 9, "Post-Decision Assessment," draws its illustrations mainly from public-sector programmatic assessment, while indicating that the underlying principles can apply to others as well.

Despite these differences of emphasis, each of the authors responds to a similar charge. Each identifies key assumptions or parameters of the tools in the category, describes characteristic tools within that category, and addresses a common set of themes:

- What questions does the tool address, and how does the tool frame these questions?
- How are answers reached? What type of knowledge is gained?
- What are the tool's strengths and limitations?
- Who uses the tool, and how do they employ its results?
- For participants in environmental decision making, what are the likely constraints on the use of the tool?

Decision-aiding tools are in a constant state of development, and the transition from tool development to tool use is being hastened by electronic means such as the Internet. Therefore, each author also was asked to consider the nonstatic nature of tools by addressing the following set of topics:

- How have tools in this category evolved?
- What types of new tools are now being developed?
- What new tools are on the horizon?
- What tools need to be developed, and what impediments slow their development?

Following each chapter, a practicing environmental decision maker provides a brief response, covering such topics as the use, misuse, and potential of tools in the category, as well as factors that may constrain the use of tools and the communication of their results and how these constraints might be alleviated. Ways to broaden the number of people using the tools are also considered, as are situations in which tools might be integrated. These practitioner comments are meant to ground the analysis of decision-aiding tools in the day-to-day reality of environmental decision making.

Reading These Chapters

As a potential user of one or more of the tools described, you will find that they vary widely in complexity, standardization, and computerization as well as in the specialized skills, data sets, and equipment that you need to use them. This variability is caused by two factors. First, the answer to the question, "What is a tool?" can vary greatly. A tool may be a theory, a means to elicit people's views, a piece of information that is particularly difficult to obtain, a method to determine the applicability of law, a monitoring system, or software for an expert system (to name only a few). Second, tool development has been influenced by the social, political, economic, institutional, and cultural realities of various environmental decision-making settings, as well as by the paradigms of various academic disciplines and the availability of comprehensive, reliable data.

Despite this variability, however, some "words of wisdom" can be identified that are common to all the tools discussed:

• Environmental decision making of the sort addressed in this book involves collective behavior, which should not be equated with individual behavior. A feasible and rational decision process for an individual may be neither feasible nor rational for a group or for groups interacting with each other.

• A balanced approach to decision making is needed; your resources should not be squandered on only one step of the process. This suggests the need for a single person or small group to coordinate and manage the total process.

- Some tools can stand alone; others necessitate companion tools. And some tools are compatible with each other while others are not. The tools to be used for the various components of a decision process should be selected with an awareness of the optimum suite of complementary tools.
- Tools often can be used for more than one function. For example, fieldwork techniques can be used for both identifying people's environmental values (Chapter 2) and characterizing the social, political, and economic setting (Chapter 4). Similarly, models can be used for characterizing the environmental setting (Chapter 3), integrating information (Chapter 6), and forecasting (Chapter 7). Although tools are described in this book on a function-by-function basis, their utility often is not limited to a particular function.
- Information-gathering and analytic tools can be simple or complex. Although our collective ability to measure physical and social phenomena and amass data has increased enormously with greater wealth and improved instruments, the "quick and dirty" approach may be preferable in some instances. In addition, as noted in Chapter 7, it may even provide more reliable results.
- In collecting and interpreting data, it is important to remember that, because changes can occur over space or time, the contextual settings of the data must be considered.
- Some mental tools should be used throughout the decision process. For example, the "gaps and blinders" techniques described in Chapter 4 are relevant to all aspects of environmental decision making.
- The time frame of the decision process, especially its urgency, is a key factor in selecting decision-aiding tools. Some of the tools described here might be ideal for highly deliberative decisions, but not feasible for decisions that need to be reached quickly; others can be tailored to the available time and resources.
- Although this book focuses on environmental decision making, many of the tools described here could apply equally well to environmental planning. In effect, planning is (or should be) deliberative, iterative, adaptive decision making.
- Environmental decision making (and environmental planning) should be a *goal*-driven process. Thus, the measurements made with tools are, like the tools themselves, just aids to the process; they should not dominate the process.
- New decision-making tools and techniques may demand new organizational structures as well as new skills. They may initially meet with resistance because they necessitate individual and organizational change.
- Post-decision assessment should not be an afterthought; it should be integral to the decision-making process. To carry it out, organizational and interorganizational continuity may need to be improved.

Looking Ahead

The final chapter focuses on tools needed to aid environmental decision making in the decade ahead. This chapter discusses the need for the development of new tools and the modification of existing tools, and general criteria for the information-gathering and analytic tools of tomorrow. In addition, because decision-aiding tools are in a state of change, information from this book is summarized and updated on the website of the National Center for Environmental Decision Making Research: http://www.ncedr.org.

Despite the ever-changing nature of decision-aiding tools, a comprehensive understanding of today's tools is essential to the improvement of environmental decision making. Taken together, the chapters of this book provide this comprehensive understanding. They offer analyses of current tools from the perspective of both researchers and practicing decision makers, and a look to the future in tool development and use.

The central argument of this book is that for people to be well-equipped to participate in the discussion and debate surrounding an environmental decision, they need access to, or at least a general understanding of, the types of tools described in the following chapters. They otherwise risk becoming marginalized as, in the coming years, increasingly complicated issues are tackled by increasingly sophisticated means. Knowledge of decision-aiding tools is not enough; other conditions, such as appropriate decision-making processes, are also crucial. Nor will the tools discussed here always lead to "good" environmental decisions; like other tools, they can be used for destructive as well as constructive purposes. But access to and a working understanding of these tools is becoming essential for meaningful, extended involvement in environmental decision making.

Key Resources

Cothern, C.R. (Ed.). 1996. *Handbook for Environmental Risk Decision Making: Values, Perceptions, and Ethics.* Boca Raton, FL: Lewis Publishers.

Katz, M. and Thorton, D. 1997. *Environmental Management Tools on the Internet: Accessing the World of Environmental Information.* Delray Beach, FL: St. Lucie Press.

Keeney, R.L. 1992. *Value-Focused Thinking: A Path to Creative Decisionmaking.* Cambridge, MA: Harvard University Press.

Kleindorfer, P.R., Kunreuther, H.C., and Shoemaker, P.J.H. 1993. *Decision Sciences: An Integrative Perspective.* New York: Cambridge University Press.

National Research Council 1996. *Understanding Risk: Informing Decisions in a Democratic Society.* Washington, D.C.: National Academy Press.

Office of Technology Assessment 1995. *Environmental Policy Tools: A User's Guide,* OTA-ENV-634. Washington, D.C.: U.S. Government Printing Office.

References

Bacharach, M. and Hurley, S. 1991. Issues and advances in the foundations of decision theory. In: M. Bacharach and S. Hurley (Eds.), *Foundations of Decision Theory: Issues and Advances*. Cambridge, MA: Basil Blackwell. Pp. 1–38.
Barry, B. and Hardin, R. (Eds.). 1982. *Rational Man and Irrational Society*. Beverly Hills, CA: Sage Publications.
Bartlett, R.A. 1995. *Troubled Waters: Champion International and the Pigeon River Controversy*. Knoxville: University of Tennessee Press.
Behn, R.D. and Vaupel, J.W. 1982. *Quick Analysis for Busy Decision Makers*. New York: Basic Books.
Boulding, K.E. 1966. The ethics of rational decision. *Management Science* 12(6):B161–169.
Bunn, D.W. 1984. *Applied Decision Analysis*. New York: McGraw-Hill.
Christensen, N.L., Bartuska, A.M., Brown, J.H., Carpenter, S.R., D'Antonio, C., Francis, R., Franklin, J.F., MacMahon, J.A., Noss, R.F., Parsons, D.J., Peterson, C.H., Turner, M.G., and Woodmansee, R.G. 1996. The report of the Ecological Society of America committee on the scientific basis for ecosystem management. *Ecological Applications* 6:665–691.
The Conservation Foundation. 1987. *State of the Environment: A View Toward the Nineties*. Washington, D.C.: The Conservation Foundation.
Diesing, P. 1962. *Reason in Society: Five Types of Decisions and Their Social Conditions*. Urbana: University of Illinois Press.
Dixit, A.K. and Nalebuff, B.J. 1991. *Thinking Strategically: The Competitive Edge in Business, Politics, and Everyday Life*. New York: W.W. Norton and Co.
Dunlap, R.E. and Mertig, A.G. (Eds.). 1992. *American Environmentalism: The U.S. Environmental Movement, 1970–1990*. Philadelphia: Taylor & Francis.
Feldman, D.L. 1997. Public participation in environmental decision-making involving the use of risk information: Results from three cases. International Association for Public Participation Annual Conference, Toronto, Sept. 6–10.
Feldman, D.L. and Nicholas, N.S. 1997. Community-based environmental decision making under changing regulatory regimes. Ecological Society of America Annual Conference, Albuquerque, N.M., Aug. 11–14.
Feldman, M.S. 1989. *Order Without Design: Information Production and Policy Making*. Stanford, CA: Stanford University Press.
Fraites, E.L. and Flanigan, F.H. 1993. Perspectives on the role of the citizen in Chesapeake Bay restoration, National Forum on Water Management Policy, American Water Resources Association, Washington, D.C., 1992 June 28-July 1.
Gelt, J. 1993. Long-awaited CAP delivers troubled waters to state. *Arroyo* 6(3):1–6; also at http://ag.arizona.edu/AZWATER/arroyo/063captr.html [accessed Nov. 5, 1997].
Hannigan, J.A. 1995. *Environmental Sociology: A Social Constructionist Perspective*. New York: Routledge.
Harsanyi, J.C. 1977. Advances in understanding rational behavior. In: R.E. Butts and J. Hintikka (Eds.), *Foundational Problems in the Special Sciences: Part Two of the Proceedings of the Fifth International Congress of Logic, Methodology and Philosophy of Science, London, Ontario, Canada, 1975*. Boston: D. Reidel Publishing. Pp. 315–343.
Heifetz, R.A. 1994. *Leadership Without Easy Answers*. Cambridge, MA: Harvard University Press.

Hofrichter, R. (Ed.). 1993. *Toxic Struggles: The Theory and Practice of Environmental Justice*. Philadelphia: New Society Publishers.

House, P.W. and Shull, R.D. 1988. *Rush to Policy*. New Brunswick, NJ: Transaction Books.

Keeney, R.L. 1992. *Value-Focused Thinking: A Path to Creative Decisionmaking*. Cambridge, MA: Harvard University Press.

Majone, G. 1989. *Evidence, Argument, and Persuasion in the Policy Process*. New Haven, CT: Yale University Press.

McKinney, M.L. and Schoch, R.M. 1996. *Environmental Science: Systems and Solutions*. St. Paul, MN: West Publishing.

Miller, G.T., Jr. 1992. *Living in the Environment: An Introduction to Environmental Science*. Belmont, CA: Wadsworth Publishing.

National Research Council. 1983. *Risk Assessment in the Federal Government: Managing the Process*. Washington, D.C.: National Academy Press.

National Research Council. 1992. *Water Transfers in the West: Efficiency, Equity, and the Environment*. Washington, D.C.: National Academy Press.

National Research Council. 1993. *Issues in Risk Assessment*. Washington, D.C.: National Academy Press.

National Research Council. 1996. *Understanding Risk: Informing Decisions in a Democratic Society*. Washington, D.C.: National Academy Press.

Office of Technology Assessment. 1995. *Environmental Policy Tools: A User's Guide*, OTA-ENV-634. Washington, D.C.: U.S. Government Printing Office.

Sen, A.K. 1977. Rational fools: A critique of the behavioral foundations of economic theory. *Philosophy and Public Affairs* 6(4):317–344.

Shklar, J.N. 1964. "Decisionism." In: C.J. Friedrich (Ed.) *Nomos VII: Rational Decision*. New York: Atherton Press. Chapter 1.

Simon, H.A. 1957. *Models of Man: Social and Rational; Mathematical Essays on Rational Human Behavior in a Social Society*. New York: John Wiley and Sons.

Stanford, J.A. and Poole, G.C. 1996. A protocol for ecosystem management. *Ecological Applications* 3:741–744.

Suter, G.W. 1993. *Ecological Risk Assessment*. Boca Raton, FL: Lewis Publishers.

Tennessee Valley Authority. 1995. *Executive Summary: Energy Vision 2020; Integrated Resource Plan/Environmental Impact Statement*. Knoxville, TN: Tennessee Valley Authority.

Tonn, B., English, M., and Travis, C. In preparation. *Environmental Decision Making: A Comprehensive View*, draft manuscript.

Tversky, A. and Kahneman, D. 1981. The framing of decisions and the psychology of choice. *Science* 211:453–458.

U.S. Department of Energy, Office of Environmental Management. 1998. *Accelerating Cleanup: Paths to Closure*. Washington, D.C.: U.S. DOE.

U.S. Environmental Protection Agency. 1992. Framework for ecological risk assessment, EPA/630/R-92/001. Cincinnati, OH: U.S. Environmental Protection Agency.

U.S. Environmental Protection Agency [web page]; http://www.epa.gov [accessed Nov. 6, 1997].

von Winterfeldt, D. and Edwards, W. 1986. *Decision Analysis and Behavioral Research*. New York: Cambridge University Press.

2
Identifying Environmental Values

ROBIN GREGORY

What people care about constitutes their values. Some of these values directly involve features of the natural environment: trees, views, animal habitats, or plant species. Other values involve related economic concerns (e.g., resource-sector jobs), social concerns (e.g., the stability of rural communities), or health and safety concerns (e.g., air pollution from emissions) that are influenced by aspects of the natural and built environment.

Actions that affect the natural environment become matters of concern whenever they create impacts that change something we care about. Some environmental changes, such as the scouring of a small streambed, are location-specific and affect only a few individuals. Others, such as climate change, are global and can affect millions of people. To the extent that we can improve our ability to identify and to define environmental values, we can do a better job in developing and implementing strategies that successfully address and satisfy these concerns.

In this chapter, I first review some of the reasons why the identification of environmental values is both important and challenging. I next look at four types of tools that are widely used to define environmental values, discussing their strengths and weaknesses as well as reviewing several illustrative applications. In the subsequent section, I examine contexts for using environmental values as part of policy decisions. Finally, I note some challenges to the existing set of tools used to identify environmental values and suggest several ideas for their improvement.

The Challenge of Identifying Environmental Values

A rich literature exists on values and the relation of values to attitudes, beliefs, opinions, and preferences (Rokeach, 1973; Crites, Fabrigar, and Petty, 1994). This chapter follows Brown (1984) in distinguishing between held values, which refer to an enduring belief about what is preferable or desirable, and assigned values, which refer to the values given to specific activities, products, or functions. In both domains, environmental values are concerned with the construction or expression of preferences by which we care about something or by which we consider one thing to be better than another.

Many conceptual approaches exist for organizing environmental values. Some of the literature connects the basis for environmental concerns to three general classes of valued objects: the self (egocentric), other people (homocentric), and nonhuman life (ecocentric) (Stern and Dietz, 1994). A complementary distinction exists between environmental values that reflect direct or indirect human uses of natural resources and mental values that do not require extractive or onsite activities, also known as non-use values. Values derived from direct human uses include consumptive activities (such as harvesting timber or hunting) and nonconsumptive activities (such as hiking or scientific study). Values reflecting indirect human uses include the scientific study of ecosystem functions (corresponding to "ecocentric" values) as well as the use of environmental stimuli as part of communication media (such as books or photographs). Non-use values derived from the natural environment include values associated with the knowledge that a natural area exists (generally a fundamental, "held" value), the desire for more information about it (e.g., to make informed choices), and values associated with the retention of future options (e.g., to visit a natural site next year).

Concern for environmental values has become far more significant to many individuals and to social policies during the past 25 years. This increase in significance is demonstrated by behavioral responses (such as the growth in outdoor-recreation activities); by the prominence of interest groups seeking to protect natural environments; and by the rise in federal legislation, such as the National Environmental Policy Act, the Endangered Species Act, and the Clean Air Act. Each of these legislative initiatives is based on an implicit or explicit set of environmental values, as are the thousands of routine, more minor regulations that cumulatively have a significant effect on the provision of environmental services. Each of these initiatives also recognizes, to a greater or lesser degree, that protection of the natural environment requires individuals and society to make decisions that acknowledge the tradeoffs between environmental and other types of values.

It is this aspect, stemming from concerns about value tradeoffs among the diverse environmental, social, economic, and health consequences of personal, corporate, or societal actions, that has proved most controversial and has led to difficulties in the acceptance and implementation of environmental policies. In the absence of tradeoffs, nearly everyone would favor a more healthy environment, although disagreements would remain about exactly what constitutes environmental health. Value tradeoffs do exist, however, and they lead to conflicts in assessing the consequences of an action. For example, an individual who wishes to preserve a nearby wilderness area for low-impact recreational camping must recognize that the range of commercial forest products available from the site will be different than if timber harvesting were allowed. A corporation seeking to meet tough new environmental regulations needs to realize that resources must be allocated to these efforts to ensure compliance. A government body wishing to preserve air quality in a fast-developing area must also recognize that this protection may result in lower employment and tax revenues, alter support among local voters, or influence regional trade accounts.

Assessment of these different environmental values requires the comparison of information derived from numerous sources. The process of selecting this information includes both technical questions—What information is currently available? What resources are at hand to gather new data?—and philosophical questions—How does one define the term value? Can alternative information sources be trusted? For example, economic markets provide useful information on the prices of many environmental commodities, such as timber, fish, and minerals. Although care must be taken to understand the role of subsidies, controls, and industry structures on the accuracy of market-based information, many people believe that these prices reveal something important about the relative values that society holds for different environmental goods. Others would disagree, citing philosophical differences with the willingness-to-pay paradigm of economics and pointing instead to ecologically or spiritually based value systems that may present a dramatically different picture of the relative importance of an environmental change or the associated impacts under consideration.

The focus of this chapter is on tools that can help identify the many environmental values that are not well reflected in market transactions. This emphasis on assessing nonmonetary impacts underscores a major shift in federal environmental policy guidelines during the past decade. These guidelines now require, as part of environmental assessments, explicit evaluations of human-health, ecological, and social impacts that may have economic implications, but are not typically sold in markets or measured in dollar terms. Such nonmarket impacts include improvements in visibility or aesthetics, protection of threatened plant or animal habitats, and health benefits associated with water-quality improvements. As noted above, even non-user benefits (which include the value of simply knowing that a species or wild area exists) can legitimately be included. Although this broadening of target values is both reasonable and widely supported, policy evaluations are now challenged by the difficulties of their identification and assessment.

One of the reasons for this challenge to policy creation and assessment stems from characteristics and limitations of the tools that are currently available for identifying and measuring these values. (This topic forms the subject of the next sections.) Another reason stems from a change in the context for environmental decision making, marked by a shift from analysis by experts to analysis by multiple stakeholders. This shift reflects a new emphasis on public perspectives and stakeholder consultation as well as a new emphasis on process, reflecting a focus on *how* decisions are made rather than on just *what* is decided (Simon, 1978).

This new context for environmental decision making has elevated the significance of negotiated settlements and increased the importance of procedural factors that highlight stakeholder concerns, such as *who* is involved in decision making, the *meaningfulness* and *openness* of that involvement, and the role of process considerations such as *trust* and *equity* (English et al., 1993). This enlarged context adds complexity but also insight for analysts because it helps to "unpack" complex decisions by clarifying important linkages and relationships among different stakeholders and their environmental values. These linkages range from the inclusion of economic externalities in policy decisions to an increased appreciation of how income, culture, and gender differences influence perceptions of environmental benefits, risks, and costs (Flynn, Slovic, and Mertz, 1994).

As noted in Chapter 1, a further challenge to the identification of environmental values stems from the nature of the preferences at issue. Many environmental actions evaluated as part of public-policy decisions are complex, unfamiliar, and richly multidimensional, involving a broad range of scientific, aesthetic, life-support, ecological, religious, recreational, and economic values. Research on human judgment and decision making clearly shows that, when asked to make judgments about complex matters, individuals often adopt simplifying cognitive strategies, such as searching their memories for similar situations or comparing alternatives based on a single, most important dimension (Tversky, Sattath, and Slovic, 1988). The use of such rules-of-thumb, or heuristics, in identifying complex environmental values has led to questions among many researchers about the validity of participants' responses and concerns regarding the use of their assessments in resource-management decisions.

In addition, experiments by behavioral decision researchers show that preferences for unfamiliar choices do not exist full blown in people's minds but are constructed during the decision-making process. The construction relies heavily on the available cues and the method of elicitation (Payne, Bettman, and Johnson, 1992). The phenomenon of preference reversals (Slovic, 1995) provides one of the best-known examples of constructed preferences: Although object A is preferred over object B under one method of measurement, B is clearly preferred under a different, but formally equivalent, measurement procedure. Other evidence for constructed preferences comes from empirical studies demonstrating the striking effects that can be produced by changing the frame of a valuation question from emphasizing gains to emphasizing losses. One oft-cited example is Tversky and Kahneman's demonstration of reversals of preference when the description of two public-health interventions is shifted from a "lives-saved" frame to a "lives-lost" frame (Tversky and Kahneman, 1981) (Sidebar 2.1).

Sidebar 2.1
Different Valuation of Gains and Losses

The conventional assumption of environmental-policy analysis is that valuations of gains and losses are, for most practical purposes, equivalent: A gain of $10 and a loss of $10 will leave an individual's welfare (or satisfaction) unchanged. Extensive experimental work in psychology (Tversky and Kahneman, 1981) and behavioral economics (Knetsch, 1995) has demonstrated that, in fact, people experience losses far more strongly than formally equivalent gains. This means that losses matter more to people than do gains, and that reductions in losses will be considered more valuable than will forgone gains.

This robust finding has important implications for the identification of environmental values and for their use in constructing acceptable environmental-policy initiatives. In particular, it suggests a strong link between the choice of a measure and the resulting valuation of an environmental action. For example, suppose that an initiative is contemplated that would change water-quality levels in a river from the current moderately degraded state (e.g., no swimming and fishing for carp and suckers only) to a higher level of quality (e.g., swimming allowed and fishing for species such as trout). If this change is viewed as a gain, it will be valued at some level X. If the same change is viewed instead as the restoration of a loss, for example from historical levels, then it will be valued at an even higher level (Gregory, Lichtenstein, and MacGregor, 1993). This change will result in a different benefit-cost ratio or, if the proposed change is put on the ballot, a different level of voter acceptance. Because the valuation difference typically is a factor of at least two or three, it means that the outcome of this framing choice (whether an environmental initiative is presented as a gain or as the restoration of a loss) can have an effect on the perceived value of the proposed action that is far larger than generally appreciated.

Another implication of the asymmetrical valuation of gains and losses is whether the economic value of a proposed change is measured in terms of people's willingness to pay for the new state or their willingness to accept compensation to move from the current to the new state. The issue is particularly important in the context of economic assessments of environmental damages. Consider the famous case of the Exxon Valdez oil spill. One highly publicized contingent-valuation study that followed the spill asked citizens their maximum willingness to pay to avoid another such disaster; however, because losses are valued more highly than gains, the resulting measure of loss was seriously understated compared to the assessment that would

> have followed from asking people their minimum willingness to accept another spill. In a similar way, a host of routine activities with negative environmental impacts will be unduly encouraged if a willingness-to-pay measure is used because their true adverse impacts will be understated. Mitigation initiatives, designed to prevent further losses, also will be measured incorrectly if viewed as gains and will therefore be valued relatively lower than if they were viewed as addressing a prior loss.

These challenges to understanding the nature of expressed values should serve to increase the modesty of those engaged in identifying and assessing the environmental effects of actions. Nevertheless, the interdisciplinary study of environmental decision making contains a variety of powerful tools for identifying stakeholder values and for clarifying significant environmental relationships.

Tools for Identifying Environmental Values

A multitude of approaches exists for identifying and comparing the values of environmental assets. Four principle value-identification approaches are noted here (and summarized in Table 2.1). The reader can find additional insights into their use and limitations in other chapters of this book.

Economic Markets

One approach to identifying what matters in the environment is to look for market-based clues, wherein values for environmental decision making are revealed through individuals' decisions about goods that they purchase or sell. As with other goods or assets, environmental resources have economic value to the extent that people are willing to make sacrifices of other things to acquire them or to prevent their loss. If a person is willing to pay $10, for example, to enjoy a day of fishing, then the experience is valued at this much or more because the individual is willing to give up the other things that this $10 could buy to acquire one day's angling. Similarly, if a person would take no less than $20 to accept the loss of access to a park, then he or she would be willing to give up what this $20 would buy; this willingness to accept compensation is a measure of the person's economic valuation of the park. Either way, it does not matter if money actually is exchanged so long as we are sure that the payments would be made or accepted if necessary.

In some instances, an action may lead to an environmental disruption that can be remedied by *replacement* or *restoration*, and implementation

TABLE 2.1. Tools for identifying environmental values.

Tool	Use	Strengths	Weaknesses
Economic measures			
Restoration/ replacement costs	Assigns economic cost to environmental damages	Estimates costs directly related to the damaged resource	Some resources irreplaceable; ignores loss of use before replacement; measures costs rather than values
Travel costs	Assigns economic value to resource based on visitation	Works well when distance to site is key for estimating benefits	Trips often have multiple objectives; confuses payments (expenditures) with value
Hedonic pricing	Assigns economic worth to component of resource values	Can expand market prices to nonmarket environmental amenities	Difficult to identify contributions of various nonmarket factors; reflects market prices rather than values
Damage schedules	Estimates the relative seriousness of adverse impacts	Facilitates quick response and saves transaction costs; reflects community concerns	Provides relative rather than absolute values; difficult to anticipate all types of possible losses
Ecological relationships			
Health	Relates ecosystem quality to the performance of key indicators	Provides useful summary measures to gauge impacts of changes over time	Hard to link cause and effect in ecological relationships; choice of indicators may be controversial
Integrity	Focuses on synergistic and system relationships	Recognizes system-wide characteristics of complex ecosystems	Definitions can vary greatly across experts; human vs. nonhuman factors problematic
Resilience	Assesses the long-term viability of a resource	Captures threats to future environmental quality based on past events and ecosystem response	Difficult to measure; translation into comparable policy terms can be controversial
Carrying capacity	Relates fundamental qualities of ecosystem value to productivity	Tracks key threats to future resource use and availability	Relation of productivity to value may be contested; choice of impact baseline difficult
Expressed-preference surveys			
Attitudinal and opinion surveys	Gathers information about ecological understanding and support for policies	Viewed as egalitarian and democratic; can be closely targeted to issues or population	Subject to strategic and motivational biases; may encourage superficial responses
Contingent valuation	Places an economic value on a resource not	Derives numbers that can be compared to	Value estimates subject to biases; measures gains

TABLE 2.1. Continued

Tool	Use	Strengths	Weaknesses
	Expressed-preference surveys		
	sold in conventional markets	other economic valuations	only; confuses economic and other motives
Constructed preference	Elicits values used in making decisions about environmental choices	Attempts to reflect actual decision processes and the key tradeoffs of stakeholders	Responses may be difficult to integrate into cost-benefit framework
Image	Assesses affective and psychological reactions to scenarios or events	Incorporates perceptions and beliefs associated with a proposed action	Stimulus-response characteristics tough to anticipate; high geographic variability in responses
Narrative and affect	Elicits concerns of stakeholders through dialogue and conversation	Can yield compelling stories; methods grounded in familiar feelings and emotions	Subject to bias via small-sample selection; coding of responses is problematic
Referenda	Asks individuals to vote for or against a specific proposed action	Provides familiar method for gauging opinions of diverse stakeholders	Knowledge level of participants can vary widely; responses sensitive to framing of questions
	Small-group elicitations		
Focus groups	Elicits responses to proposed action through informal small-group discussions	Inexpensive; directly targets question of concern; uses insights from diverse populations	Sessions can be dominated by one point of view; values remain implicit, and conflicts are difficult to address
Advisory committees	Develops a broad perspective on an issue; involves interested and knowledgeable representatives	Allows for open discussion; can increase trust in agency and empower local citizens	Objectives and powers of committee may be unclear; diversity of viewpoints easily suppressed
Multi-attribute elicitations	Structures the objectives and tradeoffs of participants vis-à-vis policy alternatives	Structures problem and improves understanding of stakeholders' values; distinguishes ends and intermediate goals	May appear overly quantitative; difficult for participants opposed to problem decomposition

costs may then serve as a useful indicator of economic value (Kopp and Smith, 1993). Another frequently used tool for identifying and estimating economic values for environmental assets is the *travel-cost method*. Using the relationship between the number of people who visit a site and the travel costs they incur, this technique derives an estimate of how much visitors would pay over and above their cost of travel to gain access to the site. *Hedonic* (or characteristics-based) methods of economic valuation build on the recognition that many unpriced environmental values are captured in the prices of marketed goods, such as the price of land or houses. The economic value of an unpriced amenity thus can be measured by the difference in prices between assets that are otherwise similar (e.g., two houses that are identical except that one has a view and the other does not [Freeman, 1993]).

An alternative economic approach to identifying environmental losses employs the concept of a damage schedule that provides scaled rankings of the relative importance of various environmental harms. The rankings reflect relative damages (of which people typically are more certain) rather than absolute values (of which people are far more uncertain). All environmental impacts listed on the damage schedule are thereby acknowledged as legitimate, thus forming the basis for regulatory and other controls or for the setting of damage awards in much the same way that schedules now are used to settle worker's compensation claims and establish workplace safety regulations (Gregory, Brown, and Knetsch, 1996).

The strength of these economic methods is their relation to the familiar market system and their widespread acceptance by both policymakers and citizens. The principles of welfare economics that underlie comparisons of the monetary costs and benefits of an environmental change are used throughout the world and establish a common currency for dialogue and decision making. The weakness of economic methods is underscored by the equally widespread dissatisfaction with their ability to capture what really matters to people. This list of omissions includes many important ecological, cultural, and spiritual aspects of the natural environment that lie outside the scope of most economic analyses because they are not typically bought, sold, or exchanged. Although some propose that the economic paradigm of revealed preferences should be expanded to include these other sources of environmental value, many are looking instead to alternative tools that focus on ecological relationships or more direct, multidimensional expressions of individual and group preference.

Ecological Relationships

A second approach to identifying values for environmental decision making is to employ ecological techniques that focus on the expression and modeling of ecosystem functions. Although a detailed discussion of ecological indicators is outside the domain of this chapter (and the expertise of its author), a diverse set of tools has been developed by ecological scientists to understand the constituents of natural ecosystems and the forces that operate on them over time. The relevant environmental functions may be expressed in summary terms (including concepts such as ecosystem health, integrity, resilience, or carrying capacity) or in more detailed terms, such as the identification of nutrient dynamics, critical thresholds and synergisms, the need for migratory pathways, or major external threats to stability

(Mangel et al., 1996). In some cases, developing integrated assessment frameworks is useful. Such frameworks can combine various levels of ecosystem indicators, resulting in the designation of certain areas as critical habitat or in the construction of multivariable indices of effects at the level of organisms, populations, or ecosystems (e.g., energy flow, material cycling, or biodiversity indexes) (Suter, 1993).

A key to the identification and understanding of many ecologically based environmental values is the role of uncertainty and variability in natural systems (Costanza and Cornwell, 1992). Probabilistic analyses can be used to provide insights about either event-based uncertainty, which refers to a lack of data about key ecological relationships, or knowledge-based uncertainty, which refers to a lack of knowledge (or agreement) among experts regarding the ecological impacts of a proposed action. Recognition of ecological uncertainty can lead to values associated with precautionary (conservative) strategies (O'Riordan and Jordan, 1995) or it can enhance the usefulness of additional information used to help determine values for, and priorities among, possible impacts on the environment.

This chapter includes ecological functions alongside other component environmental values because stakeholders involved in policy decisions frequently are willing to trade economic, social, or health objectives to retain ecosystem characteristics, such as resilience or carrying capacity. Education efforts by ecologists and economists have done a great deal to demonstrate the worth of ecosystem functions. For example, marshlands are now widely recognized as being valuable, and tradeoffs are made routinely between economic development and ecological factors. The use of ecological relationships as tools for identifying environmental values provides a clear link between many ecological functions and human welfare. But this link also poses a dilemma, because not all ecological relationships are so readily translated (e.g., some estuaries may yield clear economic benefits, but what are the economic benefits of biodiversity?) or so well understood (e.g., does the species in question play an essential role in the environment?). In addition, specific components of the natural environment may function as ecological indicators, which means that their well-being provides a signal for the quality of a larger habitat or group of species. Valuation of such indicator species is therefore problematic, an issue that sits squarely in the midst of proposed changes to the Endangered Species Act.

The translation of ecological relationships provides another dilemma: Many ecologists are uncomfortable with the adoption of human-centered perspectives as the basis for environmental valuation. At a minimum, the relation of ecological (ecocentric) to human-based (homocentric) values is complex. Some argue that all values are human based; after all, anything we care about is in reference to our roles as humans, even if these values include recognizing the equal rights of all animals or plants to coexist. Others argue that ecological functions have a status beyond or outside human-based values, extending to an "inherent worth" as depicted by a deep ecology or spiritual perspective. The associated promise and challenge of this debate can be heard in many of today's significant environmental controversies, which place economists, ecologists, fisheries biologists, and engineers in heated discussions with ethicists, environmental philosophers, First Nation elders, and community activists.

Ecological risk assessment also offers a framework in which the question of ecological values must be addressed (Environmental Protection Agency, 1992). The risk-assessment paradigm structures a problem so that the values must be considered up front in the decision-making process. Because this paradigm has feedback loops, however, the values can be revisited. The same need for continuing reevaluation of ecological values is part of an adaptive-management strategy (Christensen et al., 1996; Walters, 1986), which encourages ecological experimentation as a basis for learning. This approach explicitly recognizes that management practices and goals may need to change over time in response to changes in information or environmental conditions.

Expressed-Preference Surveys

A third approach to identifying environmental values is to ask people about their preferences and to use the answers as an indicator of their values. As discussed in Chapter 3, such expressed-preference surveys can rely on a wide range of approaches and measures of value.

One important group of expressed-preference tools is opinion and attitudinal surveys, which are being increasingly used as the basis for information about environmental values and opinions. Opinion surveys are seen as a relatively inexpensive and user-friendly mechanism for developing a broad-based understanding of public views about environmental policies, and for testing the acceptability of specific proposed policy actions. For example, Dunlap (1991) used an opinion questionnaire to compare the relative importance of environmental values in industrialized nations to values in developing nations. Axelrod (1994) used an attitudinal survey to examine how subjects balance specific economic, social, and personal needs with their desire for increased levels of environmental preservation.

Opinion surveys can be designed to provide information from a broad cross section of the public so that results are not limited to advocate, opponent, or special-interest perspectives. They can also be structured to compare the values of different segments of the population. In addition, attitudinal surveys can ask questions about nearly anything, eliciting general expressions of interest or support as well as expressions of understanding about specific environmental issues or conflicts (Kempton, Boster, and Hartley, 1995). These strengths of opinion surveys are countered by several well-known weaknesses: Responses may be hypothetical and not reliable, respondents are susceptible to manipulative contexts or biased questions, and tradeoffs may be unaddressed (leading, in many cases, to inconsistent answers). In addition, response rates may be low. Although a carefully designed opinion survey can go a long way toward addressing many of these methodological concerns, other questions (for example, the level of information and understanding required for defensible policy-relevant responses) lack clear answers and remain topics of heated debate among practitioners.

Because dollar payments are often used as an indicator of value, another type of survey approach expresses environmental values in dollar terms. If such monetary expression is successful, integration of these values with other economic impacts is straightforward (e.g., as part of a cost-benefit analysis of project or program options).

The most versatile of the economic-survey methods are contingent-valuation (CV) techniques (Mitchell and Carson, 1989). These tools posit a hypothetical market for an unpriced environmental item and ask people to state the maximum price they would be willing to pay to obtain more of the item (if good) or to reduce or avoid the item (if bad). These surveys use samples of as many as several thousand people, and the results are taken as indicators of the value placed by society on environmental goods. Contingent-valuation methods now have been employed for a wide range of environmental policy issues; a recent bibliography lists more than 1,600 studies. In addition, CV methods have been granted substantial authority by the popular press (following the State of Alaska's use of CV techniques to estimate damages caused by the 1989 Exxon Valdez oil spill) as well as by academics, legislatures, and the courts (Smith, 1996).

Nevertheless, the use of CV methods has attracted many critics from within and outside the evaluation community (Hausman, 1993). Most notably, their accuracy has been called into question by evidence demonstrating that minor variations in the information provided to participants or the way in which valuation questions are asked (e.g., their context, wording, and order) can have large effects on the magnitude of respondents' answers (e.g., Kahneman and Knetsch, 1992). Practitioners of CV have responded to these concerns by making numerous design changes in elicitation procedures. They have used citizen groups to understand the range of environmental values at issue and have employed multiple information or payment strategies in an attempt to overcome cognitive and emotional responses to particular aspects of the survey. Most CV professionals, however, view these new design options as providing refinements to current approaches and as fine-tuning to address problems of survey design and response bias. They assume that people's true values for environmental assets are being distorted by imperfect, but steadily improving, monetary measurement methods.

Research in behavioral decision making has led to a different approach to survey development, one based on the assumption that true values for environmental assets do not exist beforehand but instead are constructed as part of the selected survey (or other) elicitation process (Fischhoff, 1991; Gregory, Lichtenstein, and Slovic, 1993). This alternative approach to the expression of values typically provides extensive supplementary help to participants in thinking about their concerns and priorities. The respondents, after all, are being asked about complex resource-management tradeoffs about which they may not have thought deeply in the course of their everyday lives. Such constructive approaches typically use multiple scales for identifying and comparing values in the belief that many of the effects of environmental policy actions are not cognitively represented in monetary terms. The rationale is that a reliance on monetary responses will, at the least, place an additional burden on respondents and imply that ecological, cultural, health, or other aspects of a decision are considered less important. At the extreme, people may consider the monetization of a valued cultural or ecological impact to be impossible or immoral and refuse to participate.

Several new survey and questionnaire approaches to the elicitation of complex environmental preferences are now being proposed. One technique uses conjoint analysis to build up an understanding of an individual's environmental values by asking a participant to make a series of structured, pair-wise comparisons (Opaluch et al., 1993). Another approach is to explore the positive and negative images that are associated with various policy options and their possible consequences (Slovic et al., 1991). Images are often particularly useful indicators of value in the context of environmental risks because they easily extend to the fears, hopes, and perceptions that individuals associate with the anticipated possible effects of an action. As a result, image-based techniques are useful when values are poorly formed or reflect strong affect, such as dread or worry about a proposed action or the use of a particular management technique.

Other survey techniques use the insights of multi-attribute utility theory (Keeney and Raiffa, 1976) to elicit values for the individual characteristics of an environmental resource and to combine these attributes as part of the creation, ranking, or rating of a set of proposed resource-management alternatives. Known by various names, such as public-value surveys and value-integration surveys, constructive multi-attribute methods address the cognitive complexity of environmental values by decomposing the decision problem and then recombining these component parts. This ability to organize complex information around both values and facts is an appealing strength of multi-attribute techniques, and the approach rests on a strong axiomatic foundation. However, the reliance on multiple measures of value leads to a quite different decision-making process than the single-measure approach favored under cost-benefit analysis. At this time, the adaptation of multi-attribute approaches to environmental surveys remains experimental, but their potential is high for creating new insights about environmental choices and, in particular, the value tradeoffs made by individuals or groups.

A related experimental constructive technique, called a decision-pathway survey, attempts to draw out participants' reasoning by providing a set of linked questions that encourage participants to self-select a response pathway that reflects their thinking about an environmental policy option (Gregory et al., 1997). The questions emphasize reasons why alternative policy options might matter and encourage respondents to address potential value conflicts, thereby defining more clearly the relative benefits, costs, or risks associated with selected policies. For example, the province of Ontario used the results of a decision-pathway survey of the environmental values of the general public, forest professionals, and residents of timber-dependent communities to develop new policies for managing the growth of unwanted forest vegetation (Sidebar 2.2) (Ontario Forest Research Institute, 1995).

Sidebar 2.2
Ontario Vegetation Management

Early in 1994, Decision Research in Eugene, Oregon, was asked by the Ministry of Natural Resources in Ontario, Canada, to design a survey that would inform provincial resource managers about public attitudes and opinions concerning alternative forest-vegetation management policies. Issues relating to vegetation management are important in Ontario because the province's extensive forests provide a major source of income and employment. They are also controversial because public attitudes toward environmental management have shifted since the 1950s from a pro-industry perspective to one that acknowledges, and at times favors, ecological concerns. Vegetation-management issues, such as the spraying of herbicides from airplanes, the use of tractors or other machinery to control unwanted growth, and the introduction of genetically altered plants, are therefore important issues for provincial decision makers.

The Decision Research team (Jim Flynn, myself, Steve Johnson, C.K. Mertz, Terre Satterfield, and Paul Slovic) approached the problem by designing a survey within a survey. The larger survey asked questions about the perceived risks of vegetation-management actions and obtained opinions concerning the benefits, costs, and risks of specific vegetation-management options under consideration within the province (Ontario Ministry of Natural Resources, 1995). The team also included questions to determine where respondents obtained information about forest policies and to ascertain their degree of trust and confidence in these sources. A second, experimental survey used "decision-pathway" questions to probe respondents' reasoning behind their support for, or opposition to, specific management options (Gregory et al., 1997). The surveys were administered in the fall of 1994 and included stratified random samples drawn for three populations: a general-population sample of 1,500 Ontario residents 18 years of age and older; residents of timber-dependent communities; and professional foresters. Questions were asked over

> the telephone with computer-assisted telephone-interview techniques that permit the interviewer to record a sequence of questions and to select questions based on a participant's previous answer(s).
>
> The results showed strong support for environmental values across both the general public and members of timber-dependent communities. Private-industry foresters tended to be less supportive of environmental values than any other group. Not surprisingly, timber-dependent respondents were more supportive of job creation in the woods industry, and forestry professionals were less supportive of recreation as a management goal. All samples supported active vegetation management, although public support was much lower for herbicide-based control approaches (particularly aerial spraying, as compared to ground-based applications) than for other biological or manual options. Forestry professionals were more comfortable with the use of herbicides as a vegetation-management tool, in part because professional foresters believe that they have more control over risks to their health and are less likely to believe in the goal of a risk-free environment.
>
> Participants in the decision-pathway experiment tended to choose one of five paths (from a total offering of 13). Important information provided to policymakers included the large differences in the pathway choices selected by the different samples and the implications of these choices in terms of the underlying reasoning and decision-making processes of survey participants.

Other expressed-preference tools rely on the narrative of individuals or groups to understand the relationships among environmental values, ethics, and emotions. Callicott (1984) and other environmental ethicists have argued that environmental values are grounded in human feelings and emotions, suggesting that expressions of support or opposition may reveal important values that often are omitted from more quantitative modes of analysis. This approach to value identification typically uses in-depth individual interviews to explore participants' emotional responses to a real or an imagined scenario and to examine their explanations and justifications for the feelings they express or the emotional responses they make. For example, Satterfield (1996) used affective expressions to document the environmental values of loggers, environmentalists, and community residents affected by changes in harvest practices on old-growth forests in the Pacific Northwest.

A final, widely used expressed-preference approach to identifying environmental values involves the use of referenda to elicit from voters direct expressions of support or opposition to proposed policies. Voting-based structures often are used as part of contingent-valuation surveys, with respondents being asked if they would vote yes or no for a policy option that has been described in terms of its anticipated effects and cost. Referenda also have been used as part of structured, decision-analysis evaluations of environmental policy options. In these referenda, the value-identification task begins with a clear decision structure, moves to explicit tradeoffs among objectives important to voters, and culminates in a choice among alternative policy options. For example, McDaniels (1996) used a structured value referendum to assist a regional government in British Columbia to select among wastewater treatment alternatives for managing potential environmental risks from the disposal of municipal sewage. Although voting has the advantage of being a natural and familiar method for public decision making, the questions asked of voters are notorious for being partial or slanted to favor certain interests and perspectives. A voting procedure also has to distinguish clearly whether its purpose is to make a decision (i.e., the winning option will be put in place) or to inform decision makers (i.e., the results of the vote will be used as an input to a larger decision-making process). Unless this distinction is clear at the outset, both sides are likely to end the day disgruntled and upset.

Small-Group Elicitations

A final tool for identifying environmental values is to work interactively with small groups, eliciting expressions of value from participants and using a variety of approaches for further defining and understanding these interests. Perhaps the best known of small-group approaches is the focus group, which typically involves between 6 and 10 participants in a structured conversation (a group interview) about a proposed project or policy option. Focus groups typically are facilitated by a moderator working from a prepared script, with the desired outcome being a better understanding of the participants' reasoning about a proposed environmental action based on statements of their key concerns, expectations, and worries (Morgan, 1988). Focus-group results also can include rankings of the relative importance of the various environmental objectives raised by participants, which can then be used as a guide to understanding how to make the translation from concerns to values.

A second, increasingly popular approach to identifying environmental values in small groups is to work with citizen advisory committees to highlight key concerns and to suggest effective strategies for dealing with controversial aspects of environmental initiatives. Such committees typically involve 10 to 15 citizens, drawn from diverse backgrounds and neighborhoods, as part of a process lasting several months and combining presentations by experts with extensive group discussions. For example, many representative citizen committees have been formed to advise electric utilities on preferred approaches for reducing worry and risks about electromagnetic fields from transmission or distribution lines.

Both focus groups and citizen advisory committees provide a comfortable forum for identifying and talking about the concerns of participants, and they often result in useful insights to guide the design and dissemination of environmental products, programs, or technologies. Yet this comfort comes at a price: information derived from focus groups and advisory committees lacks statistical rigor and often is poorly defined, which leaves it open to reinterpretation (and, possibly, misinterpretation) at the hands of analysts, interest groups, or decision makers.

Other approaches to small-group elicitations therefore use tools that attempt to develop more structured information on environmental values. One important technique involves the use of multi-attribute utility theory and the methods of decision analysis to structure the objectives and tradeoffs of participants as a means of selecting a favored policy alternative (Keeney, von Winterfeldt, and Eppel, 1990). The tools of decision analysis are used to probe stakeholders' environmental values, to distinguish between means objectives and ends objectives, and to identify measures to describe what fundamentally matters in the decision being faced. Decision-analysis methods have been used to compare options for endangered grizzly bear populations (Maguire and Servheen, 1992), to search for novel alternatives that satisfy both economic-development and preservation values (Gregory and Keeney, 1994), and to lend insight to a wide variety of other environmental policy problems. In addition, decision-analysis techniques have been used in small-group settings to help understand why stakeholders may differ in their assessments of the likelihood of anticipated environmental impacts. For example, Morgan and Keith (1995) used decision-analysis techniques to obtain probabilistic judgments from experts about the relative contributions of various factors to uncertainty in climate-change estimates.

Supporters argue that small-group multi-attribute techniques provide a preferred prescriptive approach to clarifying environmental values that: (1) allows access to the relevant information (to remind respondents of values they might otherwise overlook); (2) asks for responses to parts of the problem (to avoid cognitive overload); (3) uses natural metrics (instead of dollars, except for naturally monetary aspects); and (4) helps respondents to combine the parts into a single whole (to facilitate an overall assessment of expressed value). Critics argue that the formal requirements for eliciting and combining measures may appear unduly quantitative or may be difficult for participants who are philosophically opposed to decomposition strategies or to what is perceived as a focus on outcome comparisons.

The Policy Context for Identifying Environmental Values

Selecting the correct tool for identifying environmental values is only the first step in assessing environmental policies. To make the value-identification process useful to further deliberations, the policy context facing decision makers must be linked with the selected value-identification tools. Three types of policy contexts are considered in this section. Once a context for a value-identification exercise is recognized explicitly, it will have the effect of highlighting or deemphasizing aspects of the selected tool.

Creating Better Project Alternatives

A key role for information about environmental values is to aid in the development of preferred project (or program) alternatives. The preferred option is the one that does the best job of satisfying the underlying values that will be affected. This common-sense statement hides a wealth of problems that may arise in deciding about:

- The preferred stakeholders (*whose* values)
- The legitimacy of their concerns (*what* values)
- The relevant scope of impacts (over *how long* a time period or *how large* a geographic area).

Until recently, information about environmental values typically was used to aid in the selection of a preferred alternative from among a small set of possible projects. For example, the environmental impact statement (EIS) process created in the 1970s required the presentation of project alternatives, but only rarely was information about stakeholders' environmental values used to broaden the project set beyond two or three options (frequently composed of a middle, clearly preferred alternative and two other dominated options). One of the reasons given for this presentation was the difficulty associated with providing the necessary data for consideration of a broader set of alternatives. Several of the newer value-identification tools, including constructive surveys and small-group multi-attribute elicitations, can help organize the massive amounts of data collected as part of an EIS and thereby facilitate the search for better project or program options.

An increased appreciation of the negotiation-based context for environmental decision making, including enhanced interest in both creativity-based tools (such as brainstorming) and formal techniques of negotiation analysis (Sebenius, 1992), has led to the recognition that the differences in values among groups often lead to better alternatives, based on trades across objectives. Role-playing exercises and case-study examples are now widely used to explore differences in participants' values and to enhance their understanding of the dynamics of environmental-dispute and -resolution processes. Numerous examples demonstrate the gains that are possible from cultivating shared interests and from converting environmental-value differences into mutually beneficial exchanges (Sidebar 2.3).

A related role of value-identification strategies is to help defuse the "we versus them" framing that is often encouraged by both the media and the

Sidebar 2.3
The Alouette River Stakeholder Committee

In the spring of 1996, Tim McDaniels (University of British Columbia) and I were asked by B.C.Hydro to lead a stakeholder group in a consultative process designed to develop and implement a revised operating plan for the Alouette River in southwestern British Columbia. The process was contentious because, for many years, B.C.Hydro had operated a dam on the river that provided electric power to the city of Vancouver and surrounding areas, but in so doing, sharply decreased the flow of water to the south Alouette River. As a result, fish populations (principally, migratory salmon) had declined sharply, and recreational opportunities on the river (e.g., boating and swimming) were curtailed.

A prior decision by the provincial government required that the new operating plan incorporate stakeholder interests, as well as the results of ongoing technical studies. Following extensive interviews with interested parties, the final Alouette River Stakeholder Committee (ARSC) included representatives from the local community, riparian homeowners, B.C.Hydro, the provincial and federal governments (e.g., representing fishery resources, regional development, and revenue), and local First Nations. The group was called upon to explore alternative operating conditions, to evaluate their implications and underlying tradeoffs, and (if possible) to provide consensus suggestions to B.C.Hydro about the operation of its Alouette River facilities. Despite this broad mandate, the official role of ARSC was purely advisory; final determination of the new operating regime was to be made by the provincial Water Comptroller.

Meetings of ARSC were held two or three times per month for six months. The basic steps in the consultative process followed a constructive, multi-attribute approach. We began by identifying values that could be affected by a new operating plan. After discussion, five objectives were agreed upon, thereby establishing the focus for subsequent discussions and negotiations. They were:

- Promote the health and biological productivity of the river and reservoir
- Avoid adverse effects from flooding
- Promote recreational activities
- Avoid cost increases to provincial residents
- Promote flexibility, learning, and adaptive management for the Alouette system.

The middle months of the process were spent in identifying factual impacts associated with alternative plans that addressed each of these objectives. For example, much time was spent discussing the issue of flushing flows, which are higher-than-usual releases from the dam for short periods of time (e.g., two to four days) designed to clean out the river system and mimic the ecological benefits of natural flooding. However, costs are incurred because of lost power generation (because extra water is released from the reservoir) and, if flushing flows are too high, damages could be experienced by homeowners or by recreationists. Although the obvious focus was on demands for accurate quantitative analysis, the subtext of these discussions included issues of trust. For example, would numbers developed by B.C.Hydro be believed by their long-time adversaries in the community? Would promises really be kept in the implementation stage? Another topic addressed was cultural diversity: How would the aboriginal rights of First Nations be considered alongside the needs of downstream communities? Input to the discussions of environmental impacts came from many of the tools discussed in this chapter, including:

- Economic studies of restoration/replacement costs and travel-cost studies of park use
- Ecological studies of the effects of alternative water releases on the health and carrying capacity of the river system, including sampling and modeling efforts
- Prior survey results showing residents' beliefs and desired end-states for the area
- Small-group elicitations, including breakout groups that worked independently to settle particularly contentious issues and then bring a recommendation back to ARSC.

Because of the extensive discussions and opportunity for interactive dialogue, both among Committee members and with outside experts, consensus was reached on all substantive content and process issues. The recommendations of the committee were accepted in their entirety by B.C.Hydro. A key factor influencing this acceptance was agreement on an ongoing Management Committee, comprised of key stakeholders and established for the life of the water license. This Management Committee holds strong powers to implement ARSC recommendations in light of changing environmental and economic knowledge and conditions, which addresses the "learning and flexibility" objective cited above.

litigation process. This framing implies that negotiations among key stakeholders are likely to fail because the groups hold different, and fundamentally incompatible, objectives. Yet in many cases, in-depth value elicitations can facilitate agreements by showing that the values held by different stakeholders are actually quite similar and that disagreements about the choice of a preferred plan are caused by differences in the priorities (or weights) assigned to these values or by factual differences in beliefs about how specific alternatives measure up in terms of the objectives. Fact-based differences, in particular, generally are far easier for stakeholders to address (e.g., by testing alternative models or bringing in additional experts) than are fundamental differences in what is wanted.

Integrating Across Environmental Disciplines

One reason why many environmental problems are challenging is because input must be obtained from many different disciplines. Valueidentification tools can usefully define the issues of concern to the different disciplinary interests and help bring these concerns to the bargaining table in a common language and format. One aspect of this process is to reach across semantic or disciplinary differences to develop a shared terminology. Another involves identifying areas of misinformation or topics where additional data would be helpful.

One example of this integration is the use of tools from the decision sciences (discussed in the section of this chapter on "Small-Group Elicitations") to characterize ecological indicators of environmental value. Tools such as influence diagrams can be used to visually depict the relationships among constituent elements of the environment and thereby demonstrate the importance of cause-and-effect relationships or nutrient and food-chain pathways. A variety of scales can be developed to express ecological values, including straightforward measures (such as average wind speed or the number of species) or more complex, constructed indexes that are based on several attribute considerations (e.g., a biodiversity measure that includes the number of species, the health of populations, and their spatial extent). Weighting procedures can be used to assist in clarifying the relative significance of predicted impacts on different environmental species, areas, or processes. In addition, the tools of probabilistic risk analysis and assessment are useful for clarifying environmental exposure and effects pathways, and for understanding their influence on ecological concerns (Morgan et al., 1984).

Communicating Environmental Values and Choices

Information about environmental values is often used as part of the communication strategies of government agencies, corporations, members of the public, and stakeholder or interest groups. In recent years, researchers interested in understanding how people think about and respond to environmental risks have begun using focus groups to obtain insights into what people already know, the additional information they would like to receive, and how this information will be processed in terms of underlying values. In this context, focus groups have been used to examine the basis for people's risk perceptions, to pretest risk-communication materials, and to design more effective risk-mitigation policies that can be tailored to the special interests and needs of particular stakeholders (Desvousges and Smith, 1988).

The communication of environmental values has become a topic of great interest to the media and, as a result, the level of coverage given to environmental issues is now quite high. This increased attention has obvious benefits: it enhances the ability of the general public to learn about environmentally significant actions and it encourages dialogue about environmental values and issues (Wilkins and Patterson, 1991). At the same time, the characteristics that make one event attractive to the media (e.g., its salient or unusual nature, a potentially catastrophic outcome, or the possibility that an identified party may be "blamed") can obscure other, equally significant environmental events. Those same characteristics can result in a mismatch in environmental understanding between the public and the technical experts, because they emphasize events that are more sensational—and thus more "newsworthy"—rather than those that are significant from an ecological point of view. Many of these same factors lie behind the amplification of certain risk-related events or processes, so that their significance (e.g., in terms of the effects on a company's sales and revenues) is increased well beyond what might have been predicted (Kasperson et al., 1988). Both private and public-sector organizations are now recognizing this potential for the stigmatization of environmentally sensitive products or technologies and are attempting to use value-identification tools as the basis for developing communication strategies that can be fine-tuned to listen, and speak, to each of several audiences.

The identification of values also expands the bounds of what can, and should, be communicated about an environmental event or process. Public reactions to many recent controversial environmental decisions have hinged on questions of trust, equity, fairness, history, or cultural effects that lie outside the domain of what agencies or corporations typically have communicated regarding project effects. However, these factors matter and are a priority for many stakeholders. Properly designed value-identification processes can highlight the role played by these concerns and can also aid in deciphering the more fundamental issues that may lie beneath those initially expressed. In some cases, developing mitigation or compensation measures to address these process and distributional concerns may be possible. In other cases, the only resolution may be an explicit recognition of differences and an acknowledgment of the limits of assessment and analysis to address certain issues of fundamental spiritual, ecological, or cultural significance.

Improving Tools for Identifying Environmental Values

Significant advances have been made during the past decade in all the tools for identifying environmental values discussed in this chapter. This rapid progress has fueled a perception on the part of many decision makers that the introduction of environmental values to decision processes is now straightforward. However, much additional work remains to be done, both on the value-identification techniques themselves and on the interpretation and use of their results.

Encouragement for continuing this improvement is found in the progress made by researchers and in the frustration experienced by practitioners, including:

- Decision makers who are frustrated by the existence of controversy and disagreement among ecological, economic, or safety experts
- Stakeholder participants in multiparty environmental decisions who are frustrated by the strong role played by implicit values and political forces
- Technical experts who are annoyed by the gap between public and science-based environmental concerns, a gap that often remains even after lengthy discussions of impacts or extended deliberations about values.

One of the major challenges to the use of tools for identifying environmental values is understanding which techniques will work best for which problems and for which stakeholders. In most cases, multiple problems require the use of multiple tools. Adding to this complexity is the frequent presence of multiple decision makers, each of whom will bring their own style, training, and perceptions to the table. Questions about how tools should be applied and sequenced, as well as how they can be disseminated in a politically attractive manner while remaining rigorous and theoretically justified, raise a host of issues to challenge those involved in today's environmental decisions.

However, the fact that these frustrations and issues are now out on the table (and in the literature, media, and courts) is a hopeful sign of the maturation of environmental-values-identification processes. As I write, many high-visibility pieces of values-based legislation (including regulations and proposed laws concerning global climate change, species diversity, and other key environmental issues) are currently being debated before the legislatures, the courts, and the public. This dialogue provides an unusual opportunity for implementing policies that take account of people's underlying environmental values and for using these policies as a means for creating better individual, corporate, and public actions.

Capturing these opportunities will require movement across disciplinary boundaries to integrate more successfully the environmental values of technical experts (economists, sociologists, engineers, and ecologists) and public or interest-group participants than has been done in the past. It will require extensions of value-structuring approaches to create more meaningful definitions of terms such as "ecosystem management" or "public consultation" in order to facilitate broadly acceptable agreements on environmental actions. At the same time, progress will require an acknowledgment of the limits of value-identification strategies and new insights about decision-making approaches that embrace the ethical, poetic, and spiritual implications of policies as fully as they embrace the analytic or scientific dimensions.

This task will require that we address fundamental questions related to how environmental values and preferences are formed and expressed. It also asks us to reexamine the place of nature in our lives and to revisit our understanding of basic activities and concepts, such as what it means to improve (or harm) the natural environment. This is a tall order, but it holds the potential for allowing environmental value-identification strategies to play a leading role in redefining the ways in which we view and behave in the natural world.

Acknowledgments. The author gratefully acknowledges the support of the NSF/EPA program on Valuation for Environmental Policy under Award No. SBR 95-25582. Helpful comments on an earlier draft were received from Tom Brown, Mary English, Terre Satterfield, and Howard Kunreuther, as well as several participants at the October 1996 Knoxville Tennessee workshop sponsored by the National Center for Environmental Decision-Making Research. Any opinions, findings, conclusions, or recommendations expressed in this article are those of the author and do not necessarily reflect the views of the National Science Foundation or the Environmental Protection Agency.

Key Resources

Freeman, R. 1993. *The Measurement of Environmental and Resource Values.* Resources for the Future, Washington, D.C. (An excellent overview of the use of economic techniques in resource management by a superb theoretician and experienced practitioner.)

Keeney, R.L. 1992. *Value-Focused Thinking.* Cambridge, MA: Harvard University Press. (An insightful look into the definition of values and their use as a guide to creative decision making.)

Fisher, R. and Ury, W. 1991. *Getting to Yes.* New York: Penguin Books. (An easily readable guide to making group and negotiated decisions based on the identification of underlying interests and concerns.)

Plous, S. 1993. *The Psychology of Judgment and Decision Making*. New York: McGraw-Hill. (An accessible overview to the science and art of decision making, written in a way that neatly blends philosophy with practical advice.)

Payne, J., Bettman, J., and Johnson, E. 1993. *The Adaptive Decision Maker*. New York: Cambridge University Press. (A revealing introduction to knowing how people adopt strategies for decision making based on a constructive analysis of their tradeoffs across accuracy, effort, and concerns.)

References

Axelrod, L. 1994. Balancing personal needs with environmental preservation: Identifying the values that guide decisions in ecological dilemmas. *Journal of Social Issues* 50:85–104.

Brown, T. 1984. The concept of value in resource allocation. *Land Economics* 60:231–246.

Callicott, J.B. 1984. Non-anthropocentric value theory and environmental ethics. *American Philosophical Quarterly* 21:299–309.

Carpenter, R.A. 1996. Ecology should apply to ecosystem management: A comment. *Ecological Applications* 6:1373–1377.

Christensen, N.L., Bartuska, A.M., Brown, J.H., Carpenter, S.R., D'Antonio, C., Francis, R., Franklin, J.F., MacMahon, J.A., Noss, R.F., Parsons, D.J., Peterson, C.H., Turner, M.G., and Woodmansee, R.G. 1996. The report of the Ecological Society of America committee on the scientific basis for ecosystem management. *Ecological Applications* 6:665–691.

Costanza, R. and Cornwell, L. 1992. The 4P approach to dealing with scientific uncertainty. *Environment* 34:12–20.

Crites, S., Fabrigar, L., and Petty, R. 1994. Measuring the affective and cognitive properties of attitudes: Conceptual and methodological issues. *Personality and Social Psychology Bulletin* 20(6):619–634.

Desvousges, W. and Smith, V.K. 1988. Focus groups and risk communication: The "science" of listening to data. *Risk Analysis* 8:479–484.

Dunlap, R. 1991. Public opinion in the 1980s: Clear consensus, ambiguous commitment. *Environment* 33:10–15, 32–37.

English, M., Gibson, A., Feldman, D., and Tonn, B. 1993. *Stakeholder Involvement: Open Processes for Reaching Decisions about the Future Uses of Contaminated Sites, Final Report*. Knoxville: The University of Tennessee.

Fischhoff, B. 1991. Value elicitation: Is there anything in there? *American Psychologist* 46:835–847.

Flynn, J., Slovic, P., and Mertz, C.K. 1994. Gender, race, and perception of environmental health risks. *Risk Analysis* 14:1101–1108.

Freeman, R. 1993. *The Measurement of Environmental and Resource Values*. Washington, D.C.: Resources for the Future.

Gregory, R. and Keeney, R. 1994. Creating policy alternatives using stakeholder values. *Management Science* 40:1035–1048.

Gregory, R., Lichtenstein, S., and Slovic, P. 1993. Valuing environmental resources: A constructive approach. *Journal of Risk and Uncertainty* 7:177–197.

Gregory, R., Brown, T., and Knetsch, J. 1996. Valuing risks to the environment. In: H. Kunreuther and P. Slovic (Eds.). *The Annals of the American Academy of*

Political and Social Science, Vol. 545. Thousand Oaks, CA: Sage Publications. Pp. 54–63.

Gregory, R., Flynn, J., Johnson, S., Satterfield, T., Slovic, P., and Wagner, R. 1997. Decision pathway surveys: A tool for resource managers. *Land Economics* 73(2):240–254.

Hausman, J. 1993. *Contingent Valuation: A Critical Assessment*. Amsterdam: North-Holland.

Kahneman, D. and Knetsch, J. 1992. Valuing public goods: The purchase of moral satisfaction. *Journal of Environmental Economics and Management* 22:57–70.

Kasperson, R., Renn, O., Slovic, P., Brown, H., Emel, J., Goble, R., Kasperson, J., and Ratick, S. 1988. The social amplification of risk: A conceptual framework. *Risk Analysis* 8:177–187.

Keeney, R. 1992. *Value-Focused Thinking: A Path to Creative Decisionmaking*. Cambridge, MA: Harvard University Press.

Keeney, R. and Raiffa, H. 1976. *Decisions with Multiple Objectives*. New York: John Wiley and Sons.

Keeney, R., von Winterfeldt, D., and Eppel, T. 1990. Eliciting public values for complex policy decisions. *Management Science* 36:1011–1030.

Kempton, W., Boster, J., and Hartley, J. 1995. *Environmental Values in American Culture*. Cambridge, MA: MIT Press.

Kopp, R. and Smith, V.K. 1993. *Valuing Natural Assets*. Washington, D.C.: Resources for the Future.

Maguire, L. and Servheen, C. 1992. Integrating biological and sociological concerns in endangered species management: Augmentation of grizzly bear populations. *Conservation Biology* 6(3):426–434.

Mangel, M., et al. 1996. Principles for the conservation of wild living resources. *Ecological Applications* 6:338–362.

McDaniels, T. 1996. The structured value referendum: Eliciting preferences for environmental policy alternatives. *Journal of Policy Analysis and Management* 15:227–251.

Mitchell, R. and Carson, R. 1989. *Using Surveys to Value Public Goods: The Contingent Valuation Method*. Washington, D.C.: Resources for the Future.

Morgan, D. 1988. *Focus Groups as Qualitative Research*. Newbury Park, CA: Sage Publications.

Morgan, G., Morris, S., Henrion, M., Amaral, D., and Rish, W. 1984. Technical uncertainty in quantitative policy analysis: A sulphur air pollution example. *Risk Analysis* 4:201–216.

Morgan, G. and Keith, D. 1995. Subjective judgments by climate experts. *Environmental Science & Technology* 29:468–476.

Ontario Forest Research Institute. 1995. *Vegetation Management in Ontario's Forests: Survey Research of Public and Professional Perspectives*. Sault Ste. Marie, Ontario, Canada: Ontario Forest Research Institute.

Opaluch, J., Swallow, S., Weaver, T., Wessells, C., and Wichelns, D. 1993. Evaluating impacts from noxious facilities: Including public preferences in current siting mechanisms. *Journal of Environmental Economics and Management* 24:41–59.

O'Riordan, J. and Jordan, A. 1995. The precautionary principle in contemporary environmental politics. *Environmental Values* 4:191–212.

Payne, J., Bettman, J., and Johnson, E. 1992. Behavioral decision research: A constructive processing perspective. *Annual Review of Psychology* 43:87–132.

Raiffa, H. 1982. *The Art and Science of Negotiation*. Cambridge, MA: Harvard University Press.

Rokeach, M. 1973. *The Nature of Human Values*. New York: The Free Press.

Satterfield, T. 1996. Pawns, victims, or heroes: The negotiation of stigma and the plight of Oregon's loggers. *Journal of Social Issues* 52:71–83.

Sebenius, J. 1992. Negotiation analysis: A characterization and review. *Management Science* 18:18–38.

Simon, H. 1978. Rationality as process and as product of thought. *American Economic Review* 68:1–16.

Slovic, P. 1995. The construction of preference. *American Psychologist* 50:364–371.

Slovic, P., Layman, M., Kraus, N., Flynn, J., Chalmers, J., and Gesell, G. 1991. Perceived risk, stigma, and potential economic impacts of a high-level nuclear waste repository in Nevada. *Risk Analysis* 11:683–696.

Smith, V.K. 1996. Can contingent valuation distinguish economic values for different public goods? *Land Economics* 72:139–151.

Stern, P. and Dietz, T. 1994. The value basis of environmental concerns. *Journal of Social Issues* 50:65–84.

Suter, G. 1993. *Ecological Risk Assessment*. Chelsea, MI: Lewis Publishers.

Tversky, A., Sattath, S., and Slovic, P. 1988. Contingent weighting in judgment and choice. *Psychological Review* 95:371–384.

Tversky, A. and Kahneman, D. 1981. The framing of decisions and the psychology of choice. *Science* 211:453–458.

U.S. Environmental Protection Agency (EPA). 1992. *Framework for Ecological Risk Assessment*, EPA/630/R-92/001. Cincinnati, OH: U.S. Environmental Protection Agency.

Walters, C. 1986. *Adaptive Management of Renewable Resources*. New York: MacMillan.

Wilkins, L. and Patterson, P. 1991. *Risky Business: Communicating Issues of Science, Risk, and Public Policy*. Westport, CT: Greenwood Press.

Decision-Maker Response

JOSEPH W. LEWIS

Robin Gregory has done an excellent job of describing the various techniques that have been used by analysts to assess environmentally related values. His coverage of the topic is comprehensive, and his discussion of the strengths and weaknesses of each technique is thorough and fair-handed. I could find little to quarrel with, so I will just expound on some general categories, perhaps from a different perspective.

The Concept of Value

There is no denying that "value" is a subjective concept, a concept with moral, ethical, and/or economic overtones. Some moral/ethical values (e.g., belief in God, the Golden Rule) need not have any economic aspects for most people. Some economic values, such as buying groceries, need not have any moral/ethical aspects for most people. However, for some people, religion has economic aspects; some people find the selection of groceries to be an ethical matter. When we speak of environmental values, it is difficult to separate the moral, ethical, and economic elements incorporated in that concept. When we try to objectively analyze something so complex and subjective, we know we are in for a difficult time.

Analyzing Trade-Offs

Policymakers cannot escape the need to compare trade-offs. This task is easier for them if the consequences of all options can be expressed in common units. The preferred common denominator is a monetary unit (dollars in the United States). If apples are valued at $.75 each, and oranges are $.50, then one can comfortably conclude that two apples are equivalent to three oranges. Policymakers do not like to compare apples to oranges, but they must; and the most objective way is by translation to dollars.

Making Decisions

People in high-level positions often say they make important decisions based on intuition or gut feelings. That is not to say these decisions are arbitrary. But they are often based on experience and knowledge rather than on empirical data or formal analyses. What is more, many of these executive-level policymakers are proud of this. Scientists, on the other hand, need lots of empirical data, covering long periods, before they are willing to conclude anything. Policymakers and scientists live in different worlds. Both would like to make use of scientifically based data, but policymakers cannot wait that long. And then there are analysts, who try to integrate science and policymaking to arrive at sound decisions. It is evident from Gregory's assessment that no evaluation tool is without flaws. Whatever analytical technique the analyst can muster will be criticized from one angle or another. Is there one technique that will meet the needs of policymakers, scientists, and analysts simultaneously?

With/Without

The answer is no! In theory, maybe, but in practice, no. However, the contentiousness can be minimized if the analysis is carefully structured to accurately describe the "with" and "without" scenarios. (I am referring to contentiousness among policymakers, scientists, and analysts; Gregory covered the need to include all viable alternatives and to emphasize compromise to deal with contentions among stakeholders.)

Addressing all viable alternatives is critical, but in many cases, the analysis boils down to taking a proposed action (with) compared with not taking that action (without). Too often, the "without" case is not adequately examined. Sometimes the "without" case is a straw man set up to make the proposed action look good. More often, it is a depiction of undesirable effects on intangible values. For example, say the quality of a town's water supply is threatened by an eroding watershed that requires restoration to maintain the safety of the water supply. What is the benefit of restoring the watershed? That depends on how the "without restoration" scenario is formulated. If the "without" situation assumes that the town's water supply becomes contaminated and causes a percentage of the population to get sick (some of whom might die), then the benefit of restoration would be avoidance of sickness and death. However, if the "without" situation assumes that the town would not accept a contaminated water supply, but would build a treatment facility, then the benefit of restoration is avoidance of the cost to construct and operate a treatment facility. The latter scenario is not only the more realistic choice; it is also easier to quantify in both physical and monetary terms.

Here is another example: Suppose you had 10 beautiful trees in your yard. You value these trees because they provide shade, a home for birds, and scenic beauty. The trees are threatened by a disease that will kill them all if you don't spend $600 on a preventive treatment. What would your "without" scenario be? Would you live with a yard full of dead trees, or would you cut down the dead trees and replant new, smaller trees in their place? In the former case, the benefit of disease prevention is avoiding the loss of shade, avian habitat, and scenic beauty (difficult to quantify). In the latter case, the benefit would be avoidance of tree removal and replacement costs (not difficult to quantify). Of course, the latter "without" case still results in diminished shade and scenic beauty because the replacement trees are likely to be small. The point to be made is that the most realistic "without" scenario is often also the one most easily quantified.

Careful attention to the "without" scenario may be the key to providing an analysis that is perceived to be reasonably objective by policymakers, scientists, and analysts.

3
Tools to Characterize the Environmental Setting

Virginia H. Dale and Robert V. O'Neill

Many tools are used to characterize environmental conditions. People often use these devices without even thinking of them as tools. Simple examples include binoculars, hand lenses, thermometers, or other instruments that enhance human senses and are a part of many biologists' tool kits. At the other extreme are sophisticated tools, such as computer simulation models, laboratory analyses, or statistics, which require considerable training or auxiliary equipment to use. Decision makers who use environmental data are sometimes not aware of the diversity of tools available or the assumptions involved in their use.

The purpose of this chapter is to describe tools that are used by industry, government agencies, and citizens or citizen groups to characterize the environmental setting and to present the strengths and limitations of these tools. The tools are applicable to the range of conditions found in built environments, as well as natural situations. The chapter considers the information needed for environmental decision making, constraints in obtaining the information, and how the constraints interface with the scientific approach to problem solving. The final section of the paper discusses ways to make the tools both useful and used in communicating information.

Constraints on Information Needed to Characterize the Environment

The goals and values, socioeconomic conditions, and appropriate regulations for the issue at hand determine the type and quality of information needed by the decision maker (see Figure 3.1). A private citizen may be most interested in the aesthetics, recreational use, or cultural aspects of a site; a business executive may be most concerned about a site's consumptive use; and a government agency may promote multiple uses of a site. The information pertinent to these diverse interests would be different. In reality, of course, there is typically some overlap of goals.

3. Tools to Characterize the Environmental Setting   63

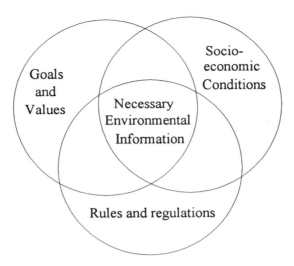

FIGURE 3.1. Factors that determine what information is needed to characterize the environmental setting.

Chapter 2 describes how tools can be used to identify environmental goals and values for an issue although, as noted, there may be other types of relevant goals and values (e.g., profit, reputation, and distributional justice).

The regulatory context in which the decision process takes place constrains the type of information that is needed. Laws set the lower limits for measurement of environmental information. These limits deal with spatial or temporal resolution, the degree of biological complexity, or the precision of the data. For example, regulations and legal requirements may set scheduling deadlines. Also, a regulation may require the use of a specific tool, such as the use of the Environmental Protection Agency (EPA) standard analytical methods.

Socioeconomic conditions determine the availability of resources for gathering information relevant to the decision and the timing of the decision (as discussed in Chapter 4). In some cases, the availability of environmental information itself jeopardizes the social or economic conditions of people. For example, coal miners may not support studies of black lung disease because this information can jeopardize their jobs. Social mores also constrain the type of environmental information that is collected. An example of this impact is the lack of debate about population control in the United States because moral and religious values impede discussion of the options. Thus, the socioeconomic setting must be considered in determining what environmental information can and should be obtained.

The Scientific Approach to Characterizing the Environmental Setting

Given these constraints, the environmental scientist typically relies on the scientific approach to determine what kind of information is needed and what kind of tools will be used in its collection. This approach consists of defining the issue; reviewing existing information; forming testable hypotheses of possible cause-and-effect relationships; collecting evidence; drawing conclusions; reformulating the hypotheses; and then repeating the sequence. The process used by the scientist to determine what information is needed differs from that used by the public or high-level managers to set the goals or values that are addressed by the decision (see Chapter 2).

Specific questions must be considered to characterize the environment when a particular decision is at hand:

• *What do we need to know about the environment to make the decision?* This question is important because one can spend a lifetime and many millions of dollars and still not acquire the appropriate information. The issue at hand (i.e., the impending decision) determines the type and extent of environmental information collected, the tools used in the analysis, and the types of conditions that one hopes to achieve or maintain (e.g., aesthetics and health).

• *What are the relevant background conditions for the particular site and decision?* This information includes previous land use; prevailing weather conditions; atmospheric-pollution levels; traffic patterns; current expenditures to maintain widely shared values; and estimates of how these conditions may change in the future. Such background information is an important part of deciding which environmental data must be obtained.

• *At what scale or scales will the decision impact the environment?* A management decision may affect a number of temporal and spatial scales (as discussed in Chapter 1). For example, many people in the United States are concerned about the clearing of the Amazonian rain forest because of the potential effects that it may have on atmospheric CO_2 and global climate change. However, the people living in the Amazon forest are most concerned about how rapid deforestation affects soil fertility and thus their ability to make a living by farming. Some land-management decisions provide a solution to one of these concerns, but other decisions exacerbate both trends. Being aware that these two scales of issues exist is an important aspect of the decision being considered.

• *How might proposed changes influence future conditions?* To answer this question, one must project future conditions under different management scenarios, including no action at all. In addition, one must know not only what environmental conditions will be affected, but also what secondary effects might occur.

Addressing these questions raises issues about the scale of the data, availability of existing information, and relevance of data in view of the natural conditions. Decision makers who are aware of these issues make more-informed choices and interpretations, hence they can make appropriate use of the available information.

Spatial and Temporal Scale of Data

A common issue of environmental information relates to the spatial and temporal resolution of the data. Often, environmental data are collected in a small plot and at only one time. Yet, decisions usually address changes that occur over a large area and that span months, years, or decades. For example, when the location of a pipeline across the state of Washington was considered, the consultants performed a very cursory survey, reportedly because of cost limitations. The sampling scheme monitored plant diversity in one-meter-by-one-meter plots at two times of the year in a few places on the Olympic Peninsula. These samples were too small and inadequate to measure the vegetation diversity and complexity across this area, but published studies on the vegetation of the Peninsula supplemented the field measures.

On the other hand, some model projections or remote-sensing data are at too broad a resolution for the situation at hand. For example, global circulation models project climate changes for very large resolutions (e.g., larger than the entire state of Maryland). It is difficult for decision makers to grasp the impact that such projections have for their constituents who may occupy only a portion of the projected area. As another example, the most widely available remote-sensing data set is Landsat Thematic Mapper, which typically provides a resolution of 30 meters by 30 meters. However, even this scale is too broad for some decisions, such as siting a road relative to the location of a wetland.

Temporal resolution of the data is another common issue. Biologists are frequently asked to collect field data in a very short window of time. These short time frames are characteristic of environmental impact statements, which are often prepared on strict schedules. Because populations and ecosystems are dynamic, environmental conditions may change drastically from season to season and year to year. Short time frames frequently do not allow appropriate characterization of the field situation. In the best impact statements, continuous monitoring occurs to fully evaluate impacts (Bernard et al., 1993; Draggan et al., 1987).

To some extent, the scale of the data is determined by the scale at which the environmental decisions are being made. At the township scale, decisions focus on air and water quality, waste management, zoning, and critical-area protection. The decisions deal with relatively small spatial scales and can depend on locally gathered field data. But many decisions have long-term effects, and thus appropriate temporal scaling of the information remains important. At regional scales, authority extends over complete airsheds and river systems. Decisions that deal with broader ranges of environmental issues require larger scales of information.

Frequently, decisions become problematic because of scale issues. For example, at the local scale, each small dam on the Columbia–Snake River system is responsible only for local impacts. Yet the Federal Power Commission is faced with the challenge of evaluating the cumulative impacts of all of the dams in the system and requires a much larger perspective and larger-scale data to make informed choices.

Conflicts of scale are common occurrences in environmental decision making. Often, a decision purported to affect a site influences a larger area than what is being considered. For example, a Texas state senator protested a federal decision to prohibit the building of an industrial complex because the complex would impact a population of fish that was on the endangered species list. To paraphrase the senator's comments: "Nobody cares! It's a damn trash fish; you can't eat it or nothing! They got some more of them over there in Nevada. I say we go catch ours and release them over there and build the complex!" The argument was probably compelling to a majority of the audience. The federal decision, however, was based on information about the distribution of the fish and its habitat across the entire United States. More frequently, data are available for large-scale entities but are not appropriate or helpful for small-scale concerns.

Such conflicts of authority are a fact of life to decision makers. The important point here is that this same conflict of scale carries over into the environmental data used in decisions. Small-scale data only reflect local conditions and can seriously bias the decision maker's characterization of the environmental trends.

Using Available Information Versus Collecting New Data

The issue of scale leads us directly to a second topic: Using existing data or collecting new data. In many cases, available data cover large spatial and temporal scales but are less focused on the present environmental issue than data collected to specifically address the concern. New data might be immediately relevant, but would need to be collected during the short time frame or at the small spatial resolution and extent permitted by available resources. This issue frequently arises in concert with the question of modeling. Sometimes it is more cost effective to produce a model that extrapolates the existing information than to collect new information that would be woefully deficient in scale.

The decision to rely on available data and modeling is a common resolution of this issue. As a result, encyclopedic collections of environmental data have been compiled (Jorgensen, 1979; Boden et al., 1994; Golden et al., 1980; Gross and Pake, SAMAB, 1996; Eco-Inforum, 1996; Brown et al., 1993). More and more of this environmental information is becoming available over the Internet. When using extant data, the quality of any data must be evaluated and considered. The decision maker must be cognizant, however, that although the physical aspects of the environment tend to be relatively stable and available data are often reliable, biological systems are more complex and less stable, and important changes may have occurred since the available data were gathered. Therefore, some checking of field conditions may be necessary before relying on available data.

Biological Data and Life Cycles

In addition to the need to collect environmental data at the appropriate scale, it is important to recognize that biological data are relevant only when they are collected during the proper periods of an organism's life cycle. The most common concern is that only the adults of a biological species can be easily observed. As a result, quantifying the impact of an environmental decision on the reproductive success of the population is frequently very difficult. For example, biologists studying Cerulean warblers, a rare and endangered species, have a very difficult time locating the nests, which the birds build high in the canopy (Robbins et al., 1989). Because these nests are not typically observed, it is difficult to relate specific environmental-management actions to the reproductive success of this rare species.

Some organisms experience cycles in their life-history patterns that may be predictable or, conversely, may depend on an unpredictable event, such as seasonal rainfall or wildfire disturbances. For example, for most conifers in the United States, seed production occurs en masse every seven years; thus, it would be inappropriate to assess the effect of an environmental change on seed production until the mast year had occurred. Effects of seed production of lodgepole pine may be even harder to determine, for it requires fire to release seeds from the cones. Thus, it is very difficult to get an accurate measure of how changes in the environment affect the reproductive success of these organisms. In large or long-lived organisms (e.g., elephants), assessing how an environmental stress affects statistics on birth, reproduction, and death is even more challenging.

Separating Impacts from Natural Variability

Simply stated, all change is not equal. The observation of a change in the environment, no matter how reliable the information, does not necessarily mean that human agents are the cause of the change. Many biotic populations undergo cyclic changes in numbers from year to year. Although the most radical examples are found in regions with seasonal changes in the environment, the phenomenon of fluctuations in population size is universal. Understanding the magnitude of natural variability is critical to determining whether impacts have occurred. In highly variable natural systems, it may be difficult or impossible to demonstrate an impact unequivocally, as was the case of bird mortality in the wake of the Exxon Valdez oil spill (Wiens, 1996). It may, in fact, be impossible to prove an effect until the impact has proceeded to irretrievable limits. Our society brought about the extinction of many rare species before any clear demonstration of the impending loss could be proven (e.g., the woolly mammoth and the carrier pigeon).

A principle from decision theory, known as mini-max, may need to be called into play when dealing with variable biotic communities. This principle recommends that decisions minimize the probability of the maximum, or worst, outcome. The principle logically recommends a cautious and conservative approach to collecting data from systems that have a high degree of environmental variability.

Dealing with Keystone Effects in Communities

Impacts on some species in a community do not have much influence on the rest of the environmental system. However, changes in certain other species, known as keystone species (Paine, 1980), have severe effects on the entire system. For example, Paine demonstrated that removal of the pisaster starfish in the intertidal zone of the Pacific coast has a major effect on the ecosystem, while removal of other organisms is not so serious. The starfish scours the rock surfaces while feeding, opening areas for the establishment of a wide variety of organisms. When the starfish is removed, algal growth covers the rocks, other organisms cannot establish a presence, and the entire system is changed. Similarly, coral organisms form the substrate for thousands of other species. Pollutant impacts on the sensitive coral polyps will indirectly alter the entire ecosystem (Jackson et al., 1989). As another example, red-cockaded woodpeckers is the only species of bird in the southeastern United States that creates cavities in living trees. These cavities serve as nests for the red-cockaded woodpeckers, but also provide homes for more than 25 other species (Dennis, 1971). Thus, the threatened and endangered status of red-cockaded woodpeckers jeopardizes many other species.

Sources of Environmental Information

Environmental data provide the information to determine if there is a trend in environmental conditions and what may cause the trend. Data are the most basic tool available for characterizing the environmental setting. Environmental data sometimes provide the basis to suggest the reason for observed trends. Ideally, the data should establish the direction, magnitude, and extent of any trend; the causes of changes in environmental conditions;

and appropriate junctures in space and time at which decisions or interventions might alter the trend.

To the naive observer accustomed to weather reports with morning coffee and the evening newspaper, the ability to report environmental change may seem woefully inadequate. But the magnitude of the task must always be kept before our eyes. There are more than 4,000,000 chemicals registered with the Chemical Abstracts Service, and 43,000 of these, excluding pesticides, are listed by the EPA as subject to the Toxic Substances Control Act (Council on Environmental Quality, 1979). For the moment, it seems out of reach to consider the development of a reporting system similar to the weather bureau for conveying the status of toxins and other impacts to the water, soil, biota, or built environments. Without a central source of information, environmental decision makers must begin by considering the types, scale, and quality of the available data.

Risk-assessment approaches help define data needs; these approaches are discussed in Chapters 1, 2, and 8. This definition of data needs is such an important part of the decision-making process that the EPA developed a guide book just for establishing data-quality objectives (USEPA Quality Assurance Management Staff, 1994). A planning approach completes the seven steps required to specify the appropriate data:

1. Clearly define the situation including reviewing existing information.
2. Specify the questions relevant to the decision.
3. Lay out information needs sufficient to answer these questions.
4. Set the temporal and spatial boundaries of the study.
5. Specify a decision rule; this step includes defining the statistical parameter of interest and the action level.
6. Determine the tolerance limits on decision errors.
7. Set forth the optimal design for collecting needed data.

Major sources of data used in environmental decision making include citizen observation, field data, laboratory data, maps, remote imagery, and simulation models. Ordinarily, information is available from some of these sources and must be integrated into the decision-making process. Standard approaches have been established for collecting these data (Clarke, 1986). See Sidebar 3.1 for a more detailed look at one component of the risk-assessment approach.

Citizen Observation

Most environmental trends are identified by relatively informal observation. For example, the first sign of an environmental concern may be trout disappearing from the stream, bass missing from the lake, or the lack of eagles in the sky. Lifelong observation of their environs has often meant that the farmer, hiker, and hunter can provide a sensitive system for environmental monitoring.

Sidebar 3.1

Dose-response functions (Suter 1993, particularly Chapters 1 and 8) refer to the relationship between the level of a stressor (e.g., the concentration of a pollutant) and its effect on biota. These algebraic relationships quantify a biotic response, such as increased mortality or decreased fecundity, across all possible levels of stress. Ideally, the relationships are derived from controlled studies in the field or laboratory. But developing these relationships is difficult because of the many features of ecosystems.

A dose-response relationship generalizes an organism's responses to a specific pressure. Once the relationship is determined and graphed, the response to any level of the stress can be read directly from the curve, permitting extrapolation to a wide variety of situations. Whenever such a relationship can be established, the need for further laboratory data may be eliminated, depending on the universality of the conditions under which the relationship was established. However, actually applying this approach is very difficult for a variety of circumstances.

Mixed stressors. When environmental conditions involve a number of simultaneous stressors, biotic responses may not be the sum of the responses to the individual stressors. Standard dose-response relationships cannot be extrapolated in this situation. Further, responses to a mix of stressors are difficult to mimic in a laboratory, and it may be impossible even to establish a dose-response relationship for a specific mixture.

Complex life histories. When large or long-lived organisms are involved, the stress does not affect all stages of the life of the organisms equally. To use a general dose-response approach would require information on each life stage separately, as well as data on various combinations of impacts to various combinations of life stages.

Prior conditions. Organisms may become more susceptible to a given impact by previous conditions. In the most common situation, organisms are weakened by an unrelated additional stress. For example, Fraser fir trees in the high elevations of the Great Smoky Mountains appeared to be relatively resistant to damage from air pollution. However, following an outbreak of a pest—the balsam woolly adelgid—the dose-response relationship for air pollutants has changed, and the trees appear more susceptible (Hain and Arthur, 1985).

Complex communities. Dose-response relationships are established in the laboratory with individual organisms. In nature, the organisms are also involved in a complex network of interactions. The dose-response relationship, established under laboratory conditions, may not extrapolate well to the complex web of interactions in the natural world. Importantly, all of the interacting organisms are being impacted, and toxic impacts on food organisms or predators may be more important than direct toxic effects to the organism.

Spatial complexity. Laboratory data must be extrapolated to the natural landscape with caution. Not all organisms on the landscape receive the same dose because of topography, spatial isolation, etc. The effects of a given dose may be intensified or ameliorated depending upon time of year, availability of refuges, and numerous other factors. It is difficult to argue that each individual organism on a complex landscape receives the same average stress and that population response can be characterized by the dose-response relationship appropriate to the average organism.

Difficult observation. A dose-response approach is difficult to apply to some critical parts of the environment. For example, measurements of groundwater pollution are difficult to obtain, making it problematic to determine the dosage of a chemical delivered to the ecosystem from this source.

Changing conditions. The dose-response relationship assumes that an organism is in a stable environment. However, change is an integral part of the natural system. Therefore, extrapolation of the laboratory relationship may be difficult because the organism in the field is subjected to seasonal changes, life-cycle changes, and a variety of other alterations in its background environs. Many of these natural changes can be expected to alter susceptibility to imposed stress.

Of course, the quality of the information provided by citizens leaves much to be desired. Citizens are *sensitive* observers of change, but are seldom *critical* observers. Thus, bias is a big concern with using this kind of information. Also, casual observation cannot distinguish between random variations and significant trends. Nevertheless, the citizenry provides the most extensive observation network possible.

Preliminary information is sometimes available through recreational clubs (e.g., hunting, fishing, or hiking organizations), high school biology classes, and special-interest groups (e.g., bird watchers or wildflower enthusiasts). Such organizations often keep long-term records that establish background variability and help determine whether a real trend might be occurring. Members are often in personal contact with others in the region with common interests. These contacts help establish an initial hypothesis about the spatial extent of change.

However, beyond serving as an extensive trend detector, citizen observations can do little more than alert the public or the decision maker to the *possibility* of an environmental concern. Uncritical observation cannot prove environmental damage. The chief limitation of citizenry information is that it usually *cannot establish* the reality of environmental damage because natural variations are frequently not considered and the information cannot be linked to cause-and-effect relationships.

Field Data

Field data are collected for the purpose of monitoring the environment or to characterize a particular issue. Both purposes require a systematic and critical approach. It may become necessary to prove, scientifically, that the trend is real, demonstrate its intensity, and establish its spatial extent. Field data are typically gathered according to established protocols. Data may be collected from air, land, water, or biota, or from natural or built environments. Land samples can include measures either on the ground or below, although the collection of below-ground measurements, such as the flow rate of groundwater and the contaminants in it, are typically expensive and labor intensive.

Field data ideally include a variety of information. The data should involve a detailed physical description of the scene, such as atmospheric conditions, soil texture, moisture, geologic conditions, water quality, slope, exposure, runoff, buildings, roads, etc. Such information is invaluable in determining the type of system in which trends are observed and what physical conditions amplify or mitigate the effects. Early in the field investigation of acid-rain effects, for example, it was discovered that mature, alkaline lakes showed little impact, while younger, glacial lakes were seriously affected. Adequate data on the physical aspects of the environment were needed to interpret the confusing field record of effects in some places and no effects in other instances.

Field data also consider the biological organisms that are components of the systems. The actual measurements depend upon the environmental system and the issue being investigated. Sampling techniques have been developed for a great variety of biotic components. For example, many wildlife species are difficult to observe directly. A combination of secretive habits and mobility make direct counting impractical. Census techniques based on trapping may endanger the very organisms we seek to protect. In such cases, vegetation measures provide sensitive habitat indicators. Background studies demonstrate significant correlations between the vegetation structure and the wildlife that occupy the area. In other cases, the presence of biological organisms may be related to built structures. For example, the extensive search for the cause of Legionnaire's disease was resolved when researchers determined that the air-conditioning system of a hotel was distributing the disease agent.

To provide reliable field measures, field-sampling methods are an area of active research. The effort has been aimed at developing techniques that are both efficient and statistically valid. Consider, for example, the method of *mark and recapture* for mobile animals (Jolly, 1965). During an initial period of live trapping, a sample of individuals are collected harmlessly, permanently marked (perhaps with an ear tag), and released. For illustration, let us assume that 10 individuals are captured and tagged. The released animals are assumed to mix back into the general population. Then a second collection is made. This time, 10 individuals are captured and only one is from the original tagged group. Simple logic dictates that the original tagged group was approximately one tenth of the total, and the entire population contains 100 individuals. But even this clever method has limitations because it assumes that the tagged individuals mixed uniformly with the rest of the population and that the sampling is random. The method also ignores the fact that the original tagged group has discovered it can eat the bait (or be otherwise trapped) and escape unharmed. As a result, that group is less likely to be caught during the second collection period. Furthermore, the approach does not consider immigration or emigration of the population. Thus, even well-tested methods for field-data collection require considerable judgment in interpretation. For this reason, field sampling and its interpretation is typically a job for experienced professionals in resource agencies and universities. However, citizen groups can train their members to provide reliable field measures (e.g., the Christmas bird count comes from both amateur bird watchers and professional ornithologists who locate and enumerate bird species every Christmas day).

Laboratory Data

Having established that an environmental concern exists, attention usually turns to a scientific understanding of the mechanisms involved and the causes. This information will be critical to develop strategies for appropriate decisions and will often entail laboratory measurements and experimentation.

Laboratory studies may be required to establish a specific mechanism linking a cause to the observed environmental effect. It was not possible from field work alone, for example, to establish a link between pollutants and the disappearance of large raptor birds. Laboratory studies revealed that pesticides make the egg shells thin and brittle so that few eggs survived the nesting period and hatched. Biota possess a great variety of mechanisms to detoxify chemicals. In some cases the material is sequestered and accumulated in fatty tissue and only has a significant effect when the organism is consumed by a predator. The predator then receives a more concentrated dose than had previously been contained in its normal diet. Because of the diversity of biochemical mechanisms, laboratory studies may be needed to establish the toxicity of a particular chemical and its tendency to be concentrated in the food chain.

Laboratory data, therefore, provide critical information to the decision-making process. A scientific demonstration of the specific cause of the environmental trend may be needed to establish credibility, develop public understanding and backing, and justify legal actions to alter the trend. However, laboratory studies can be expensive. Furthermore, the laboratory study must be properly designed to have a sufficient sample size and to address the appropriate question.

The decision maker, therefore, is often faced with the prospect of relying on available studies. Typically, laboratory studies are designed with a specific question in mind. The results are specific to an area, a soil type, a toxicant, type of biota, type of structure, etc. Therefore, extreme caution is required in extrapolating the information to a new situation.

In spite of the concerns associated with gaining access to laboratory data pertinent to a specific issue, the information can be critical in monitoring conditions, establishing causality, and developing a plan for corrective action. As with the collection of field data, the interpretation and extrapolation of the available data require expertise, and the solution is often to call upon local resource agencies and universities to tap the requisite expertise.

Maps

Some environmental data are best viewed and interpreted in a spatial context. The most accessible means for identifying the location of monitoring information is provided by maps. For example, an on-ground observation of forest defoliation by insects is often followed by examining a map of regional forests and checking susceptible locations to establish whether the outbreak is widespread. Under normal conditions, defoliating insects strike local areas at infrequent intervals, and there is little cause for alarm. Maps can show the spatial scope of the outbreak and illustrate whether point sources of contaminants can be correlated to the pattern of the insect damage. Mapped information depicts the location of linear features (such as roads and underground pipes or wires) or point information (such as the location of a specific building).

Mapped information also illustrates the spatial *extent* of an issue. The spatial extent will often differentiate local, possibly transient, events from more widespread trends. Information on the spatial extent will also be required to narrow the identification of potential causes. For example, observed changes in the chemistry along a particular stretch of a river system might lead one to suspect a local source of water pollution. Finally, the spatial extent of the trend will help determine the social units and legal jurisdictions that might need to be involved in the decision-making process.

Remote-Sensing Data

More and more commonly today, remote-sensing data are being used to inform environmental decisions when the spatial context is vital (e.g., Iverson et al., 1994). Historically, the data were based on aerial photography, but now satellite imagery is becoming available that has the appropriate resolution for environmental questions (Westman, 1985). Available satellite imagery contains information about the absorption and reflectance patterns of the objects being observed on the ground (i.e., how brightly the sunlight is being reflected). The information is collected in individual wavelength bands, most often in the visible and near-visible (i.e., the infrared and ultraviolet) spectra. Radar is also useful because it obtains unique reflectance characteristics at different wavelengths. The imagery, whether light or radar derived, is interpreted with information on how surface features reflect each wavelength. For example, live vegetation has a characteristically large ratio of infrared-to-red reflectance and is readily identified (Tucker, 1979). The fundamental information contained in the imagery is spatial extent and patterning on the Earth's surface. The specific environmental characteristics that can be studied in this manner depend on the satellite sensor, and such instrumentation is the subject of active research.

Once the spatial extent of various land covers is known, it may be possible to hypothesize the cause of an observed pattern in environmental conditions (e.g., a proximity to industrial or urban areas). There are well-established correlations, for example, between land use on a watershed and the quality of the water flowing into the stream draining the watershed. Roads crossing a stream, extensive agriculture, or industrial complexes along a stream bank are possible causes of water-quality trends that can be identified from remote imagery.

The spatial configuration of land-cover types may reveal that the landscape has become fragmented by human activities. This fragmentation may impact the ability of wildlife to access resources in their habitat and result in changes in population numbers (Andren, 1994; Dale et al., 1994a). Similarly, fragmentation may affect the ability to move machinery between patches in a cost-effective manner, for example, moving harvesting equipment from field to field within the short time that the crops are ready. Thus, an observed reduction in wildlife or loss of crop may have little or nothing to do with direct pollution effects, and remote imagery may be needed to determine if the cause is fragmentation or some other influence.

The type of information that can be extracted from the remote imagery depends on the correlations that can be established between reflectance patterns and biological and physical processes in the environment. In current research, training sites are used where the actual ground information is known, correlations are established with reflectance, and then this relationship is used to interpret the imagery. Insect attacks and disease processes that change the appearance of vegetation show promise of being remotely detectable. As technology for remote sensing advances, so will the reliability, resolution, and appropriateness of the data available for environmental decisions.

Model Projections

In many cases, the field, laboratory, and environmental data are not available or not appropriate to the decision being made. In these cases, results from simulation models are important sources of environmental information. These model results may be needed to complement existing information or to relate extant data to the conditions at hand. However, even when extensive data are available, the complexity of the situation may require a model for interpreting interactions. (The use of models as a tool rather than the projections that are derived from them is discussed in the next section of this chapter.)

Basically, the results from a simulation model summarize our understanding of the interrelationships, interactions, and correlations known to exist in the natural world. Properly considered, the model results do not mimic data from the real world so much as they reveal our current understanding of the environment. They can provide information regarding what the real world might and could do, but not necessarily what it *will* do. In addition, the model results *always* contain uncertainties because they are based on current understanding of interactions and field and laboratory studies. That is why we call model results *projections* (estimates of future possibilities) rather than *predictions* (something that is declared in advance). Therefore, great caution is required in basing decisions solely on model results. Models produce approximations to real situations and are only as good as the assumptions on which they are based. Until information is available to validate the model, its results should be viewed with caution. Nevertheless, the model results are the logical implications of existing data produced in a manner that assimilates and applies what we do understand.

The caution required in interpreting model calculations is perhaps best illustrated by an example documented in Van Winkle (1977). Under the scrutiny of legal proceedings, two computer simulation models were developed to determine the potential impact of a power plant on fish populations. One model, emphasizing one body of understanding, concluded that there would be little impact and that changes in the fish population could be explained by natural factors. The second model, relying on a different understanding of how fish populations interact with their environment, concluded that significant impacts would occur. Both models were subjected to intense scrutiny, but the difference in interpretation remained. The simple fact is our current level of understanding of complex environmental systems, as reflected in the model, will not always be adequate to provide simple answers to environmental questions. In spite of the limitations to our understanding of complex environmental systems, model projections remain our best source of information for extrapolating limited field and laboratory data to the real-world decision arena.

As a final note, it is important to recognize that the environmental data used in decision making are almost always a combination of types and sources of information. For example, information from remote sensing, field, laboratory, and model simulations can be combined to estimate habitat types, potential impacts, groundwater conditions, or other information that is very difficult to obtain by direct measurement.

Categories of Tools for Characterizing the Environment

Up to this stage, the chapter has discussed environmental information as a tool for characterizing the environment and issues involved in the use of that data. Now we consider the variety of approaches required to obtain the necessary information. For example, the decision may require apparently simple information, such as changes in tree size over time. However, a number of attributes describe the size of the tree, including diameter, rooting configuration, leaf area, crown height, and canopy structure. Other features of the environment are important to estimating future tree size, such as soil conditions, competing species, light levels, air quality, and potential for pathogen or insect attack. For example, observations suggest that trees grow faster under certain light and soil conditions. Diameter tapes can be used to measure the size of a tree as it responds to changing environmental conditions. Computer models can integrate all of the factors into a single projection of tree growth in the face of environmental changes over time. Statistical methods can be used to characterize the variation or mean tendency in tree size or to test if changes in the conditions would produce statistically different tree sizes. Thus, the type of information that is available influences the ability to address some pertinent questions. For example, estimating the effects of a decision to harvest trees of a particular size depends on the ability to project how the prevailing climate affects tree growth.

Types of tools used in characterizing the environment include:

- Devices to enhance human senses
- Measurement instruments
- Models
- Statistics

The strengths and limitations of these tools are summarized in Table 3.1.

Tools Used to Supplement Human Senses

Human senses are important for characterizing the environment, but are limited in their range and ability to discriminate. Tools that aid the senses include binoculars or telescopes, which enhance seeing; amplifiers, which enhance hearing; and odorants (such as those added to natural gas so leaks

TABLE 3.1. Strengths and limitations of characterization tools.

Tool	Strengths	Limitations
Devices to enhance senses	• Everyone understands them • Can produce a long-term record of sensual observations	• Difficult to compare or repeat • Some important features are out of the range of sensing
Measurement devices	• Quantitative • Repeatable	• Sophisticated devices may lead to discomfort or mistrust • May be costly, be time consuming, or require special expertise
Models	• Quantitative • Repeatable • Can summarize known information • Can be tuned to appropriate spatial and temporal resolution	• May deter people from questioning results • May be costly, be time consuming, or require special expertise • Require validation
Statistics	• Quantitative • Repeatable • Provides both average values and variations	• Public is more interested in individual effect rather than average effect • Sophistication may lead to discomfort/trust • May be costly, time consuming. or require special expertise

can be easily detected), which enhance smelling. A concern in relying on human senses is that the measure may not be repeatable. For example, early observations of air pollution were made by individuals who noted their decreasing ability to see distant mountains. However, the decline of visibility may have been caused by a loss of visual acuity with age. Sensory information has the advantage of widespread comprehension and the potential for a long-term record. The biggest limitation is that some important features are out of range of the senses even with enhancement devices.

Measurement Devices

The second type of tool consists of the large number of devices used to take measurements. Basic measurements are length, mass, time, and temperature. From these features, numerous derived measures can be attained. For example, the chemical constituents of an object can be measured by a spectrometer, which spreads the light emanating from the object into a spectrum, allowing the observation of emission or absorption lines at specific wavelengths of the light. The patterns of those lines are determined by the electronic structure of the atoms and molecules being analyzed. Therefore, the patterns can be used to identify the different types of atoms and molecules present in the sample being analyzed.

Measurement devices range from simple, direct mechanisms (such as a tape measure) to sophisticated analytical methods. The advantage of measurement devices is that observations are quantitative and may be repeated. However, application of some of these devices leads to discomfort with their interpretation and use, as is indicated by the Toutle River example in Sidebar 3.2. Some sophisticated devices lead to inappropriate confidence in the data, even when the measures are not well understood. As with models and statistics, these devices may be expensive, be time consuming, or require unavailable expertise.

Sidebar 3.2

When Mount St. Helens erupted on May 18, 1980, it created the largest landslide in history. This landslide flowed down into the North Fork of the Toutle River Valley for 23 kilometers and had an average depth of 45 meters. It reshaped Spirit Lake and created two new lakes, Castle Lake and Coldwater Lake. The unknown stability of the new material gave great concern to the towns downstream. Sudden loss of the debris avalanche material would result in a failure in the impoundment of one or more of the lakes, a torrential mudflow downstream, and an inundation of the streamside homes and businesses. Therefore, an effort was made to monitor the status of the impoundment material and, at the same time, to reduce the water level of Spirit Lake. This reduction was being done in two phases. The long-term solution was drilling a tunnel through hard rock into a reservoir on the east side of the mountain. While the tunnel was being bored, seven barges in Spirit Lake pumped water into the eastern reservoir.

While the barges were pumping the material into the eastward reservoir, the situation was explained to the local people at a town meeting. It took an hour and a half and great elaboration for officials to explain that the avalanche debris may be unstable and that extensive steps were being taken to characterize the material and to quickly notify people living adjacent to the river of any imminent impoundment failure. A laser was aimed at the impoundment so that changes in the structure would be detected instantly. Any indication of an approaching failure would be conveyed via satellite to Reston, Virginia, and transmitted over the phone lines to Seattle, Washington. The information would then be transmitted by telephone from Seattle to Toutle, Washington, at which point, helicopters would be sent up and down the river valley to warn people of the impending disaster.

After this lengthy explanation, the county sheriff stood up, rubbed his beard, and said that he did not know about the rest of the folks in the audience, but he was going to be looking for those barges floating

> downstream, and as soon as he saw them, he was going to head for the hills! This example illustrates that even though extremely sophisticated tools may be used to monitor the environment, the understanding of the situation and of the proposed means of dealing with it are of the utmost importance if the people involved are to trust the decisions made.

Environmental measurements may require specific instruments and procedures and are usually formally recorded. For example, monitoring the air temperature in a building requires that thermometers be placed in a well-ventilated location away from drafts or heat sources, such as lights. Selecting the number of temperature gauges and the time interval over which the temperature is recorded should be based on the building layout and use. (For example, does it contain a lobby where people are usually wearing coats? Is the building used at nights and/or on the weekends?)

Advances in measurement devices are increasing their accuracy and portability as well as adding new metrics to the repertoire of measurements that can be made. For example, field gas chromatographs now allow the rapid identification of chemicals and avoid the storage of samples from which chemical reactions might occur. Global-positioning satellites permit rapid and accurate identification of almost any location on Earth.

The selected approach for measuring environmental conditions should address appropriate questions and, to distinguish between alternatives, should hypothesize cause-and-effect relationships. It is not always easy to design approaches to collect information because all the pertinent information is seldom available. A hundred years ago when scientists were first examining the characteristics of uranium, they thought they were measuring a fluorescent material and designed experiments that would measure luminescence. Later studies showed that these scientists had inaccurately interpreted the darkening of photographic plates used in these experiments because they were not aware of radiation as a form of energy.

Models

Models are tools that simulate some, but not all, of the essential features of the environment. Modeling may be used to mimic the natural condition, and simulations may be done with the model to examine potential impacts of a decision. The ability to project effects is only as good as the model itself. Some sort of validation is useful to determine if the model produces realistic projections.

There are at least three types of models: heuristic, physical, and mathematical. Heuristic models are fairly simple and help to emphasize the underlying relationships of the system. They can be expressed as pictures, diagrams, words, or mathematical relationships. Sometimes scientists call these "back of the envelope" models because they can be explained in simple terms. They are appealing in that they are generally simple and relatively easy to understand. However, their simplicity may mean that some of the important interactions in the system are not adequately characterized.

Physical models are scaled-down versions of the real world. They include wind tunnels (used to examine aerodynamic properties of airplanes, cars, and seeds) and model houses (used in fire studies). An interesting example is the scale model of the Ocoee River used to design a kayaking course for the 1996 Olympics. In this model, quantities of water were allowed to run past the simulated boulders and cascades. Model kayaks were put on the miniature course so that engineers could observe potential hazards that the Olympian athletes might encounter. The physical model characterized streambed constraints on kayakers' safety and mobility and greatly reduced the cost and adverse environmental consequences of modifying the stream. Similarly, a scale model of San Francisco Bay is used to understand and predict the effects of tides and river influences on pollution dispersal, salinity, and navigation in the bay. Although the Ocoee River and San Francisco Bay models are primarily designed for use by decision makers, they are both on display for public viewing.

Mathematical models depict relationships with equations. Part of the art of modeling is making sure that the relationships are appropriate and the assumptions are realistic for the situation. The challenge of presenting the results from a mathematical model in a courtroom demonstrates the general need to effectively communicate the assumptions, form, and outcomes of the model and to choose appropriate equations for the question at hand. Swartzman (1996) discusses the use of a population-dynamics model to predict the future population size of colonial seabirds subsequent to an oil spill. The model and its projections became embroiled in litigation arising from an oil spill from a barge off the coast of California. Several lessons emerged from this experience:

• An accepted model may not make common sense. For example, a Leslie matrix model (Leslie, 1945) is commonly used to analyze population dynamics, but can project the bird population as eventually going to infinity. To avoid this unbelievable possibility being discussed in the courtroom, Swartzman introduced a density-dependent fecundity term into the model.
• A model must be simple enough for the judges, lawyers, and jury members to understand.
• Jargon must be avoided.
• The model and its projections must be clearly described; simple illustrative graphics are helpful.

These lessons are general enough to be applicable to all environmental decisions to which models might be applied.

Models have the advantage that they are quantitative and, when run in a deterministic mode, are repeatable. They are able to integrate known information from a number of different sources. They can also be tuned to the desired spatial and temporal resolution (e.g., a particular locality). However, the sophistication of models often leads to a false sense of confidence and may inhibit people from questioning the results. In addition, the use of models may be costly, time consuming, or require special expertise. Models need to be validated by comparing projections to field data; yet such a comparison is not always done and may be infeasible. Backcasting and comparing model results to historical conditions offers a useful way to validate a model (e.g., Dale and Gardner, 1987).

Many mathematical models exist, and they deal with all media (land, water, and air) and associated biotas. Space is not available in this chapter to review even a small portion of these models. Table 3.2 provides some insight into the range of available ecological models, but the table is far from comprehensive. The application of models is discussed in greater detail in Jorgensen (1994), Botkin (1992), Stalnaker (1993), and McKelvey and Hull (1996). These sources should be consulted for further information and insights. Modeling textbooks (e.g., Swartzman and Kaluzny, 1987; Bossel, 1994; and Haefner, 1996) provide introductory explanations of many aspects of ecological modeling. The textbooks are particularly helpful because of their extensive examples.

In addition to particular models to address a set management questions or to describe a particular system, tools are available for developing models. They come in the form of software packages that provide a brief introduction to modeling and to the types of models that can be implemented with the software, and they provide the analytical tools to develop a spectrum of models. For example, ModelMaker^{TM} provides the tools for developing simulation-type models (information about this software can be located on the Internet at www.cherwell.com). Another set of models is RAMAS^{TM}, which is a software library for building ecological models of age and stage structure, ecotoxicology, metapopulation dynamics, and spatial relations. The RAMAS models are described at www.ramas.com. STELLA^{TM} is a computer package for building models to simulate dynamic systems and processes (see www.hps-inc.com). Using a simple set of building block icons in STELLA, you can construct a map of a process or issue, and the program produces equations that are used to make graphs, tables, diagrams, animations, or movies. The advantage of these software libraries is that they come with manuals that present the basics of modeling as well as examples; the codes are designed to be user friendly. The limitation of this approach is that these software packages are restricted to the circumstances for which they were created. Furthermore, these models are best used when one understands the mathematical assumptions behind these approaches and the basic theory of modeling.

TABLE 3.2. Some existing environmental models with potential application to decision making.

Issue	Management situations to which models are applied
Deforestation	Tropical forests but adaptable to temperate regions[9,14,19,24]
Water pollution	Fresh water and estuaries[20,21,33]
Air pollution	Impacts of pollutants on forests[31] and grasslands[16]
Disturbance spread	Fire or pests as influenced by land management[18,28,34]
Risk assessment	Estimating values of environmental risk[3,32]
Ecosystem productivity and nutrient cycling	Effects of harvesting, grazing, irrigation, fertilization, animal production, and chemical deposition in: Forests[2,5,10,33,36] Grasslands[2,15,25,33] Deserts[23,33] Tundra[6,23,33] Tropical forests[11,23,33] Salt marshes[33] Marine plankton[33]
Fisheries	Regulating catches[33] and impacts of power plants[35]
Forest management	Strategies for maximizing productivity[1,4,33] and pest control[30]
Range management	Strategies for efficient grazing[15,22,33]
Watershed hydrology	Water and nutrient cycles within a watershed[17,27,33,7,30]
Water resource management	Impact of dams, diversions, and reservoir operations on aquatic systems[33,26]
Climate change	Impacts on land and water resources[2,8,12,29,36]
Land use	Projecting large-scale land-use changes[13]

1. Aber et al., 1982
2. Agren et al., 1991
3. Bartell et al., 1992
4. Belcher et al., 1982
5. Botkin, 1992
6. Banal and Sculler, 1975
7. Change et al., 1992
8. Dale and Rauscher, 1994
9. Dale et al., 1994b
10. Dale and Gardner, 1987
11. Doyle, 1981
12. Emanuel et al., 1985
13. Esser, 1989
14. Grainger, 1990
15. Hanson et al., 1985
16. Heasley et al., 1984
17. Huff et al., 1977
18. Johnson and Gutsell, 1994
19. Jones and O'Neill, 1992
20. Jorgensen et al., 1991
21. Jorgensen, 1994
22. Joyce and Kickert, 1987
23. King and DeAngelis, 1986
24. Lambin, 1994
25. Lauenroth et al., 1986
26. Leavesley et al., 1996
27. Luxmoore, 1983
28. O'Neill et al., 1992
29. Pastor and Post, 1988
30. Rose et al., 1991
31. Schaefer et al., 1988
32. Suter, 1993
33. Swartzman and Kaluzny, 1987
34. Turner et al., 1989
35. Van Winkle, 1977
36. VEMAP, 1995

The use of computer models can be expensive, yet their value can be well worth the cost. Unfortunately, this fact became apparent in the winter of 1997 when the state of Florida cut back on funding the model used to project freeze events. Because farmers were not warned of a possible freeze, they did not take protective actions and subsequently lost millions of dollars in crops.

Sometimes mathematical models are available that project environmental conditions, such as the distribution and type of emissions from a coal-fired power plant set in a particular locality. However, the average decision maker or private citizen may not have the skills to use these sophisticated models. Model interfaces need to be developed so that the lay public can understand the assumptions and projections from models. Ideally, such tools could be used in a gaming mode, where options for decisions are explored and the repercussions examined. For example, SimCity™ is a computer game that can project patterns and densities of urban development. As the game progresses, a need for expanded infrastructure (roads, railroads, fire or police service, power, parks or recreational facilities) becomes clear. The player who serves as city manager has the choice to spend financial resources on these various infrastructure components, and his popularity among the citizenry is estimated on the basis of how well their needs are met. In playing the game, one learns that the spatial arrangement and amount of infrastructure determines whether the city can grow or can withstand the periodic disturbances, such as fires, that occur.

Besides developing user-friendly interfaces, workshops can be held or examples of the model use can be put on the Internet that demonstrate the assumptions of a model and detail ways to interpret the projections. Conveying how models should be used in decision making becomes particularly important when the citizens perceive that they could be at risk as a result of a decision (e.g., their air quality may degrade).

Statistics

Statistics are an essential tool for characterizing the environment. The use of statistics helps to identify trends in even highly variable information and to specify the probability of occurrence of particular events (Ott, 1995). One of the ways statistics are used is to infer generalities from specific observations. Descriptive statistics can summarize large data sets and specify the range of behavior of the environmental factors over space or time, including the average, minimum, and extreme values. However, the public is often more interested in the individual-level impact of these values than the average effect. In some cases, a decision is not considered important to an individual unless it is perceived to affect the individual's own lifestyle or health. Then, the "not in my backyard" (NIMBY) philosophy may come into play.

Statistics are also important in accepting or rejecting hypotheses of the cause and effect of a phenomenon. Numerous statistical tests are available to compare observations. Each test has an explicit set of assumptions. A concern with some environmental data is that the assumptions of the test may be violated. For example, many environmental trends do not follow normal distributions, and nonparametric tests (which do not assume normality) should be used in those cases.

Finally, statistics provide a method for determining how and when data should be collected and how many observations need to be obtained (e.g., Green, 1979; Greig-Smith, 1983; Hurlbert, 1984; Dale et al., 1991). Sampling and experimental approaches are even available for environmental issues that involve several factors at the same time.

Bayesian statistical inference offers some advantages to decision makers. Instead of testing a hypothesis with a large sample, Bayesian statistics uses knowledge of prior probabilities, usually based on a large amount of valid data. As with classical statistics, Bayesian approaches can estimate environmentally meaningful parameters and provide a measure of the uncertainty in parameter estimates (Ellison, 1996). Bayesian decision-theoretic techniques can use prior opinions to formalize possible consequences of decisions (Wolfson et al., 1996). Use of prior opinions may be particularly important for environmental decision making where uncertainties can be high. However, care should be taken not to use prior information unless it is considered reliable by all concerned (Edwards, 1996).

Making Tools Both Used and Useful and Communicating Their Results

It is important to consider who uses tools to characterize the environment in order to evaluate their effectiveness. The range of potential users is wide. Decision makers or people who provide information to them might be trained industry or government staff who are very knowledgeable about sophisticated tools. The other actors in the environmental-decision context generally are private citizens who may not have formal education, but who are aware of how decisions about the use of the environment can affect their livelihoods and aspects of the environment that they value. These people may not be familiar with sophisticated tools, such as models or satellite imagery, but they certainly understand both the short- and long-term implications of decisions. The challenge is to use tools to characterize the environment in a way that meets the needs of all these people.

Decision makers often find themselves as the communication link between sophisticated tools and private citizens. This function requires the ability to speak the language of science as well as to speak the language of local people. Decision makers vary greatly in their success at such outreach activities. Because there are so many types of users, a variety of tools must be considered, and communication of the information gleaned by these tools is of utmost importance.

Conclusions

The complexity of the natural and built systems and our meager comprehension of that complexity make the job of environmental decision making difficult. The available sources of environmental information will vary from issue to issue, and the decision maker must focus on balancing the advantages and limitations of each of these data sources and the tools used to collect the information. An understanding of the basic characteristics of the data and tools is critical to weighing the information used in the decision-making process.

Acknowledgments. Discussions with Glenn Suter were extremely insightful. Reviews of the paper by John Beauchamps, Rebecca Efroymson, Robert Washington-Allen, Mary English, Howard Kunreuther and Charlie Van Sickle are appreciated. Bob Cushman, Linda Mann, Dick Olson, and Mike Sale pointed the authors to appropriate references for some of the information. The development of the paper was supported by an NSF grant to the National Center for Environmental Decision-Making Research (NCEDR). This material is based upon work supported by the National Science Foundation under Grant SBR-9513010. Any opinions, findings, and conclusions or recommendations expressed in this material are those of the authors and do not necessarily reflect the views of the National Science Foundation. Oak Ridge National Laboratory is managed by Lockheed Martin Energy Research Corp. for the Department of Energy under contract DE-AC05-96OR22464. This is Environmental Sciences Division publication number 4696.

Key Resources

Ways to obtain measurements of environmental conditions are summarized in books by Greig-Smith 1983 and Hildebrand and Cannon 1993. Statistical methods for environmental applications is the focus of books by Green 1979 and Ott 1995. Numerous books on ecological modeling are referenced within the chapter, but the most appropriate introductory text would be the books by Jorgensen 1994 and Swartzman and Kuluzny 1987.

References

Aber, J.D., Melillo, J.M., and Federer, C.A. 1982. Predicting the effects of rotation length, harvest intensity, and fertilization on fiber yield from northern hardwood forests in New England. *Forest Science* 28:31–45.

Agren, G.I., McMurtie, R.E., Parton, W.J., Pastor, J., and Shugart, H.H. 1991. State-of-the-art of models for production-decomposition linkages in conifer and grassland ecosystems. *Ecological Applications* 1:118–138.

Andren, H. 1994. Effects of habitat fragmentation on birds and mammals in landscapes with different proportions of suitable habitat: A review. *Oikos* 71:355–366.

Bartell, S.M., Gardner, R.H., and O'Neill, R.V. 1992. *Ecological Risk Estimation.* Ann Arbor, MI Lewis Publishers.

Belcher, D.W., Holdaway, M.R., and Brand, G.J. 1982. *A Description of Stems, the Stand and Tree Evaluation and Modeling System.* General Technical Report NC-79. ST. Paul, MN: U.S. Department of Agriculture, Forest Service, North Central Forest Experiment Station.

Bernard, D.P., Hunsaker, D.B., Jr., and Marmorek, D.R. 1993. Tools for improving predictive capabilities of environmental impact assessments, structured hypotheses, audits and monitoring. In: S.G. Hildebrand and J.B. Cannon (Eds.). *Environmental Analysis: The NEPA Experience*. Boca Raton, FL: Lewis Publishers. Pp. 547–564.

Boden, T.A., Kaiser, D.P., Sepanski, R.J., and Stoss, F.W. 1994. *Trends '93: A Compendium of Data on Global Change*, ORNL/CDIAC-65. Oak Ridge, TN: Oak Ridge National Laboratory.

Bossel, H. 1994. *Modeling and Simulation*. Wellesley, MA: A.K. Peters.

Botkin, D.B. 1992. *Forest Dynamics: An Ecological Model*. Oxford: Oxford University Press.

Brown, D.R., Kane, H., and Acres, E. 1993. *Vital Signs 1993: The Trends That Are Shaping Our Future*. New York: W.W. Norton & Company.

Banal, F.L. and Sculler, K.A. 1975. ABISKO II: A computer simulation model of carbon flux in tundra ecosystems. In: T. Rosswall and O.W. Heal (Eds.). *Structure and Function of Tundra Ecosystems*, Ecological Bulletin 20. Stockholm: Swedish National Science Research Council. Pp. 425–448.

Chang, L.H., Draves, J.D., and Hunsaker, C.T. 1992. *Climate Change and Water Supply, Management, and Use: A Literature Review*, ORNL/CDIAC-52. Oak Ridge, TN: Oak Ridge National Laboratory.

Clarke, R. 1986. *The Handbook of Ecological Monitoring*. Oxford, England: Clarendon Press.

Council on Environmental Quality. 1979. *Environmental Quality 1979: Tenth Annual Report of the Council on Environmental Quality*. Washington, D.C: USGPO.

Dale, V.H., Franklin, R.L.A., Post, W.M., and Gardner, R.H. 1991. Sampling ecological information: Choice of sample size. *Ecological Modelling* 57:1–10.

Dale, V.H. and Gardner, R. 1987. Assessing regional impacts of growth declines using a forest succession model. *Journal of Environmental Management* 24:83–93.

Dale, V.H., Offerman, H., Pearson, S., and O'Neill, R.V. 1994a. Effects of forest fragmentation on neotropical fauna. *Conservation Biology* 8:1027–1036.

Dale, V.H., O'Neill, R.V., Southworth, F., and Pedlowski, M.A. 1994b. Modeling effects of land management in the Brazilian settlement of Rondônia. *Conservation Biology* 8:196–206.

Dale, V.H. and Rauscher, H.M. 1994. Assessing impacts of climate change on forests: The state of biological modeling. *Climatic Change* 28:65–90.

Dennis, J.V. 1971. Species using red-cockaded woodpecker holes in northeastern South Carolina. *Bird Banding* 42:79–87.

Doyle, T. 1981. The role of disturbance in the gap dynamics of a montane rain forest: An application of a tropical forest succession model. In: D.C. West, H.H. Shugart, and D.B. Botkin (Eds.). *Forest Succession: Concepts and Applications*. New York: Springer-Verlag. Pp. 56–73.

Draggan, S., Cohrssen, J.J., and Morrison, R.E. 1987. *Environmental Monitoring, Assessment and Management*. New York: Praeger Press.

Eco-Inforum. 1996. *Proceedings of Eco-Informa '96: Global Networks for Environmental Information, Nov. 4–7, 1996, Lake Buena Vista, FL.*, Ann Arbor, MI: Environmental Research Institute of Michigan.

Edwards, D. 1996. Comment: The first data analysis should be journalistic. *Ecological Applications* 6:1090–1094.

Ellison, A.M. 1996. An introduction to Bayesian inference for ecological research and environmental decision-making. *Ecological Applications* 6:1036–1046.

Emanuel, W.R., Shugart, H.H., and Stevenson, M.P. 1985. Climate change and the broad scale distribution of terrestrial ecosystem complexes. *Climate Change* 7:29–43.

Esser, G. 1989. Global land use changes from 1860 to 1980 and future projections to 2,500. *Ecological Modelling* 44:307–316.

Golden, J., Oullette, R.P., Saari, S., and Cheremisinoff, P.N. 1980. *Environmental Impact Data Book*. Ann Arbor, MI: Ann Arbor, Science.

Grainger, A. 1990. Modelling deforestation in the humid tropics. In: M. Palo and G. Mery (Eds.). *Deforestation or Development in the Third World?* Bulletin 349, Vol. III. Helsinki, Finland: Finnish Forest Research Institute. Pp. 51–67.

Green, R.H. 1979. *Sampling Design and Statistical Methods for Environmental Biologists*. New York: John Wiley and Sons.

Greig-Smith, P. 1983. *Quantitative Plant Ecology* (3rd ed.). New York: Plenum Press.

Gross, K.L. and Pake, C.E. 1996. *Final Report of the Ecological Society of America Committee on the Future of Long-Term Ecological Data (FLED)*. Washington, D.C.: Ecological Society of America.

Haefner, J.W. 1996. *Modeling Biological Systems: Principles and Applications*. New York: Chapman and Hall.

Hain, F.P. and Arthur, F.N. 1985. The role of atmospheric deposition in the latitudinal variation of fraser fir mortality caused by the balsam woolly adelgid, *Adelges piceae* (Ratz.) (Hemipt., Adelgidae): A hypothesis. *Zeitschrift fuer Angewandte Entomologie* 99:145–152.

Hanson, J.D., Parton, W.J., and Innis, G.S. 1985. Plant growth and production of grassland ecosystems: A comparison of models. *Ecological Modelling* 29:131–144.

Heasley, J.E., Lauenroth, W.K., and Yorks, T.P. 1984. Simulation of SO_2 impacts. In: W.K. Lauenroth and E.M. Preston (Eds.). *The Effects of SO_2 on a Grassland*. New York: Springer-Verlag. Pp. 161–184.

Hildebrand, S.G. and Cannon, J.B. (Eds.). 1993. *Environmental Analysis: The NEPA Experience*. Boca Raton, FL: Lewis Publishers.

Huff, D.D., Luxmoore, R.J., Mankin, J.B., and Begovich C.L. 1977. *TEHM: A Terrestrial Ecosystem Hydrology Model*, EDFB/IBP-76/8. Oak Ridge, TN: Oak Ridge National Laboratory.

Hurlbert, S.H. 1984. Pseudoreplication and the design of ecological field experiments. *Ecological Monographs* 54:187–211.

Iverson, L.R., Graham, R.L., and Cook, E.A. 1994. Regional forest cover estimation via remote sensing: The calibration center concept. *Landscape Ecology* 9:159–174.

Jackson, J.B.C. et al. 1989. Ecological effects of a major oil spill on Panamanian Coastal Marine Communities. *Science* 243:37–44.

Johnson, E.A. and Gutsell, S.L. 1994. Fire frequency models, methods, and interpretations, *Advances in Ecological Research* 25:239–279.

Jolly, G.M. 1965. Explicit estimates from capture-recapture data with both death and immigration: Stochastic model. *Biometrika* 52:225–247.

Jones, D.W. and O'Neill, R.V. 1992. Endogenous environmental degradation and land conservation: Agricultural land use in a large region. *Ecological Economics* 6:79–101.

Jorgensen, S.E. 1979. *Handbook of Environmental Data and Ecological Parameters.* Oxford, England: Pergamon Press.

Jorgensen, S.E., Nielsen, S.N., and Jorgensen, L.A. 1991. *Handbook of Ecological Parameters and Ecotoxicology.* Amsterdam: Elsevier.

Jorgensen, S.E. 1994. *Fundamentals of Ecological Modeling.* Amsterdam: Elsevier.

Joyce, L.A. and Kickert, R.N. 1987. Applied plant growth models for grazing lands, forest, and crops. In: K. Wisiol and J.D. Hesketh (Eds.). *Plant Growth Modeling for Resource Management. Volume I: Current Models and Methods.* Boca Raton, FL: CRC Press. Pp. 17–55.

King, A.W. and DeAngelis, D.L. 1986. *Site-Specific Seasonal Models of Carbon Fluxes in Terrestrial Biomes,* ORNL/TM-9749. Oak Ridge. TN: Oak Ridge National Laboratory.

Lambin, E.F. 1994. *Modelling Deforestation Processes. Trees,* Series B, Report 1. Luxembourg: Office for Official Publications of the European Community.

Lauenroth, W.K., Hunt, H.W., Swift, D.M., and Singh, J.S. 1986. Estimating aboveground net primary production in grasslands: A simulation approach. *Ecological Modelling* 33:297–314.

Leavesley, G.H., Markstrom, S.L., Brewer, M.S., and Viger, R.J. 1996. The modular modeling system (MMS): The physical process modeling component of a database-centered decision support system for water and power management. *Water, Air, and Soil Pollution* 90:303–311.

Leslie, P.H. 1945. On the use of matrices in certain populations mathematics. *Biometrika* 33:183–212.

Luxmoore, R.J. 1983. Water budget of an eastern deciduous forest stand. *Soil Science Society of America Journal* 47:785–791.

McKelvey, R. and Hull, V. (Eds.). 1996. Special issue: Ecological resource modeling. *Ecological Modelling* 92.

National Research Council. 1993. *Issues in Risk Assessment.* Washington, D.C.: National Academy Press.

O'Neill, R.V., Gardner, R.H., Turner, M.G., and Romme, W.H. 1992. Epidemiology theory and disturbance spread on landscapes. *Landscape Ecology* 7:19–26.

Ott, W.R. 1995. *Environmental Statistics and Data Analysis.* Boca Raton, FL: Lewis Publishers.

Paine, R.T. 1980. Food webs: Linkage, interaction strength, and community infrastructure. *Journal of Animal Ecology* 49:667–685.

Pastor, J. and Post, W.M. 1988. Response of northern forests to CO_2 induced climate change. *Nature* 334:55–58.

Robbins, C.S., Fitzpatrick, J.W., and Hamel, P.B. 1989. A warbler in trouble: *Dendroica cerulea.* In: J.M. Hagan, III, and D.W. Johnston (Eds.). *Ecology and Conservation of Neotropical Land Birds.* Washington, D.C.: Manomet Bird Observatory and Smithsonian Institution Press. Pp. 549–562.

Rose, K.A., Brenkert, A.L., Cook, R.B., Gardner, R.H., and Hettelingh, J.P. 1991. Systematic comparison of ILAWS, MAGIC, and ETD watershed acidification models. 2. Monte Carlo analysis under regional variability. *Water Resources Research* 27:2591–2063.

Schaefer, H., Bossel, H., Krieger, H., and Trost, N. 1988. Modeling the responses of mature forest trees to air pollution. *GeoJournal* 17:279–287.

Southern Appalachian man and the Biosphere Cooperative (SAMAB). 1996. *The Southern Appalachian Assessment,* five volumes. Washington, D.C.: U.S. Department of Agriculture.

Stalnaker, C.B. 1993. Fish habitat evaluation models in environmental assessments. In: S.G. Hildebrand and J.B. Cannon (Eds.). *Environmental Analysis: The NEPA Experience*. Boca Raton, FL: Lewis Publishers. Pp. 140–162.

Suter, G.W. 1993. *Ecological Risk Assessment*. Boca Raton, FL: Lewis Publishers.

Swartzman, G.L. and Kaluzny, S.P. 1987. *Ecological Simulation Primer*. New York: MacMillan.

Swartzman, G. 1996. Resource modeling moves into the courtroom. *Ecological Modelling* 92:277–288.

Tucker, C.J. 1979. Red and photographic infrared linear combinations monitoring vegetation. *Remote Sensing of Environment* 8:127–150.

Turner, M.G., Gardner, R.H., Dale, V.H., and O'Neill, R.V. 1989. Predicting the spread of disturbances across heterogeneous landscapes. *Oikos* 55:121–129.

U.S. Environmental Protection Agency (USEPA). 1992. *Framework for Ecological Risk Assessment*, EPA/630/R-92/001. Washington, D.C.: U.S. Environmental Protection Agency and Risk Assessment Forum.

USEPA Quality Assurance Management Staff. 1994. *Guidance for the Data Quality Objectives Process*, EPA QA/G-4. Washington, D.C.: U.S. Environmental Protection Agency.

Van Winkle, W. (Ed.). 1977. *Assessing the Effects of Power-Plant-Induced Mortality on Fish Populations*. New York: Pergamon Press.

VEMAP. 1995. Vegetation/ecosystem modeling and analysis project: Comparing biogeography and biogeochemistry models in a continental-scale study of terrestrial ecosystem responses to climate change and CO_2 doubling. *Global Biogeochemical Cycles* 9:407–437.

Westman, W.E. 1985. *Ecology, Impact Assessment, and Environmental Planning*. New York: John Wiley and Sons.

Wiens, J. 1996. Oil, seabirds, and science: The effects of the Exxon Valdez oil spill. *BioScience* 46:587–597.

Wolfson, L.J., Kadane, J.B., and Small, M.J. 1996. Bayesian environmental policy decisions: Two case studies. *Ecological Applications* 6:1056–1066.

Decision-Maker Response

WILLIAM R. MILLER III

The collection, analysis, synthesis, interpretation, and presentation of environmental data is perhaps one of the more challenging areas of scientific inquiry. All aspects of societal interaction are involved in the use of environmental data and how decisions are made based on the data at hand.

By way of a humorous example, let me recount a true story that occurred during the initial site assessment in 1985 at the new home of Saturn in Spring Hill, Tennessee. As you may recall, there was a lot of interest in July 1985 when General Motors announced that the small, rural town of Spring Hill was going to be the home of the new $2-billion Saturn plant. One of the initial tasks we had when we took over 13 working farms was to remove and dispose of numerous, problematic farm chemicals that were no longer needed. We ran into Toxaphene, DDT, 2,4-D, 2,4,5-T, used oil, waste gasoline, and many other hazardous materials. On one occasion, we discovered a 55-gallon drum on the second floor of a two-story dairy barn. The drum was a waste profiler's nightmare. The bung on the drum was gone, a thick black liquid was oozing out of the top, and the drum was obviously swollen. It had all the indications of a drum that was going to require special handling and care. Our waste-handling consultant had someone suit up in Level A containment (a fully encapsulated space suit) to take a sample for analysis so we could properly assess the correct handling and disposal methods. We knew that it would take about a week to get the results back. In the meantime, I had mentioned the drum to a local farmer who indicated he might know what the "mystery drum" was and how to handle it. We crept up the ladder to the second floor of the dairy barn and approached the drum with caution. I told the farmer that samples had been taken to determine the drum's contents; and before I knew it, the farmer walked over to the drum, stuck his finger in the black ooze, and licked it. I nearly went ballistic thinking how great it was going to be that one of my first acts of community environmental interaction was going to be the death of a prominent local citizen. Fortunately for me, the farmer turned around after he licked his finger, and said, "It's molasses for the cattle feed." Well, as you might expect, the analysis came back high in unknown carbohydrates, and the disposal method of choice was incineration in a hazardous waste incinerator at $750 for the drum. Ultimately, we gave the drum to the farmer who had identified it. He blended it with oats and fed it to his cattle. I relay this story to help point out the complexities we face sometimes as environmental professionals in dealing with complex data sets.

We are often faced with very complicated environmental issues and data that must be communicated to others, most often the public, in a straightforward and uncomplicated fashion. The intent of this chapter is to provide an overview of available tools for characterizing the environmental setting and for providing useful data to aid in decision making. The authors have done an excellent job in summing up a complex topic while simultaneously providing enough breadth for the practitioner who deals with these issues every day.

Let me start by identifying the customers of environmental data (i.e., who needs to know what was collected, what was found, and what it means):

- The public
- The regulatory community
- Upper management
- Fellow workers
- Neighbors

Even though you may be working with a common data set, the way you communicate to each of these customers is quite different. The public's interest may vary from one of no concern to outrage, depending on the perceived risk of the environmental situation under study.

Explaining environmental data to the public is further complicated by the inherent mistrust that sometimes accompanies the communication of complex, scientific information. Regulators are familiar with dealing with complex information on a routine basis; that is the basis for much of what they do. However, regulators are frequently in the unenviable position of balancing the public good against the prevailing politics of the time, positions that at times may be diametrically opposed. Upper management within a company or industry has historically been very nervous about environmental issues. Typically, the only time it would ever hear about an environmental issue at its location would be when something went wrong (e.g., an underground tank leaked into a neighbor's well and the neighbor was going to sue; a key process was out of compliance with an environmental limit and production would suffer; or a new law was passed that increased environmental penalties and also made it possible for a senior manager to be found criminally negligent). Fellow workers at a company are normally concerned about environmental data from a number of perspectives. They usually want their company to do well, and environmental performance is often a key indicator of how attentive a company is to detail and corporate responsibility. Often there is a link between an environmental issue and health in the workplace that is of paramount concern to many. Finally, neighbors are a key customer of the methods and techniques used to communicate environmental decision-making results. I separate neighbors from the public because frequently neighbors are most severely impacted when something goes wrong, and therefore they are the most important and often most difficult customers to deal with.

A number of issues must be addressed in selecting an environmental tool and the method(s) of communicating decisions to a diverse customer base. These issues include:

- Why was the data collected in the first place? Was it a regulatory requirement? Is it to avoid problems in the future? Is it in support of anticipated litigation? Is it pure research?
- Who are the potential customers for the results?
- Will the customers trust the data collectors?
- Will the data support a popular political position?
- How will the data be displayed: graphically, geographically, numerically, narratively, publicly, privately, or otherwise?
- How much money was spent to collect the data?

The tool(s) chosen to communicate environmental decisions vary with the intended customer(s) and objectives for the study. This chapter provides the reader with a method for choosing the right tool given the intent of the study.

Finally, to broaden participation by diverse customers, a greater understanding is needed of how to communicate environmental decisions. Historically, many barriers have hindered accomplishing this task. Those barriers include mistrust, risk communication, lack of shared risk and reward, suspicion, hidden agendas, complexity, politics, and resources. Through dialogue like that established through the research done by groups like the National Center for Environmental Decision-Making Research (NCEDR), we can begin not only to overcome these barriers, but also to build trust across a diverse customer base.

4
Tools for Understanding the Socioeconomic and Political Settings for Environmental Decision Making

WILLIAM R. FREUDENBURG

Under a broad definition, the number of tools for characterizing the socioeconomic and political settings for environmental decision making is nearly infinite, including the accumulated techniques of all the social sciences. I will focus on the tools used most often by social scientists in environmental decisions or, in some cases, actually developed for such decisions.

Most policy-level decision makers have only a minor interest in the full complexity of the human environment; instead, interest generally centers around understanding the ways in which decisions might influence how humans value and interact with the environment. Much of the relevant social-science work, accordingly, is found in the interdisciplinary specialty known as social (or socioeconomic) impact assessment (SIA).[1] Fortunately,

---

[1] While social and behavioral scientists have played a variety of roles in environmental management, the level of involvement was greatly increased by the National Environmental Policy Act, or NEPA (P.L. 91–190, 42 U.S.C. 4371 *et seq.*). Along with its requirement that federal agencies prepare environmental impact statements (EISs) for "major federal actions significantly affecting the quality of the human environment," Sect. 102(2)(A) of the act requires federal agencies to make "integrated use of the natural *and social sciences* . . . in decision making which may have an impact on man's environment" [emphasis added]. Section 1508.8 of the official *Regulations for Implementing NEPA* (U.S. Council on Environmental Quality, 1978; 40 CFR 1500 *et seq.*) notes that EISs need to consider direct and indirect social and cultural impacts, as well as physical and biological impacts. Section 1508.14 of the *Regulations* notes that, while social and economic effects by themselves do not require preparation of an EIS, "When an environmental impact statement is prepared" because of physical environmental impacts, and when the social and the bioenvironmental impacts are interrelated, "then the environmental impact statement will discuss all of these effects upon the human environment" (Council on Environmental Quality, 1978:29; see also Savatsky, 1974; Meidinger and Freudenburg, 1983; Jordan, 1984; Freudenburg and Keating, 1985; Llewellyn and Freudenburg, 1990).

Probably the key definition of "the human environment" is the one provided by the President's Council on Environmental Quality (CEQ), which had lead responsibility for overseeing implementation of the law. CEQ's Regulations for Implementing the Procedural Provisions of the NEPA (40 CFR 1508.14) note that

SIA has developed to the point that the major social-science professional associations have been able to compile a set of consensus guidelines and principles for such assessments (see Interorganizational Committee on Guidelines and Principles for Social Impact Assessment 1994, hereafter referred to as Interorganizational Committee on SIA 1994). These guidelines and principles will be drawn upon in several portions of this chapter, and a relatively typical list of the variables that need to be considered is provided in Table 4.1.

Those accustomed to thinking about the environment in biological terms find both similarities and differences when it comes to *Homo sapiens*. Among the similarities is the need to consider the *distributions* as well as the *severity* of the impacts a given decision might produce. In addition, the extent to which impacts might be *cumulative* and/or *reversible* must be considered. For example, when a scientist is attempting to understand the biophysical environment, concerns will generally extend to a wide range of species; ignoring the potential impacts on one group of species is not considered appropriate, even if those impacts would aid other species. If a proposed development would change the temperature and chemical composition of a lake's water, posing threats to a specific species of fish, it would be inappropriate to ignore those threats even if the overall level of biomass in the lake were to be increased. Similarly, overlooking the impacts on specific segments of the human population in SIAs is inappropriate, even if other segments are likely to benefit (e.g., if an area's overall income would be increased). Just as biologists need to devote particular attention to species having special vulnerabilities, socioeconomic analyses need to devote special attention to the impacts on particularly vulnerable segments of the human population, such as the poor; the elderly; adolescents; unemployed women; minorities; or those occupational, cultural, or political groups for whom a given community, region, or use is particularly important.

At the same time, major differences exist between the impacts on the biophysical environment and those on the human environment. The most

---

the "human environment" is to be "interpreted comprehensively," to include "the natural and physical environment and the relationship of people with that environment." Agencies need to assess not just so-called "direct" impacts or effects, but also "aesthetic, historic, cultural, economic, social, or health" impacts, "whether direct, indirect, or cumulative" (40 CFR 1508.8). Perhaps the clearest congressional statement on the meaning of "the human environment" in a statute involving natural-resource development is provided by a different law, the Outer Continental Shelf Lands Act, as amended (43 U.S.C. 1331 *et seq.*). In that law, the term is given at least as expansive a definition: "The term 'human environment' means the physical, social, and economic components, conditions, and factors which interactively determine the state, condition, and quality of living conditions, employment, and health of those affected, directly or indirectly" by the resource-development activities in question (43 U.S.C. 1331 (i)).

TABLE 4.1. Variables to be considered in the assessment of the social impacts of a proposed policy or project.

Population
Will the population of the affected area be changed?
Will the area's ethnic, racial, or cultural mix be altered?
Will populations be relocated?
Will the area experience influxes and outflows of temporary workers?
Will the population change seasonally?

Community and Institutional Structures
How stable are the extant residential communities?
What is the current density of acquaintanceship and how will it change?
What effects will debates about the project's opportunities and threats have on the community?
How will the activities and staffing of the community's voluntary organizations be affected?
What interest-group activities might be anticipated?
Will the size and structure of the local government need to be changed?
Has the community experienced relevant changes in the past, and how did it deal with those changes?
Will any new local, regional, or national linkages need to be established?
Are adequate planning and zoning activities in place?

Political and Social Resources
How will the distribution of power and authority be affected?
Who are the stakeholders in this situation?
Who are the interested and affected publics?
Can the local organizations, institutions, and social structures muster the required leadership capabilities and characteristics?

Individuals and Families
Are any risks to health and safety perceived?
Are there any concerns about displacement or relocation?
How strong is the citizenry's trust in the local political and social institutions?
What are the popular attitudes toward the policy or project that is under consideration?
How tight and enduring are the current family and friendship networks?
What concerns does the populace have about the social well-being of the community?

Community Resources
Will the diversity of industrial and commercial enterprises in the community be affected?
What changes might be effected or required in the community's infrastructure?
What levels of employment and compensation will result?
Will minorities be equitably employed?
Will Native American tribes be affected?
How might land-use patterns be altered?
What impacts might be made on cultural, historical, and archaeological resources?

obvious differences include the sheer pervasiveness of human impacts on the global ecosystem and the fact that other species are not likely to read a scientific report about themselves and challenge its conclusions in court. Perhaps the key difference, however, has to do with the distribution of impacts over time. In the physical or biological sciences, impacts generally do not take place until a project actually alters the physical or biological conditions. In the human environment, though, observable and measurable

4. Understanding the Socioeconomic and Political Settings 97

impacts can take place as soon as there are changes in *socioeconomic* conditions, which often means from the time of the earliest announcements or rumors about a project. In the words of Freudenburg and Gramling (1992, p. 941):

Speculators buy property, politicians maneuver for position, interest groups form or redirect their energies, stresses mount, and a variety of other social and economic impacts take place, particularly in the case of facilities that are large, controversial, risky, or otherwise out of the range of ordinary experiences for the local community. These changes have sometimes been called "pre-development" or "anticipatory" impacts, but they are far more real and measurable than such terminology might imply. Even the earliest acts of speculators, for example, can drive up the *real* costs of real estate.

For example, when buildings were identified as being in the path of future highways, landlords were found to cut back on maintenance and repairs, even if years were to pass before any highway construction took place (Llewellyn et al., 1981; Llewellyn, 1981). A more recent example is the case of a proposed storage facility for low-level radioactive waste in New York State. The "pre-development" impacts were ignored in decision-making documents because they were thought to be "too speculative" or "not real." When the plans became known, citizens set up round-the-clock lookouts at county boundaries, chained themselves to bridges, and took other actions that ultimately prevented the state's Siting Commission from setting foot on a single site. The net effect was that none of the supposedly "predictable" changes to the physical environment ever took place, while the supposedly "unpredictable" public hostility proved to be so real that it led to the downfall of the siting process (National Academy of Sciences/National Research Council, 1996). In fact, as social scientists had attempted to explain to the Siting Commission even before the public explosion took place, any behavior pattern that is essentially universal should prove relatively simple to predict.

As the New York State Siting Commission and other organizations have learned, these are not merely "perceived" or "anticipated" impacts. They are real, and they carry real consequences. The technical literature, accordingly, has begun to refer to the earliest stages of development as the "opportunity-threat phase" (National Academy of Sciences, 1996; Interorganizational Committee on SIA, 1994). The terminology reflects the fact that the socioeconomic impacts characterizing this phase of development result predominantly from the efforts of interested parties to define and to respond to the implications of development, whether as opportunities (for those who see the changes as positive) or as threats (for those who feel otherwise). The process of attempting to shape both the development and the way in which it is perceived can play a key role in determining the social and economic impacts, not only in cases where no facility is ultimately constructed, but also in cases where construction and development actually take place. In the latter case, however, additional impacts can be expected. The impacts that emerge are shaped by the characteristics of the social negotiation process, such as fairness and openness (Creighton, 1980; Howell and Olsen, 1981) as well as by the nature of the proposal under consideration.

Another difference between the human and biophysical environments is that, in the case of the human environment, the policy world has at times shown a greater interest in the *tools* than in the *findings* of socioeconomic analyses. In a depressing number of cases, the same agencies that have been willing to invest significant resources in producing another how-to manual have proved unwilling to invest in improving the empirical database. As a result, dozens of manuals spell out "how to" carry out social or socioeconomic impact analyses (for a recent overview, see Interorganizational Committee on SIA, 1994) while the available base of empirical findings is much weaker than it might have been. Unfortunately, despite an apparent precision and sophistication, the models and other tools can be no better than the data and the assumptions that go into them. In many cases, the needed information is lacking, and the models rely on assumptions that will make sense only as long as the system has a high level of continuity.

Still, it is necessary to avoid not just unduly optimistic expectations, but also overly pessimistic conclusions. After all, no law of nature prevents us from using the scientific method just because questions of human behavior are involved. In addition, there are many encouraging signs that the gap between the social sciences and the biological and physical sciences is beginning to be bridged. For example, social scientists are showing interest in the complex yet vital interrelationships between human beings and their biophysical surroundings; and biophysical scientists and resource managers are recognizing the need to understand the same interrelationships. As one colleague put it, "We call ourselves 'fisheries managers,' but we don't really manage fisheries as much as we try to manage people; and we need more help from you social scientists to do our job right." Such insights, of course, are only an initial step. A further step is the need to recognize that human beings are often quite resistant to being "managed," particularly by those who fail to understand the complexities of the systems within which human behaviors take place. Given the need to obtain an improved understanding of the complexities of the interrelationships between *Homo sapiens* and the rest of the ecosystem, however, the tools reviewed in this chapter offer a reasonable way in which to start.

Characteristic Tools

Information on the socioeconomic and political settings for environmental decisions can be obtained with three characteristic tools. The first involves the use of existing data that are at least potentially quantifiable, whether in the form of publicly available data sets, archival information, agency records, or other sources. Given that these data sources are compiled by someone other than the researcher in question, this characteristic tool is often seen as involving secondary data or archival techniques. The second characteristic tool requires original or first-hand data collection through what are often called fieldwork techniques, which can require research time in the field for gathering data. The third characteristic tool includes techniques for identifying and dealing with the gaps and blinders that so often bedevil environmental decisions. The gaps involve information that is directly relevant to a decision but that is not traditionally included within the disciplines that have been consulted for that decision. The blinders involve matters that might actually be within the traditional concerns of one or more of the disciplines but that fail to receive adequate attention.

Archival/Secondary Techniques

Standard Information Modules

Each of the three characteristic tools is made up of an array of specific, relevant techniques. For archival and/or secondary data, the more specific techniques can be grouped within three major subcategories.

The first subcategory involves the systematic effort to identify socioeconomic environments and/or decisions that are to some degree similar to the one in question. Such an exercise is sometimes relatively straightforward, but at other times it is far more of an art than of a science, requiring a substantial level of judgment. The appropriateness of a comparable situation will often depend on the uniqueness or sensitivity of a given socioeconomic setting and on the nature of the decision being considered. In some cases (e.g., if one merely needs to make some reasonable assumptions about the levels of formal education or the availability of electricity-distribution systems), it may be sufficient to know whether the decision involves a relatively developed country such as the United States or a less-developed one such as Bangladesh. In other cases, higher levels of resolution will be important, such as knowing whether the area is predominantly urban or rural; whether a rural region is more or less affluent; whether a less affluent rural area is predominantly white or includes more persons of other races; or whether distinctive cultural characteristics of the predominantly white residents of a less-affluent rural area within the United States come into play. An example might be the importance of knowing whether the residents of a region are predominantly Mormon (a characteristic often associated with particularly strong support for development and progress) or predominantly urban refugees, who will often view proposed developments as offering anything but "progress."

Even where the characteristics can be identified most narrowly, it is often important to recognize that the remaining variation can be truly substantial.

For example, two predominantly Mormon communities of comparable population and affluence, located within one or two hundred miles of each other in the intermountain western United States, can differ quite dramatically in how they view a proposed nuclear waste facility. The differences might spring from a community's being downwind during the above-ground nuclear-weapons testing in the 1950s, from its distance to the nearest railroads that might carry nuclear waste, from the ability of residents to reach the proposed site by road for employment, or from many other factors.

Within any given community, moreover, substantial variation can virtually be guaranteed. One group might well embrace a proposed facility, while others passionately oppose it. Differentiation is a particularly important fact of life when considering the human environment: In political elections, after all, a vote in which one candidate gains the support of 60 percent of the voters is commonly considered a landslide. In science, by contrast, a decision that effectively ignores 40 percent of the population (or in some cases, 4 percent or even .4 percent of the population) is simply unacceptable. (See, for example, the discussion in Interorganizational Committee on SIA, 1994.)

In spite of all of these limitations, information on other communities and/or decisions similar to this one will almost always be important to a decision. The best of all possible worlds, in terms of decision-making, may be one in which enough communities of a given type have actually been studied to permit the development of what Finsterbusch (1980; 1995) has termed "standard information modules." The prototypical example of standard information modules involves the Rocky Mountain energy boomtowns of the 1970s and early 1980s. Enough studies were done in enough specific communities (which in turn had enough similarities to one another) to produce an overall assessment of such experiences (Finsterbusch, 1980) and to permit a limited degree of quantitative comparison or meta-analysis to test and refine the overall assessments. (See, for example, Freudenburg and Jones, 1992. For other examples of standard information modules, see Finsterbusch, 1995; Boothroyd et al., 1995.)

For most locations, unfortunately, the existence of enough research to permit the development of anything like standard information is unlikely. Even in the case of large-scale, energy-related construction projects in the western United States, most of the relevant compilations of information are now well over a decade old. (See especially Finsterbusch, 1980; see also Freudenburg, 1986; Weber and Howell, 1982.) For a decision being faced closer to or after the year 2000, the relevance of earlier conclusions is likely to be mixed. On the one hand, the earlier conclusions about community social structure may continue to be relevant. For example, a sudden doubling of community population would still likely reduce a community's density of acquaintanceship (the proportion of people in that community who know one another), reducing the effectiveness of community watchfulness in controlling deviant behaviors and leading to an increase in crime rates. On the other hand, earlier conclusions about economic dynamics might not be relevant today because of changes in technology or changes in socioeconomic and cultural conditions. For example, present-day energy facilities might require larger or smaller construction workforces, thus altering the primary driving forces behind the boom/bust patterns that occurred in so many communities during the 1970s and early 1980s.

Economic/Demographic Data and Models

The second major subcategory of the archival/secondary tools involves the use of economic/demographic data sets and models. In some sense, the strengths and weaknesses of this second subcategory are the reverse of those in the first. At least for cases where county- or state-level information is sufficiently detailed for decision-making purposes, almost any region in the developed world will likely have at least some standardized statistical information available. In terms of substantive focus, however, the coverage of available data is likely to prove spotty. Coverage is likely to be adequate only in the topic areas that are important for societal bookkeeping (e.g., employment, income, and population). Unfortunately, the things that have already been counted are not necessarily the only things that count. As a result, much of the information needed about a community or region is not likely to be available for analysis. This problem is exacerbated by policymakers and scientists being seduced by the visible: So long as *something* has been calculated and analyzed, there can be a temptation to overlook or forget about other questions that may be equally or even more important. This problem will be noted again under gaps-and-blinders tools.

A related problem involves what is sometimes called the "tyranny of illusory precision"—cases where estimates are seductive because they appear so precise when, in fact, they are little more than quantitative guesses. A useful analogue involves the clock that no longer runs but that still reports an apparently precise time. The time will be accurate twice each day, but if people do not know the clock has stopped, they may linger over a leisurely cup of coffee, assuming that the reported time is correct. The problem is particularly dangerous in cases where it is possible to develop an apparently precise projection of what the real number might be. The temptation to treat such estimates as being valid can be very strong, but in many cases, the estimates have amounted to what Moen (1984) calls "voodoo forecasting."

The commonality of a behavioral pattern, however, does not necessarily indicate the degree to which it is reasonable or prudent. In areas where enough projections have been compiled to permit a quantitative analysis, results have scarcely been comforting. In one study, Mountain West Research (1979) showed how the official numbers and techniques in three federally sanctioned manuals on projection techniques would have led to three different estimates, even if the techniques were applied to the very same project. While the three estimates appeared to be equally reasonable and precise, they differed from one another by a ratio of more than three to one. In another study, Murdock et al. (1984; 1985) tested estimates of the employment and population changes associated with nuclear power plants and with large-scale construction projects. They found that the average absolute-value error of estimates was often more than 50 percent and sometimes greater than 100 percent. For example, a highly technical projection might estimate that there would be 1,743 employees at the peak of construction, leading to a total population increase (including spouses and children, plus the induced population working at restaurants, bars, gas stations, mobile home parks, etc.) of 3,196, a number that possessed a comforting apparent precision. The actual experience, however, might have been either 100 percent smaller (as when Exxon suddenly ceased construction of its giant oil shale project near Rifle, Colorado, bringing the work force to virtually zero in one day [Guilliford, 1989]) or more than 100 percent larger (in other projects, where schedule changes or unforeseen problems in construction or production techniques led to dramatic increases in the need for workers). To make matters worse, it is possible for the actual number of workers to be twice as high as the projected value during one calendar quarter, while dropping virtually to zero during the next.

Despite these real and significant problems, one should not be too harsh on those who prepare the economic/demographic projections. Most of these professionals try to do the best job they can and to be as straightforward as possible in pointing out the assumptions and limitations of their analyses. In addition, many of the most dramatic errors of past estimates were not actually caused by errors in the models, but by errors in the engineers' assessments of the number of construction workers that their projects would require.

Archival Research Techniques

A third subcategory includes the gathering of data that are at least potentially available but that are not in the latest Census reports or the *Survey of Current Business*. A certain level of creativity might be necessary to identify or access the necessary information, and one must be alert to variations across time and, especially, across jurisdictions in record-keeping practices. Still, many government agencies routinely compile data that can prove extremely helpful in answering questions. Businesses often keep detailed records, as well, although the proprietary and/or confidential nature of the information in those records can sometimes make access difficult to obtain. Beyond that, most cities and a surprisingly large number of rural areas will have been the focus of at least some level of earlier research and/or analysis that can be mined. To find these data, the researcher must be resourceful and persistent in combing through community plans, theses and dissertations, community histories, project reports from local school districts, utility records, booster pamphlets, and more.

4. Understanding the Socioeconomic and Political Settings   103

While all of these forms of information are potentially available, in other words, one must be realistic about the amount of effort required to locate and obtain relevant data. In one case, I had to become an "unpaid staff member" of a mental-health center, complete with an absolute confidentiality pledge to protect the sensitive content of the files. I then had to spend several months computerizing the center's records so I could compile and analyze these available records. In another case, I had to apply gentle but persistent pressure for more than 10 years (and to wait for the results of several new elections) before I was allowed access to the records from a small-town sheriff's office. These, however, are relatively extreme examples, and often it is possible to locate relevant information with just a few more days or even hours of extra effort.

Despite such challenges, moreover, the creative use of unorthodox data sources often permits access to information that is directly relevant to environmental decisions and that is not available from any other source. In a study of the differences between the coastal regions of southern Louisiana and northern California in attitudes toward offshore oil development, Freudenburg and Gramling (1993, 1994) compiled line-by-line analyses of the written and oral comments on federal environment impact statements (EIS) and measured road maps on a mile-per-mile basis. The differing patterns of comments on the EISs and the institutional affiliations of those who expressed concerns helped to illustrate the differences between the regions. The road-map measurements provided further information, showing that 70 to 90 percent of the coastline in California is served by roads, while only about 12 percent of the coast in southern Louisiana is within a mile of a road of any kind. Creativity and persistence, in short, can make data available for environmental decisions.

Primary or Fieldwork Techniques

Almost inevitably, however, environmental decision making will require different kinds of information than will already have been compiled for other reasons, and much of that information can only be gathered firsthand. In the words of one country lawyer, disdainful of the generic studies of his community done by researchers who had spent little or no time there, "Even an idiot on the spot is worth far more than a genius a thousand miles away." Of the many primary or fieldwork techniques, three have particular relevance for environmental decisions: formal surveys, focus-group and key-informant interviews, and participant-observation techniques.

Surveys

Formal surveys are the best known and most readily quantified and analyzed of the three types of fieldwork techniques. The basic idea is that representative and/or random-sampling techniques can be used to obtain information directly and systematically from the very people who know the most about the relevant attitudes, values, behaviors, and experiences: the people themselves. Such techniques are the basis of everything from basic social–science research, to formal program evaluation, to political polling, to assessing the potential markets for new products. A substantial body of research and experience has been developed with these tools, permitting a certain amount of standardization in the informal rules of the game. In addition, if questions and methods are carefully chosen, a familiarity with the surveys that have already been done in other regions can permit comparisons that will show whether attitudes, levels of satisfaction, or experiences in one community, for example, resemble or differ from those in other communities.

These characteristics often make surveys the primary-data-collection technique of choice. Surveys allow us to answer certain questions that simply cannot be answered any other way. In addition, they can permit the development of quantitative estimates with relatively well-known statistical properties. Indeed, many of the statistics routinely assumed to result from something like a census, which contacts everyone or at least every household, are actually the result instead of extrapolations based on surveys (e.g., statistics on employment and earnings).

Still, surveys have many difficulties. In particular, they cannot provide the answers to any questions that are not asked. A poorly designed survey (or even a well-designed one that does not reflect enough familiarity with the community or the environmental decision in question) can fail to provide the relevant information. Worse, surveys often provide apparently precise answers to the questions that are asked, whether those questions (and answers) are sensible or not. In addition, the voluntary response rate to surveys has been declining in recent years, as more and more organizations have started to use phony surveys for other purposes, such as fund-raising appeals.

Surveys that are well-crafted can often be quite expensive to administer. The cost problem, however, can sometimes be addressed by the use of lower-cost techniques, such as mail and telephone surveys instead of personal interviews (Dillman, 1978) or by what Finsterbusch (1976) calls "mini-surveys." Mini-surveys use very small sample sizes, on the order of 20 to 100 cases, thereby giving up larger surveys' small increases in precision (and substantial increases in the ability to draw inferences about specific population segments) in exchange for major reductions in cost.

For federal agencies, there is another problem—a perverse interpretation of legal wording that was first enacted in 1946, even before the advent of what most would consider modern-day survey research. Under this regulation, it is essentially illegal for most federal agencies to pose the same question to ten or more people if those agencies actually want to know the answer. (For a more detailed discussion, see Freudenburg 1986.) At least the research that is funded the National Science Foundation is exempted from this regulation, however, and the states and other levels of government are not bound by it.

Nonrandom Interviews

Focus groups and what are sometimes called key-informant interviews also fall into the overall data-collection tool category of fieldwork techniques. They, too, involve obtaining information directly from the people involved in a region or a decision. In a small number of cases, participants in focus groups are chosen in a statistically representative way. In general, however, the participants in focus groups and key-informant interviews are chosen not because they are statistically representative, but because they are unusually insightful, because they are expected to be particularly outspoken, or in some circumstances, simply because they are available.

Although focus groups and key-informant interviews can vary substantially, a focus group can be seen as a kind of group interview, while key informants are more often interviewed one at a time. In practice, this approach can mean that a focus group is better suited for questions that can best be answered by observing group dynamics and social interaction (sometimes with videotapes or one-way mirrors). Focus groups can also be handy for giving researchers (and their clients) a relatively quick way of gaining a feel for a given set of socioeconomic conditions. At the same time, focus groups involve an artificial situation. They are often conducted on the researcher's turf, raising the potential for answers that are more guarded or strategic (and/or more oriented toward persuading the people in the group) than would be true in a private interview. As with any group discussion, even quite articulate people may say less than will those who are merely more forceful or outgoing. The private interview, on the other hand, will often protect confidentiality and be carried out on the home turf of the person being interviewed. In addition, a group meeting at a specified time and in a strange location may be substantially more inconvenient than having a researcher show up at one's office or home at a pre-arranged time. As a result, many of the busiest people in a community (potentially those who have key information) will tend to prefer one-on-one interviews.

One-on-one interviews and focus groups tend to share two other key weaknesses. Social scientists call the first one "reactance." The very act of asking questions about certain topics can change people's behaviors, causing them to reconsider what they had done before or leading to answers that are more strategic than honest. The other weakness is elite bias: When local residents suggest "other people you should talk to," they tend to think of those who are above them in the status hierarchy, not below. They are more likely to suggest the town banker than the town drunk, or the minister of the congregation rather than a member of the flock, let alone an unemployed member of the congregation who has stopped coming to church altogether.

Participant Observation

Participant-observation techniques are often lumped together with key-informant interviews, perhaps because they are often carried out by the same researchers during the same field visits. Technically speaking, however, participant-observation techniques involve a very different researcher role, observing and participating in the daily life of an affected community, much as anthropologists originally came to understand "exotic" cultures by living within them.

Approaches to participant observation vary broadly, especially in the intensity and length of time of contact. At one extreme, a skilled and experienced researcher can often gain at least a useful set of first impressions of a community and develop some initial hypotheses in as short a time as a day or two, or even on the basis of what one of my colleagues calls "windshield work." Still, the limitations of such an approach are obvious. At the other extreme is the purist's definition, where the fieldwork would last for a year or more and would result in previously unrecognized insights into a community's culture and subcultures and into its socioeconomic, political, and broader traits.

The specific strengths and weaknesses of participant-observation techniques vary with the research being conducted. Extensive participant-observation research has a relatively high cost, if only for the amount of time and the travel involved in getting a researcher to the field, but the costs may not be nearly as high as for research in the other environmental sciences. A full-scale participant-observation research effort by a senior investigator, complete with clerical and logistical support, can cost far less than just one day of ship time for a physical oceanographer. The costs can also be lower than those for other forms of socioeconomic data collection and analysis, as in the cost of obtaining, tweaking, and running many economic models.

The more serious problems tend to be those that are tied to inexperience. Some people are simply not well suited to the role of being an unobtrusive observer; others will become such enthusiastic participants that they will "go native." Other problems can bedevil even the most experienced researcher: Many of the most important observations and insights do not lend themselves well to quantification or testing, posing potential challenges both for assessing whether one's conclusions are accurate and for convincing other people of the robustness of the data, although such problems often prove to be less severe than is often assumed by those with little contact with the technique. If one's goals are to improve one's understanding of the human environment, rather than to gather ammunition for lawsuits, outright errors in participant-observation conclusions may actually be fewer than those in demographic/economic projection techniques. The more common problems may be those that are more subtle. For example, the sheer salience of direct observation can cause researchers to be unduly swayed by their plausible, but erroneous, initial impressions, particularly when those impressions are tested only against that same researcher's later impressions. A better method is to guard against this problem by testing one's impressions against other kinds of data, such as agency statistics and survey results, that would provide better triangulation. In so doing, one would be moving toward the use of the third characteristic tool, that involving gaps-and-blinders techniques.

Gaps-and-Blinders Techniques

The third characteristic tool involves techniques that are in some senses quite old, but in another sense have only begun to be recognized as explicit decision-aiding tools that are particularly important in studying the human environment. Gaps-and-blinders techniques are important because:

- People are so effective at creating surprises
- Often, the necessary research has simply not been done
- In a relatively large number of cases, political actors and interests will try to dismiss rather than deal with the groups and concerns they find inconvenient.

The commonality among the gaps-and-blinders techniques is that they are intended not so much to develop "most likely" scenarios as to recognize ways in which such scenarios may not be so likely after all. A comparison to Russian roulette may be instructive: If a gun has one bullet and six chambers, then the maximum-likelihood estimate would be that pulling the trigger will not lead to firing a bullet. The gaps-and-blinders tools are used not to calculate the one-in-six probability to ever-greater levels of precision, but to ask what might happen if the "unexpected" outcome occurs.

The gaps-and-blinders tool is also useful in dealing with an important fact about environmental decision making that is often overlooked. Scientists are experts on questions of fact, while many real-world environmental decisions require attention to questions not only about facts, but also about values and blind spots. Questions about facts generally will be along the lines of, "How safe will that be for people and the environment?" At least in principle, a scientist might be able to answer such questions in factual terms. Unfortunately, for the subsequent questions about values—such as, "Is that safe enough?"—scientists are likely to have little expertise or claim to authority. Despite all the strengths and advantages of scientific training, once attention turns to questions of values, another word for "scientist" is "voter."

Yet scientists will often have even greater difficulties with blind-spot questions, such as, "What is it that we have overlooked?" Not without reason, experts are often assumed to be particularly good at what they do, but there is a degree of truth to the common definition of an expert as being someone who knows more and more about less and less. Charles Perrow

(1984) offers a slightly more formal definition: an expert is "a person who can solve a problem better or faster than others, but who runs a higher risk than others of posing the wrong problem." Michael Davis (1989), in turn, notes the potential for what he calls "microscopic vision," or the increased ability to see the fine points within one's own area of specialization, but often at the cost of ignoring the larger portions of the world that lie beyond the viewing range of one's microscope.

In fact, the accumulated research suggests that there may actually be two main types of disciplinary distortions of vision that need to be considered in environmental decision making. The first involves insufficient humility about what we do *not* know. Most of us with scientific training tend to be cautious about not overstepping the bounds of our own expertise, at least on questions within our own disciplines, but that caution often disappears when we deal with matters that lie outside our training and expertise. We often know enough about the various specialties within our disciplines that we also have some sense of how much we do *not* know about them, while we often have too little knowledge about other disciplines to have the same kind of awareness. We seem to have a self-image that, like the children of Lake Wobegon, we are "all above average." This self-image is often held most strongly among those whose beliefs are least constrained by a knowledge of the results of relevant research. Perhaps for this reason, the problem of disciplinary myopia tends to be particularly pernicious in cases where the matters at hand involve human behavior but the scientific training does not (Fischhoff et al., 1981; Freudenburg, 1992).

The second problem of disciplinary distortion of vision involves insufficient humility about what we believe we *do* know. We often display an overconfidence or a tendency to *under*estimate the unknowns, even within a field we think we know. The problem of overconfidence has been found in such well-developed fields as physics and in quantities that are as fundamental and as carefully measured as the speed of light. Between 1875 and 1958, there were 27 published studies of the speed of light that included formal estimates of uncertainty. Unfortunately, these studies' estimates differed from the official 1984 value by magnitudes expected to occur less than .0005 of the time according to the original estimators' stated uncertainties (Henrion and Fischhoff, 1986). In fact, the 1984 estimate of the speed of light falls entirely outside the range of standard error for *all* estimates reported between 1930 and 1970. Other examples can be reported for disciplines ranging from engineering to medicine: One study asked a group of internationally known geotechnical engineers for their 50 percent confidence bands on the height of an embankment that would cause a clay foundation to fail. When an actual embankment was built, not one of the experts' bands was broad enough to include the true failure height (Hynes and Vanmarche, 1977). Another study followed a group of patients who were diagnosed on the basis of an examination of coughs to have pneumonia. Of the group listed by physicians as having an 85 percent chance of having pneumonia, less than 20 percent actually did (Christenson-Szalanski and Bushyhead, 1982).

More broadly, studies of the ability to assess probabilities accurately (the problem of calibration) have found that calibration is unaffected by differences in intelligence or expertise, while errors are sometimes increased by the importance of a task. Overall, one would expect that only about two percent of the estimates having a confidence level of 98 percent would prove to be surprises; but in empirical studies, it is more common to find a "surprise index" on the order of 20 to 40 percent (Lichtenstein et al., 1982; Lichtenstein and Fischhoff, 1977; Sieber, 1974; Fischhoff et al., 1977; Freudenburg, 1992). In cases of real-world decision making, accordingly, the need for gaps-and-blinders analyses can be quite significant.

Researcher-Sensitivity Analyses

The first subcategory of gaps-and-blinders tools involves researcher self-checking and double-checking. The techniques in this subcategory can range from ad hoc admonitions to "think again" (even about conclusions that seem quite plausible) to sets of safeguards and double-checks that can be systematic and elaborate. In the relatively few cases where extensive quantitative evidence is available, formal sensitivity or power calculations can be performed. Unlike techniques for measuring statistical significance, which ask whether a given finding is consistent enough within a sample to be considered "real" for a larger population, tests of sensitivity or power essentially ask whether a sample is powerful enough to provide some reasonable degree of assurance of finding an effect when one is actually present. (For a more detailed discussion, see Cohen, 1988.)

While gaps and blinders may be difficult to identify, much can still be done to recognize them, starting with an awareness of indicators of potential concern. One useful strategy is to look for the impacts that have been identified elsewhere in the peer-reviewed literature (Interorganizational Committee on SIA, 1994) to make sure that these impacts were considered in the analysis under review. Another approach is to ask whether the analysis devotes little or no attention to the less powerful (and less noisy) groups in society. In decision-related work done to date, the interests of women, children, the elderly, members of racial or ethnic minority populations, those from unusual or non-mainstream cultures, and so forth, appear far more likely to be overlooked than those of business owners or white males.

A related indicator is evidence of hostility or disdain toward any such group(s) on the part of more influential or mainstream groups, particularly those whose interests are most strongly represented within policy-making circles. While overtly racist comments are far less common today than in the past, dismissive or delegitimizing comments are still commonly heard about politically inconvenient groups. These comments take the form of assurances that "those people" are "just" ignorant of the facts, ill-informed, selfish, malcontents, or obstructionists. In political terms, such assertions about enemies often prove to have a significant level of short-term effectiveness. In scientific terms, however, those assertions provide a warning that (1) the interests of the less-powerful groups are likely to be at odds with the interests of the more-powerful groups backing a given proposal; and (2) the less-powerful groups may have been given less-than-thorough consideration in the development of the policies being considered.

Perhaps the least obvious warning sign is that it is possible to have too much of a good thing, even in the realm of quantification. Sometimes this problem is produced by the tyranny of illusory precision, but often it has a less direct connection: Precise quantification can use up so much research time and other resources in answering one set of questions that other relevant questions do not get considered. As Dietz noted more than a decade ago in work for the National Academy of Sciences (1984), when decision-making resources are limited (as will almost always be the case), it is more sensible to devote those resources to *identifying* the *full* range of potential impacts on the human environment than to *quantifying* a given *subset* of those impacts.

Interdisciplinary Double-Checks

Interdisciplinary double-checks are made possible when researchers have different disciplinary backgrounds and points of view. They take advantage of something that is otherwise seen as a weakness of disciplinary training, namely that such training tends to strengthen certain capacities while ignoring others. The basic idea behind interdisciplinary double-checking is that, while all disciplines tend to have vision problems, the gaps tend to be different; the blind spots of a field biologist, for example, are likely to be quite different than those of a natural-resource economist.

While interdisciplinary double-checking can be quite formal and elaborate, the interdisciplinary nature of most environmental decision-making work often makes it a relatively simple matter for an initial set of thoughts to be batted around among persons with different backgrounds. Ironically, even while pursuing such informal safeguards, it is important to have additional, appropriate safeguards. Those safeguards should start with a well-developed sensitivity to the fact that the ultimate responsibility for judgments must remain with those who have the greatest relevant expertise; judgments should not merely reflect the opinions of team members who happen to be particularly forceful, persuasive, or powerful. In addition, the purpose of such an exercise must be to recognize possibilities that had previously been overlooked, *not* to dismiss someone else's possibilities as being unlikely. The probability of a bullet through the brain in a game of Russian roulette is, after all, also technically unlikely. The gaps-and-blinders tool should ensure not only that attention is paid to the best estimates available but also that increased sensitivity is allotted to the uncertainties inherent in those estimates—and to the consequences that may follow if even the best estimates are wrong.

Public-Involvement Techniques

Beyond the need for self-checking and for interdisciplinary double-checks, the views of those who are outside the organization also need to be considered. Many of the most important outside views will come from the people or groups who are most dependent upon a given environmental context and/or will be most strongly affected by a potential decision. Fortunately, in many cases of environmental decision making in the United States, systematic public involvement is statutorily required. Particularly in understanding the human environment, however, public involvement would be a basic analytic need even if it were not a legal requirement.

Part of the analyst's job is to provide the decision maker with accurate information on how people rely on the environment or might be affected by a proposed decision. The people in question are often more knowledgeable about themselves than any outside observer can be. In addition, in the context of identifying gaps and blinders, the people outside an organization can have the advantage of being less likely to wear the same blinders shared by those within the organization.

That last point deserves elaboration. Part of what makes it possible for an organization to function is the development of more or less distinctive views that are shared by those who work there. Such shared views can be useful to the organization by simplifying the coordination of action and contributing to an *esprit de corps*. Yet they can lead to problems, as well, particularly when those views come to be so thoroughly accepted within the organization that they take on the status of (assumed) fact.

Often, the assumptions that can create the most damage are those of which we are not even aware. An important example is what Clarke (1993) calls the "disqualification heuristic," as in the conviction that "it can't happen here." Stated more broadly, the heuristic becomes "we're right, and they're wrong," where "they" are persons outside of the agency and/or others who do not happen to share the distinctive views that "we" within the agency or company hold. Conviction, however, is a poor substitute for information. When decisions are expected to be based on an understanding of the human environment, even the most passionately held of convictions (and sometimes *especially* those convictions) can be dangerous.

The problem is worsened by two additional considerations. The first is that such convictions are not always free from political influences. As noted by Freudenburg and Gramling (1994, pp. 138–139), if the other politically relevant parties "can be convinced that an expectation is not 'reasonable' or 'legitimate,' then the expectation can usually be avoided or ignored." Claims that the opposition is "just" a matter of misinformation (or of irrationality, ignorance, or selfishness) can serve at least two functions. These claims can reinforce the belief among committed partisans that they have Truth and Justice on their side, and they challenge the legitimacy of opponents. "For purposes of political sniper fire, in fact, the central importance of such claims may have little to do with whether or not they hit the mark, in terms of accuracy, having far more to do with whether or not they keep one's opponents pinned down, forced to fight their battles in the midst of unfavorable terrain" (Freudenburg and Gramling, 1994, p. 139).

A second, longer-term problem is that, while such delegitimizing tactics can produce at least a short-term political payoff, they are not without their costs. For one thing, they tend to contribute to "the spiral of stereotypes." That pattern emerges when persons on different sides of an issue stop talking *to* one another, but persist in talking *about* one another, as in characterizing the other side as being ill-informed, self-serving, or irrational (Freudenburg and Pastor, 1992; National Academy of Sciences/National Research Council, 1996). This behavior increases polarization. In addition, it produces one of the most stressful experiences for participants in the decision-making process, namely, to find that not just "unreasonable protesters" but also they, themselves, are treated as if their most heartfelt concerns are imaginary or irrelevant (Krauss, 1989; Brown and Mikkelsen, 1990; Levine, 1982). In such cases, people *do* react with frustration, even rage, especially if they are repeatedly ignored or treated with condescension or contempt. The ironic result can be that the opponents truly can start to sound and act "emotional," but they often do so as a direct result of the ways in which they have been treated. As noted in a recent report from the National Academy of Sciences/National Research Council (1992, p. 25):

[A]gency staff members often are tempted to argue that the critics of agency policies are "emotional" or "misinformed" (Hance et al., 1988). These characterizations fail to acknowledge salient socioeconomic effects and create new ones as well. They are "guaranteed to raise the level of hostility between community members and agency representatives and ultimately stand in the way of a successful resolution of the problem" (Hance et al., 1988). Such challenges can lead people to be resistant *in principle* to matters they might otherwise be willing to consider more dispassionately.... [F]or a community to have its reality disregarded by a powerful authority is profoundly alienating; it [also] leaves no common ground on which the community and the authority can stand.

When free of such polarization, public-participation programs can fill three functions for environmental decision making:

1. Insofar as public involvement is a statutory requirement, active commitment to public input can help an agency comply with the law.
2. To the degree to which public involvement is not seen *merely* as a statutory requirement, but as a way of obtaining meaningful information about the socioeconomic and political setting of a decision, input from the

people who are likely to be affected can provide some of the evidence that is needed.

3. As a growing number of analysts have noted (Clarke, 1993; Lawless, 1993; Shrader-Frechette, 1993; Freudenburg, 1992), some of the most severe cases of the disqualification heuristic have been found where agencies have been most fully insulated from outside or public influences. It may well be that an increased institutional permeability offers the best available antidote to the errors that are otherwise likely to creep in.

Table 4.2 provides a simplified summary of the three characteristic tools that have been discussed so far. The remainder of the chapter will discuss recent developments and future challenges.

Innovations and Recent Developments

As noted at the start of this chapter, the tools for dealing with the socioeconomic and political environment have been developed over the years across a broad span of the social sciences. As such, the tools range from historians' techniques, such as archival research, to economic/demographic techniques, such as projection techniques that produce apparently precise dollars-and-bodies estimates. The archival techniques tend to be (1) nonquantitative; (2) oriented toward the past; and (3) quite useful in providing an overall understanding on the basis of a spotty data record. The economic/demographic techniques are (1) highly quantitative; (2) oriented toward the future; and (3) useful to a substantial level in their own way. Given this diversity, some of the most important of recent developments are those that bring together characteristic tools for assessing the human environment. They can be grouped in two broad categories: those that integrate insights and findings across the existing social-science techniques, and those that integrate the insights and findings from the social sciences with those from the biological and physical sciences.

Integrating Across Existing Social-Science Techniques

One potential cross-disciplinary integration technique that has received a great deal of recent attention is contingent valuation, which is covered in greater detail in Chapter 2. In essence, work in this area asks what would be the right economic value to assign to something that is not actually bought or sold on the open market, contingent on the existence of such a market. The development of this technique can be seen as a response to the criticism that, in economists' analyses over the years, only economic numbers have counted. Uses of the environment for which ready price tags are available (e.g., the lumber value of the trees in an old-growth forest) have tended to be given more weight in decisions than have other important concerns that

TABLE 4.2. Summary of tools for characterizing the human environment.

Characteristic Tool	Examples	Strengths	Weaknesses
Secondary/archival techniques	"Standard information modules" Economic/demographic models Archival sources	• Can be based on actual experience, not just guesses or expectations • Can be quite cost-effective (although often more expensive than anticipated) • Often quantitative • Reliance on peer-reviewed findings can offer quality control	• Necessary information is often not available • Gaps in knowledge often unseen or hidden • Can create "tyranny of illusory precision" • Things that can be counted are not necessarily things that count
Primary/fieldwork techniques	Formal surveys Nonrandom interviews Participant observation	• Often best (or only) source of information on affected areas • Well-done surveys have known statistical properties • Permit attention to important questions, whether or not information is already "available" as a byproduct of societal bookkeeping • Can provide on-the-ground reality checks	• Surveys cannot provide answers to questions that are not asked, and sometimes provide misleading answers to questions that are asked • Can be expensive and time-consuming • Some types of findings difficult to cross-check with other techniques
Gaps-and-blinders techniques	Researcher sensitivity analyses Interdisciplinary double-checks Serious public-involvement efforts	• As National Academy of Sciences has noted, if resources are limited, it is more sensible to identify full range of impacts than to quantify a subset of impacts precisely • Can help decision makers to anticipate how much difference it makes if even "best guesses" are wrong • Can help decision makers in integrating expertise across disciplines, and between disciplinary specialists and members of broader public • Can help to neutralize inappropriate intrusion of political interests and delegitimization of opponents in environmental decision making	• Often frustrating to those who want world to be simpler than it is • Can be difficult, requiring recognition of hitherto-unacknowledged blind spots • Politically powerful groups may want impacts on the less powerful to be ignored in decision making • Professional researchers often resist idea of recognizing ordinary citizens' views as legitimate within "professional territory"

are not normally bought and sold (e.g., the value of the same forest as a watershed for downstream cities, as habitat for endangered species, or as a way of allowing scientists and future generations to study and enjoy such environmental conditions). In a sense, contingent valuation can thus be seen as an effort to add non-economists' considerations to economists' analyses.

The goal of a good deal of research in contingent valuation is to come up with dollar (and/or utility) estimates for a broad array of values that have no obvious prices. Examples might include assessments of the value to an average citizen of visiting an old-growth forest, of simply knowing that the area exists and is being protected, or of preserving that forest for future generations.

Such techniques have often proven to be quite controversial in practice, particularly when they have been brought into litigation, as in attempts to quantify the nonmarket damages associated with a disaster such as the Exxon Valdez oil spill. Some critics have argued that the numbers tend to be too high (Hausman, 1993) while others have argued that the numbers are too low (Dietz and Stern, 1995). Important advances have been made, nonetheless, and some researchers who were initially skeptical have come to view such techniques with at least a degree of respect (Mitchell and Carson, 1989; Heberlein and Bishop, 1986). In addition, while cost-benefit analyses have fallen out of favor to a sufficient degree that they receive relatively little attention in this chapter, some variant on valuation techniques will likely remain relevant and require future attention where cost-benefit assessments are carried out.

A less radical or more evolutionary extension of past techniques is represented by the work of Norgaard and Howarth (1993). Their approach makes the simple change of assessing the costs and benefits of a given environmental decision not simply for those people who happen to live in an area at a given time, but also for future generations. Norgaard and his colleagues have sometimes found that including even a simple measure of costs and benefits for future generations can significantly increase the "economic rationality" of environmental protection. Such an approach may also help to bring the views of economists significantly closer to those expressed by biological scientists, particularly for longer-term questions, as in deciding what steps to take to control global warming in the face of less-than-perfect information.

Another approach starts with a dependent variable that is of traditional concern to a discipline other than economics and then adds economic data to a pre-existing mix of data, rather than the other way around. While environmental attitudes, for example, have been studied quite extensively (Dunlap, 1992, 1993), most of our information is at the level of the individual. We know that individuals express high levels of environmental concern in virtually all of the nations that have been studied, including the less-developed countries (Dunlap, 1993). We also know that women in the United States tend to express higher levels of concern than do men over nuclear and/or other potentially hazardous technologies, particularly at the local level, while there is no clear pattern of gender difference in concern about the environment in general (Davidson and Freudenburg, 1996).

Such findings, however, have only limited value for environmental decisions, which are often effectively made at the community or regional level, as in planning and zoning decisions, ballot initiatives, and so forth. Despite the fact that most congressional districts in the United States have comparable ratios between men and women, certain districts and regions of the country consistently elect representatives who are rated by groups such as the League of Conservation Voters as significantly more (or less) oriented toward environmental protection than the national average. In addition, given the vast amount of information that now exists on interpersonal, institutional, and cultural factors that affect people's decisions (ranging from interpersonal dynamics to economic vulnerabilities to advertising and media-coverage effects), it would be remarkable if nothing other than the additive combination of individual characteristics were involved in community-level or congressional-district election outcomes. The significance of this point is illustrated by a recent study, focusing on a three-county region of California between Los Angeles and San Francisco that has faced repeated proposals for increased coastal development. The study found that, over 16 years, the 20+ communities of the region showed enough stability in voting (across eight ballot initiatives) that a series of straightforward regression equations explained well over 70 percent of the variance in the accumulated results (Freudenburg et al., 1996). The key variables involved not just economic factors (such as community income levels and the relative importance of tourism versus oil-extracting industries), but also social and political factors (including the presence of nonreligious colleges and universities; political-party affiliations; the importance of agricultural industries; and the presence of artistic institutions, such as museums and musical organizations). While such an approach to combining economic, social, and political factors is still experimental, it appears to have good potential for future applications.

Integrating Social-, Physical-, and Biological-Science Findings

Perhaps the most notable development in integrating social- and natural-science findings is the rapid growth in geographic information systems (GISs). While a substantial amount of work remains to be done, the potential is sufficiently obvious to have generated a great deal of excitement, particularly in the context of decisions involving the environment. With GIS, it may be possible to integrate the physical and/or biological information about a given environmental setting (e.g., soil types, vegetation, existence of endangered species, and locations of water resources) with the socioeconomic and political data for the same region (e.g., occupational structure, population, income flows, predominant political orientation, and the presence of unique or at-risk populations).

Substantial work, however, remains to be done before this potential can be realized, ranging from technical challenges (such as developing better measures for spatial autocorrelation) to more epistemological challenges (such as disentangling correlation from causality). Also, the work required to study a given location is often extensive. The necessary data must be located or produced, and certain types of data will often be available only for political jurisdictions (such as counties) that may have little correspondence with the biophysical resources or issues that are the focus of concern. A more detailed explanation and assessment of GIS is presented in Chapter 6.

A new and different kind of synthesis involves one of the commonalities of past environmental disputes, the supposedly "enduring conflict" (Schnaiberg and Gould, 1994) between environmental protection and economic prosperity. Whenever a decision involves what some regard as a threat to the environment (whether the loss of local habitat for a given species or potential damage to the global atmosphere), the sources of that threat are likely to be rooted in economic pressures. Stated another way, most human-induced environmental changes result from a search for increased profit or prosperity and/or for public-sector revenues and influence, and many such proposed uses of the environment are backed with the argument that opposition would increase poverty and decrease job opportunities. While it would be premature to declare the data definitive, a growing number of empirical studies have found just the opposite.

One set of such studies found that even nations with high levels of resource-related incomes, including those from the Organization of Petroleum-Exporting Countries, have actually had less income growth than countries without such resource riches (Sachs and Warner, 1995; Passell, 1995; Corden and Neary, 1983) even during the period of soaring prices (and incomes) from petroleum products in the 1970s and 1980s. Another set of studies focused on the rural, less-developed regions of the United States and found that mining- and logging-dependent regions generally have *higher* levels of poverty than do the counties dependent on the declining industry of agriculture or on the notoriously low-wage industry of tourism. These findings are not limited to regions with a past history of extraction, such as Appalachia, but include regions having high levels of *current* employment in extraction (Cook, 1995; Drielsma, 1984; Elo and Beale, 1985; Freudenburg and Gramling, 1993; Humphrey et al., 1993; Tickamyer and Tickamyer, 1988; Krannich and Luloff, 1991; Freudenburg, 1992; Gulliford, 1989; Peluso et al., 1994; Cottrell, 1951, 1955; but see also Nord and Luloff, 1993; Overdevest and Green, 1995). A third set of studies has taken a closer look at environmental regulations; these studies have found that, rather than strangling economic growth, higher levels of environmental regulation tend to be associated with greater growth in jobs and incomes. These findings seem to hold both at the level of cross-national comparisons (Repetto, 1995) and at the level of cross-state comparisons within a given country (Freudenburg, 1991; Meyer, 1992).

This increasingly consistent pattern of "unexpected" findings may reflect the fact that the economic activities that create the most serious impacts on the environment are among those that provide the *smallest* proportions of the overall job total, not the largest. Calculations based on the Environmental Protection Agency's Toxics Release Inventory show that, even under generous accounting assumptions, the industries that produce more than 50 percent of the reported toxic releases in the United States are associated with only four to six percent of the gross national product (depending on the specific measurement techniques being used) and that they are associated with a significantly smaller fraction of the nation's jobs (Freudenburg, 1997). At an international level, Repetto (1995) found that a given level of investment in environmental protection may actually be associated with the creation of more jobs than would the same level of investment in industries commonly associated with the highest levels of environmental risk. Repetto also found that the most heavily regulated economic sectors may actually have experienced greater economic growth than those that were free of such regulation.

Anyone who has ever experienced regulatory constraints would have an easy time understanding how and why the claims of regulatory excesses seem so credible; but to repeat, the empirical track record of those regulations has been just the opposite. The data to date clearly do not prove that environmental regulations are good for the economy in all circumstances, but there is far less evidence to support the opposite argument (the argument that environmental regulations are always bad for the economy) even though this inaccurate belief is often taken as fact. The conflict between economic and environmental values, in short, appears not to be as enduring as was once assumed.

Remaining Needs

Even in light of developments that provide cause for optimism, substantial needs remain. Perhaps the key need, however, is for improving the rationality of the decisions that can be made in the face of irreducible uncertainty.

In certain cases, guidance can be offered. The accuracy of economic and demographic projections will often be the lowest at the scales that are often the most relevant for environmental decisions (e.g., future conditions in a given valley, rather than in a state or a nation). Even at broader levels, however, the track record in predicting surprises has not been good. Mainstream thought among demographers, for example, failed to anticipate both the baby boom of the 1940s to 1950s and the birth dearth of the 1970s. Economic and demographic projections have thus generally proved to be more useful and accurate in cases where conditions have remained relatively stable, permitting relatively straightforward extrapolations, than in cases where a good deal of change has taken place. This outcome is quite understandable in that, under many circumstances, our best guess is that the world will continue to operate as it has in the past.

Yet what is understandable in general or abstract terms may prove to be highly problematic in real-world decision making. Some of the most important environmental decisions are those that involve no obvious analogs or precedents of the sort that would permit confident projections of what to expect. Recent examples include efforts to project population trends and technological capabilities for the next 10,000 years to support a decision about nuclear-waste storage (Erikson, 1990) or to project the ongoing social causes and consequences of global warming (National Academy of Sciences/National Research Council, 1994). Still other complexities are introduced in cases where the decision itself may ensure that the future will *not* involve a straightforward extrapolation from past experience.

To put the matter simply, whether the focus is on the social sciences, the physical sciences, or the biological sciences, the vast majority of the effort to date has been devoted to reducing the uncertainty or improving the apparent accuracy of best guesses. This work needs to continue, but it needs to be complemented with work on improving decisions that must be made whether the uncertainty can be reduced or not.

This challenge includes the need to integrate different kinds of information (one of the topics addressed in Chapter 6), but it is much messier. It also involves integrating the combined expertise of social and biophysical scientists in cases where the relevant forms of expertise are incomplete or unsatisfactory even in isolation. Just how much are global temperatures really likely to rise over the next 50 years, for example? What are the real probabilities that political leaders five years from now would implement serious policy changes even if the global climate picture could be predicted with much greater confidence?

The most straightforward answer to both questions is, of course, "We don't know." Yet this is before we get to the really important parts of the challenge involving the *interactions* between policy actors and scientists. Politicians who are looking for an excuse to avoid the imposition of costs on business and industry can continue to raise questions about the adequacy of scientific research. And scientists are rarely (if ever) shy about asserting the need for more research, especially their own, even when the rest of the scientific community is moving toward consensus. To make the picture still more complex, any scientific conclusions would likely become ammunition in political battles, and political maneuverings would likely influence the selection of hot topics and questions (and thus the potential answers) for scientific research.

Such complexities are difficult to capture with the approaches most scientists employ. Yet these complexities present the most important challenges of all. If we are to deal successfully with such challenges, we must not only bring together the social, physical, and biological sciences, but also integrate science into political decisions and bring at least provisional answers to questions that are stubbornly and inherently unanswerable (cf. Weinberg 1972).

In all likelihood, progress in meeting these challenges may be less likely to come from the assessment, refinement, and *narrowing* of options than from the *expansion* of options. New possibilities must be developed that have not often been considered in the past, including decision-making techniques that help us to reflect on both what we know and what we know we do *not* know, and to reflect on the values that we hold. How can we best go about making decisions in cases that involve not just known unknowns, but *unknown* unknowns? The challenge is not an easy one, but the problems that are most worth wrestling with are often those that fight back.

Key Resources

Interorganizational Committee on Guidelines and Principles for Socioeconomic Impact Assessment. 1994. *Guidelines and Principles for Socioeconomic Impact Assessment*, NMFS-F/SPO-16. Washington, D.C.: U.S. National Marine Fisheries Service.

This is clearly the key source to consult. The contributors represent all the major social scientific organizations with important traditions of research on socioeconomic impacts and contexts of environmental decision making. This concise and relatively jargon-free publication not only provides a guide to the dozens of "how-to" manuals now available, but also offers scientific rather than political guidance for dealing with common problems, such as selective data availability, burdens of proof, and standards of evidence. This guide is comparable to the "Underwriter's Laboratory" guidance for fire safety of electric appliances.

Finsterbusch, K. 1980. *Understanding Social Impacts: Assessing the Effects of Public Projects*. Beverly Hills, CA: Sage.

While now somewhat dated, this volume remains the best and most carefully developed example of rigorous data compilation in the effort to produce "standard information modules." The book would clearly need to be supplemented with field research and examination of more recent empirical findings, but many of the findings (e.g., those relating to the stresses of residential relocation or extremely rapid community growth) remain relevant, as does the book's overall model of how it is possible to make decisions about the human environment on the basis of evidence rather than hunches or assertions.

Creighton, J.L. 1980. *Public Involvement Manual: Involving the Public in Water Power Resources Decisions*. Washington, D.C.: U.S. Government Printing Office.

A particularly clear and helpful guide on the "public-involvement" component of decision making on the human environment that many political leaders find both so important and so vexing.

Freudenburg, W.R. and Gramling, R. 1994. *Oil in Troubled Waters: Perceptions, Politics, and the Battle over Offshore Oil.* Albany: State University of New York Press.

A more recent, comparative case study of two regions where the same technology has led to dramatically different impacts and sociopolitical decisions. It includes chapters summarizing recent developments in socioeconomic impact analysis.

Dunlap, R.E. and Michelson, W. (Eds.). Forthcoming (expected 1999). *Handbook of Environmental Sociology.* Westport, CT: Greenwood Press.

A handbook that includes contributions from many of the leading social scientists who have dealt systematically with society-environment relationships. The book includes chapters on many of the key areas of relevant research, including risk and risk analysis, social-impact assessment, and technology assessment.

References

Boothroyd, P., Knight, N., Eberle, M., Kawaguchi, J., and Gagnon, C. 1995. "The Need for Retrospective Impact Assessment: The Megaprojects Example." *Impact Assessment Bulletin* 13(#3, Sept.):253–271.

Brown, P. and Mikkelsen, E.J. 1990. *No Safe Place: Toxic Waste, Leukemia, and Community Action.* Berkeley: University of California Press.

Christenson-Szalanski, J.J.J. and Bushyhead, J.B. 1982. Physicians' use of probabilistic information in a real clinical setting. *Journal of Experimental Psychology* 7:928–935.

Clarke, L. 1993. The disqualification heuristic: When do organizations misperceive risk? *Research in Social Problems and Public Policy* 5:289–312.

Cohen, J. 1988. *Statistical Power Analysis for the Behavioral Sciences.* Hillsdale, NJ: Lawrence Erlbaum Associates.

Cook, A.K. 1995. Increasing poverty in timber-dependent areas in western Washington. *Society and Natural Resources* 8(2):97–109.

Corden, M. and Neary, J.P. 1983. Booming sector and de-industrialization in a small open economy. *The Economic Journal* (London) 92:825–848.

Cottrell, W.F. 1951. Death by dieselization: A case study in the reaction to technological change. *American Sociological Review* 16:358–365.

Cottrell, W.F. 1955. *Energy and Society: The Relation between Energy, Social Changes, and Economic Development.* New York: McGraw-Hill.

Council on Environment Quality. 1978. *Regulation for Implementing the Procedural Provisions of the National Environmental Policy Act (40 CFR 1500–1508).* Washington, D.C.: Council on Environment Quality.

Creighton, J.L. 1980. *Public Involvement Manual: Involving the Public in Water Power Resources Decisions.* Washington, D.C.: U.S. Government Printing Office.

Davidson, D.J. and Freudenburg, W.R. 1996. Gender and environmental risk concerns: A review and analysis of available research. *Environment and Behavior* 28(3):302–339.

Davis, M. 1989. Explaining wrongdoing. *Journal of Social Philosophy* 20:74–90.

Dietz, T. and Stern, P.C. 1995. Toward a theory of choice: Socially embedded preference construction. *Journal of Socio-Economics* 24(2):261–279.

Dillman, D.A. 1978. *Mail and Telephone Surveys: The Total Design Method.* New York: John Wiley and Sons.
Drielsma, J.H. 1984. *The Influence of Forest-Based Industries on Rural Communities*, Ph.D. dissertation. New Haven, CT: Yale University.
Dunlap, R.E. 1992. Trends in public opinion toward environmental issues: 1965–1990. In: R.E. Dunlap and A.G. Mertig (Eds.). *American Environmentalism: The U.S. Environmental Movement, 1970–1990.* New York: Taylor and Francis. Pp. 89–116.
Dunlap, R.E. 1993. Gallup international survey shows global concern for the environment. *Environment, Technology, and Society* 70(Winter):1, 11.
Elo, I.T. and Beale, C.L. 1985. *Natural Resources and Rural Poverty: An Overview.* Washington, D.C.: Resources for the Future.
Erikson, K.T. 1990. Toxic reckoning: Business faces a new kind of fear. *Harvard Business Review* 119–126 (Jan.–Feb.).
Finsterbusch, K. 1976. The mini-survey: An underemployed research tool. *Social Science Research* 5(1):81–93.
Finsterbusch, K. 1980. *Understanding Social Impacts: Assessing the Effects of Public Projects.* Beverly Hills, CA: Sage.
Finsterbusch, K. 1995. In praise of SIA—A personal review of the field of social impact assessment: Feasibility, justification, history, methods, issues. *Impact Assessment* 13(3):229–252.
Fischhoff, B., Slovic, P., and Lichtenstein, S. 1977. Knowing with certainty: The appropriateness of extreme confidence. *Journal of Experimental Psychology* 3:552–564.
Fischhoff, B., Slovic, P., and Lichtenstein, S. 1981. Lay foibles and expert fables in judgments about risks. In: T. O'Riordan and R.K. Turner (Eds.). *Progress in Resource Management and Environmental Planning, Vol. 3.* New York: John Wiley and Sons. Pp. 161–202.
Freudenburg, W.R. 1986. Federal regulation and scientific research. In: L.N. Wenner (Ed.). *Social Science Information and Resource Management.* Washington, D.C.: U.S. Forest Service. Pp. 13–22.
Freudenburg, W.R. 1991. A "good business climate" as bad economic news? *Society and Natural Resources* 3:313–331.
Freudenburg, W.R. 1992. Nothing recedes like success? Risk analysis and the organizational amplification of risks. *Risk: Issues in Health and Safety* 3(1):1–35.
Freudenburg, W.R. 1996. Risky thinking: Re-examining common beliefs about risk, technology, and society. *Annals of the American Academy of Political and Social Sciences* 545(May):44–53.
Freudenburg, W.R. 1997. The double diversion: Toward a socially structured theory of resources and discourses, American Sociological Association, Toronto, 1997 August 13–17.
Freudenburg, W.R., Elliott, J., Goldberger, J., Martin, S., and Molotch, H. 1996. Attitudes toward environmental preservation at the community level: A comparison of quantitative and qualitative approaches, American Sociological Association, New York, 1996 August 16–20.
Freudenburg, W.R. and Gramling, R. 1993. Natural resources and rural poverty: A closer look. *Society and Natural Resources* 7:5–22.
Freudenburg, W.R. and Gramling, R. 1994. *Oil in Troubled Waters: Perceptions,*

Politics, and the Battle over Offshore Oil. Albany State University of New York Press.

Freudenburg, W.R. and Gramling, R. 1992. Community impacts of technological change: Toward a longitudinal perspective. *Social Forces* 70(4):937–955.

Freudenburg, W.R. and Jones, R.E. 1992. Criminal behavior and rapid community growth: Examining the evidence. *Rural Sociology* 56(4):619–645.

Freudenburg, W.R. and Keating, K.M. 1985. Applying sociology to policy: Social science and the environmental impact statement. *Rural Sociology* 50(4):578–605.

Freudenburg, W.R. and Pastor, S.K. 1992. Public responses to technological risks: Toward a sociological perspective. *Sociological Quarterly* 33(3):389–412.

Gulliford, A. 1989. *Boomtown Blues: Colorado Oil Shale, 1885–1985*. Niwot, CO: University Press of Colorado.

Hance, B.J., Chess, C., and Sandman, P.M. 1988. *Improving Dialogue with Communities: A Risk Communication Manual for Government*. New Brunswick, NJ: Rutgers University Environmental Communication Research Program.

Henrion, M. and Fischhoff, B. 1986. Assessing uncertainties in physical constants. *American Journal of Physics* 54:791–798.

Hausman, J.A. (Ed.). 1993. *Contingent Valuation: A Critical Assessment*. New York: North-Holland.

Heberlein, T.A. and Bishop, R.C. 1986. Assessing the Validity of Contingent Valuation: Three Field Experiments. *The Science of the Total Environment* 56:97–107.

Howell, R.E. and Olsen, D. 1981. *Who Will Decide? The Role of Citizen Participation in Controversial Natural Resource and Energy Decisions*, Bibliography No. 765. Monticello, IL: Vance Bibliographies.

Humphrey, C.R., Berardi, G., Carroll, M.S., Fairfax, S., Fortmann, L., Geisler, C., Johnson, T.G., Kusel, J., Lee, R.G., Macinko, S., Peluso, N.L., Schulman, M.D., and West, P.C. 1993. Theories in the study of natural resource-dependent communities and persistent rural poverty in the united states. In: Rural Sociological Society Task Force on Persistent Rural Poverty (Eds.). *Persistent Poverty in Rural America*. Boulder, CO: Westview. Pp. 134–172.

Hynes, M.E. and Vanmarche, E.H. 1977. Reliability of embankment performance predictions. In: R.N. Dubey and N.C. Lind (Eds.). *Mechanics in Engineering*. Toronto: University of Waterloo Press. Pp. 367–384.

Interorganizational Committee on Guidelines and Principles for Socioeconomic Impact Assessment. 1994. *Guidelines and Principles for Socioeconomic Impact Assessment*, NMFS-F/SPO-16. Washington, D.C.: U.S. National Marine Fisheries Service.

Jordan, W.S., III. 1984. Psychological harm after PANE: NEPA's requirements to consider psychological damage. *Harvard Environmental Law Review* 8:55–87.

Krannich, R.S. and Luloff, A.E. 1991. Problems of resource dependency in U.S. rural communities. *Progress in Rural Policy and Planning* 1:5–18.

Krauss, C. 1989. Community struggles and the shaping of democratic consciousness. *Sociological Forum* 4(2):227–239.

Lawless, W.F. 1993. Interdependence and Science: The Problem of Military Nuclear Weapons Waste Management. *Research in Social Problems and Public Policy* 5:271–288.

Levine, A.G. 1982. *Love Canal: Science, Politics, and People*. Lexington, MA: Lexington Press.

Lichtenstein, S. and Fischhoff, B. 1977. Do those who know more also know more

about how much they know? *Organizational Behavior and Human Performance* 20:159–183.
Lichtenstein, S., Fischhoff, B., and Phillips, L.D. 1982. Calibration of probabilities: The state of the art to 1980. In: D. Kahneman, P. Slovic, and A. Tversky (Eds.). *Judgment under Uncertainty: Heuristics and Biases.* New York: Cambridge University Press. Pp. 306–333.
Llewellyn, L.G. 1981. The Social Cost of Urban Transportation. In: I. Altman, J. Wohlwill, and P. Everett (Eds.). *Transportation and Behavior.* New York: Plenum. Pp. 169–202.
Llewellyn, L.G., Goodman, C., and Hare, G. (Eds.). 1981. *Social Impact Assessment: A Sourcebook for Highway Planners.* Washington, D.C.: Federal Highway Administration, Environmental Division.
Llewellyn, L.G. and Freudenburg, W.R. 1990. Legal requirements for social impact assessments: Assessing the social science fallout from three mile island. *Society and Natural Resources* 2(3):193–208.
Meidinger, E.E. and Freudenburg, W.R. 1983. The legal status of social impact assessments: Recent developments. *Environmental Sociology* 34:30–33.
Meyer, S.M. 1992. *Environmentalism and Economic Prosperity: Testing the Environmental Impact Hypothesis.* Cambridge, MA: MIT Project on Environmental Politics and Policy.
Mitchell, R.C. and Carson, R.T. 1989. *Using Surveys to Value Public Goods: The Contingent Valuation Method.* Washington, D.C.: Resources for the Future.
Moen, E.W. 1984. Voodoo forecasting: Technical, Political and ethical issues regarding the projection of local population growth. *Population Research and Policy Review* 3:1–24.
Mountain West Research, Inc. 1979. *A Guide to Methods for Impact Assessment of Western Coal/Energy Development.* Omaha, NE: Missouri River Basin Commission.
Murdock, S.H., Leistritz, F.L., Hamm, R.R., and Hwang, S. 1984. An assessment of the accuracy and utility of socioeconomic impact assessment. In: C.M. McKell, D.G. Browne, E.C. Cruze, W.R. Freudenburg, R.L. Perrine, and F. Roach (Eds.). *Paradoxes of Western Energy Development.* Boulder, CO: Westview Press. Pp. 265–296.
Murdock, S.H., Leistritz, F.L., and Hamm, R. 1985. The state of socioeconomic analysis: Limitations and opportunities for alternative futures, Annual Meeting of the Southern Association of Agricultural Scientists, Biloxi, MS, February 3–6.
National Academy of Sciences/National Research Council. 1992. *Assessment of the U.S. Outer Continental Shelf Environmental Studies Program: III. Social and Economic Studies.* Washington, D.C.: National Academy Press.
National Academy of Sciences/National Research Council. 1994. *Environmental Information for Outer Continental Shelf Oil and Gas Decisions in Alaska.* Washington, D.C.: National Academy Press.
National Academy of Sciences/National Research Council. 1996. *Review of New York State Low-Level Radioactive Waste Siting Process.* Washington, D.C.: National Academy Press.
Nord, M. and Luloff, A.E. 1993. Socioeconomic heterogeneity of mining-dependent counties. *Rural Sociology* 58:492–500.
Norgaard, R.B. and Howarth, R.B. 1993. Resolving economic and environmental

perspectives on the future. *Research in Social Problems and Public Policy* 5:225–241.
Overdevest, C. and Green, G.P. 1995. Forest-dependence and community well-being: A segmented market approach. *Society and Natural Resources* 8(2):111–131.
Passell, P. 1995. The curse of natural resources: They can keep a country poor. *New York Times*, p. C2, Sept. 21.
Peluso, N.L., Humphrey, C.R., and Fortmann, L.P. 1994. The rock, the beach, and the tidal pool: People and poverty in natural resource-dependent areas. *Society and Natural Resources* 7(1):23–38.
Perrow, C. 1984. *Normal Accidents: Living with High-Risk Technologies.* New York: Basic Books.
Repetto, R. 1995. *Jobs, Competitiveness, and Environmental Regulation: What are the Real Issues?* Washington, D.C.: World Resources Institute.
Sachs, J. and Warner, A. 1995. *Natural Resources and Economic Growth.* (Unpublished report.) Cambridge, MA: Harvard Institute for International Development.
Savatsky, P.D. 1974. A legal rationale for the sociologist's role in researching social impacts. In: C.P. Wolf (Ed.). *Social Impact Assessment.* Stroudsburg, PA: Dowden, Hutchinson & Ross. Pp. 45–47.
Schnaiberg, A. and Gould, K.A. 1994. *Environment and Society: The Enduring Conflict.* New York: St. Martin's Press.
Shrader-Frechette, K.S. 1993. *Burying Uncertainty: Risk and the Case Against Geological Disposal of Nuclear Waste.* Berkeley: University of California Press.
Sieber, J.E. 1974. Effects of decision importance on ability to generate warranted subjective uncertainty. *Journal of Personality and Social Psychology* 30(5):688–694.
Tickamyer, A.R. and Tickamyer, C.H. 1988. Gender and poverty in central Appalachia. *Social Science Quarterly* 69(4):874–891.
Weber, B.A. and Howell, R.E. 1982. *Coping with Rapid Growth in Rural Communities.* Boulder, CO: Westview.
Weinberg, A. 1972. Science and trans-science. *Minerva* 10(2):209–222.

Decision-Maker Response

Roy Silver, Elizabeth Ungar Natter, and Chetan Talwalkar

Our work regularly takes us into communities that have been forced to bear the consequences of poor decisions, some made locally but most originating from outside forces with little or no consideration given to the concerns of the affected community. Simply put, our experience has been that decision making and decision-aiding tools have not been used in ways that empower local citizens and affected communities or that bring local knowledge to bear on the decisions to be made. Rather, we have found cases where they have been used in ways that exacerbate discrimination based on race, gender, and class.

Correcting these historic deficiencies will require at least that the affected communities be brought into the decision-making process at an early enough point to actually affect the outcome. As long as the discussion is narrowly focused on tools, it has potential only to offer some moderate reforms of contemporary practices. We must change decision-making institutions so that those individuals and groups most adversely affected can be equal partners in the decision-making process or play leadership roles.

Quantitative tools, for all their apparent precision, often end up producing precisely and dramatically wrong results because of underestimation of the unknowns; overconfidence in the knowns; premature disqualification of plausible outcomes; misapplication of statistical data; or lack of necessary data. The concept of the "tyranny of illusory precision" identified by Freudenburg needs to be widely publicized and understood. Because of their apparent precision, health assessments, risk assessments, and many other best guesses are sometimes presented to communities as certainties. These results are used to encourage affected communities to accept facilities that the analysts and decision makers, who understand those uncertainties, would not live near. Thus, even when the scientists understand the limitations in their own field of knowledge, by the time the tool is "used" in the community, those uncertainties and assumptions are often glossed over.

Another example of the problem of illusory precision arises in risk assessments, a tool discussed elsewhere in this volume. Risk assessments yield a number that is often given weight far out of proportion to its accuracy. "The basic goal of a risk assessment is to evaluate the potential consequences of a decision, recognizing that much necessary information is not available and may never become available" (Montague, 1990, p. 1). Complete toxicological profiles are available for only two percent of chemicals in use. Even less is known about the consequence of exposure to the toxic soups encountered outside the lab. Research suggests that chemicals in combination can be orders of magnitude more toxic than the simple sum of risks sometimes used in risk assessment. For example, low-level mixtures of a few standard pesticides are up to 1,600 times as potent as the particular pesticides (Arnold, 1996).

Freudenburg also recognizes considerations of equity, noting that "a decision that effectively ignores 40 percent of the population is simply unacceptable. . . ." We contend that race, gender, and class need to be accounted for in determining how tools are used, whom they serve, and how well they assess environmental impacts. Tools that are apparently "neutral" or "objective" often serve to increase inequities in our society.

Tools should empower and involve affected communities by providing good information in an open and honest way. The tools that offer the most potential are: increasing public input, participant-observer research, and key-informant interviews. These tools can help "increase institutional permeability" as Freudenburg advocates, although they fall short of involving affected communities as full and equal partners in decision making. In addition, the techniques described by Freudenburg that allow the interests of future generations to be on the table need to be expressed and disseminated. Techniques that ignore the impact on future generations, or reduce them to meaninglessness through the application of an economic discount rate, should not be used without further analyses that reintroduce these considerations.

Perhaps the most important insight in the chapter is Freudenburg's "gaps and blinders" analysis. Here he calls for sensitivity analyses to identify analytical biases, interdisciplinary double-checks to recognize possibilities that have been previously overlooked, and serious public involvement efforts to make use of local knowledge. More research needs to be done to further develop these techniques. Analysts and decision makers of all types should be trained in how to apply them.

Freudenburg calls for going beyond making explicit distinctions between knowns and unknowns, to a discussion of "facts, values, and blind spots." He recognizes that science may be able to tell us how safe an activity is, but that citizens (and, we would add, those most adversely affected) must decide how safe is safe enough. He calls for going beyond disciplinary boundaries to discover these blind spots within a specific field of inquiry. His focus on the "gaps and blinders" should be a required field of inquiry for social scientists, environmental scientists, policy making, and policy practitioners.

Some of the tools Freudenburg discusses have the potential to be used in a way that explores and respects local knowledge, such as nonrandom, or "key-informant interviews," public involvement, and participant observation. Too often, techniques are applied that do not make use of local knowledge. We have seen many instances of citizens trying to explain in a public hearing about the long-term decrease in diversity of animal and plant life in their area; or about a spring their grandfather used for drinking water that is now under a Superfund site; or about past mining operations that are now under an area of current groundwater contamination. Long-term residents have a knowledge of community history that is vitally important to any decision-making process. Instead of being heard, respected, and involved in decision making, local folks who are part of cultures that fall outside the mainstream are frequently treated as eccentric. Freudenburg discusses the need to identify the many gaps within an analysis based on existing science. By using these tools in a way that respects local knowledge and value systems that fall outside the mainstream, some of the gaps and blinders could be avoided.

For public-involvement techniques to be successful, "institutional permeability" must be increased, as Freudenburg insists, not only for the collection of better data but also for less polarized decision making and responses. The full sharing of information is necessary, not just in pre-chewed tidbits, but in its entirety, especially the uncertainties and assumptions involved in the tools being used. Without such information sharing, people cannot adequately represent their interests, cannot give informed consent, and cannot provide the needed testing and gap-filling function that Freudenburg identifies.

The local knowledge and local preferences thus gathered must be used in ways that can actually affect the outcome of the process. The results of many public-input processes are unfortunately doomed to irrelevance and obscurity by a lack of information or by perfunctory public-participation exercises undertaken to fulfill statutory mandates and occurring far too late in the process to make real differences in the decision. The community must be involved up front in environmental decision making when tools for expansion of options can still be used, not at the end of the process after the agency is vested in a particular decision and is using tools for refinement and narrowing of options.

In response to pressure by the environmental justice movement, the Clinton administration has proclaimed that "each Federal agency shall make achieving environmental justice part of its mission by identifying and addressing, as appropriate, disproportionately high and adverse human health or environmental effects of its programs, policies, and activities on minority populations and low-income populations." Some of the socioeconomic tools discussed by Freudenburg could be used to help identify these injustices. More research is needed to determine which tools are the best suited for retrospectively identifying these disproportionate effects.

To avoid contributing to environmental injustices in the first place, the common assumptions of decision-making tools need to be examined for their potential to favor results that disproportionately burden low-income people and people of color. One method that avoids such problems is a tool that Freudenburg does not discuss—participatory-action research.

Participatory-action research attempts to break down the distinction between the researchers and the researched, the subjects and objects of knowledge production by the participation of the people-for-themselves in the process of gaining and creating knowledge. In the process, research is seen not only as a process of creating knowledge, but simultaneously as education and development of consciousness, and mobilization for action. (Gaventa, 1991, pp. 121–122)

Participatory-action research has seven guiding principles (Silver, 1991, pp. 10–11):

1. Try not to impose an external definition of need.
2. The community must maintain control over the process.
3. Professionals have to work in solidarity with community groups.
4. Professionals have to sustain a respect for the knowledge of the community.
5. Professionals have to learn about all the dimensions of the problem.
6. Professionals should be educated by the community.
7. Professionals should transfer their skills to the community.

These principles strive to separate the investigator and the invested. They achieve this end through the participation of those who lack information in the process of gaining and creating knowledge. Decision-making tools are then conceived not solely as a means of developing knowledge, but also as education, as a means of extending awareness, and as a way of mobilization for action.

We agree with Freudenburg that progress is more likely to come from tools that help expand options than those that assess, refine, and narrow options.

References

Arnold, S.F. et al. 1996. Synergistic activation of estrogen receptors with combinations of environmental chemicals. *Science* 272:1489–1492.

Gaventa, J. 1991. Toward a knowledge democracy: Viewpoints on participatory research in North America. In: O.F. Borda and M.A. Rahman (Eds.). *Action and Knowledge: Breaking the Monopoly with Participatory Action Research.* New York: Apex Press. Pp. 121–131.

Montague, P. 1990. *Rachel's Hazardous Waste News* (204), Oct. 24, 1990.

Silver, R. 1991. Participatory-action research: Combating the poisoning of Harlan County's Dayhoit! The Sixth Annual Conference on Appalachia: Higher Education and Appalachia, Lexington, KY.

5
Characterizing the Regulatory and Judicial Setting

MARY L. LYNDON

Environmental Law and Legal Research: Tools for Decision Making

We do not usually think of law as a tool; but ideally, that is its role. While law constrains, it also offers opportunities for creative analysis and organization of decision making. A rule or legal principle should distill and express the social considerations that ought to go into a decision, while procedural rules structure participation in law making to maximize fairness and rationality. Legal research materials and methods are tools in another sense. They record what the law is and (again, ideally) provide this information in response to appropriate inquiries. Both environmental law and related legal research are always changing, but today legal research appears to be in an especially rapid state of flux, perhaps signaling a fundamental change in the law itself.

This chapter provides a brief overview of the contents of environmental law and the chief research tools available to identify existing and emerging law. These research tools range from the traditional print materials and well-established electronic law libraries, such as LEXIS® and WESTLAW®, to the new resources available on the Internet. Suggestions are offered on how researchers can sift through these tools, select the most appropriate for the research setting, gain access to them, and use and benefit from them. The concluding section describes some emerging environmental laws and suggests some ways in which new information technologies will affect law and legal process in the future.

Environmental law has virtually exploded in its 25-year history and now is a complex specialty. It includes international treaties and conventions; national, state, and local legislation; court decisions; and agency actions. It also shares much with the law concerning occupational health, the regulation of food and drugs, and the law governing the protection of consumer rights. It has penetrated the fields of law concerned with business organizations, such as insurance, real estate, and securities law. Laws on civil rights and human rights are also invoked today in relation to environmental problems. As the legal framework of international trade develops specific environmental standards and precedents, these standards will affect national legal requirements. The discussion here focuses on environmental law itself, and particularly on the law of the United States, but the decision maker should remember that other types of law may be implicated in a particular situation.

Survey of Environmental Law

The environmental decision maker cannot find out "what the law is" by looking in one place. Environmental law is a layered system. It is made and implemented primarily by administrative agencies, which act pursuant to legislative mandate. Courts also actively make environmental law; they may review an agency performance, if this is authorized by a statute, and require the agency to change its ruling. Courts also make environmental law by deciding liability claims brought according to the common law, such as tort actions to stop a nuisance or lawsuits seeking compensation for personal injury or property damage.

Federal, state, and, sometimes, local agencies often work on the same environmental problem. Federal statutes dominate the field, but states often enact parallel requirements, which are sometimes more strict than federal law. Under the United States Constitution, the states retain the "police power," the primary authority to protect the public health and welfare. However, the Interstate Commerce and Supremacy clauses of the Constitution authorize Congress to preempt some state law, and it has done so in a variety of circumstances. Environmental law is very much a combined effort, with states both following and leading federal law, and with courts and agencies in a push-me-pull-you relationship.

Moreover, "the law" consists of more than the statutes, regulations, and court decisions that are formally recognized as binding legal requirements. It also includes the documentary paraphernalia of contemporary regulatory government. Documents relevant to an environmental matter may include plans, such as state implementation plans that are developed under the federal Clean Air Act (CAA) and local zoning regulations; executive and regulatory policy statements and directives; and findings and recommendations of government institutes and science advisory boards. Core legal requirements are shaped by this context.

Environmental law embodies familiar notions of legal philosophy, such as "rights" and "property," and also more recent economic values, such as "efficiency." When conceptual schemes clash, what principles should guide us? For instance, the social contract model of society speaks in terms of individual rights: each citizen is free to act as he will, but only as long as he does not infringe on his neighbors' rights. Thus, the traditional law of nuisance held that one may not interfere with neighboring uses of land or with common air and water supplies. The social contract model also holds that the first person to claim resources in the state of nature may do so as long as he leaves, in John Locke's words, "enough, and as good" (Locke, 1690) for others. However, some contemporary economic models of law stress the importance of maximum wealth production because greater overall wealth is thought to benefit society. Environmental law tries to accommodate different perspectives; it attempts to square our commitment to enterprise with recognition of social and ecological limits. Usually, one or another view will predominate, and decision makers need to be aware of the ways that competing frameworks yield different results.

The first of several caveats is appropriate at the start of our discussion. Legal research is usually conducted by lawyers. But everyone is affected by environmental law, and all kinds of people participate in making that law. Environmental law expressly aims to involve citizens in policy making and enforcement. Nonlawyers frequently must make decisions that are influenced by the law. But if you are relying on your own research, you must be cautious because legal documents often seem more precise than they actually are.

Environmental Statutes

Environmental statutes can be roughly categorized according to their subject matter and their basic strategies. As legislatures amend and update statutes, they incorporate newer provisions and philosophical approaches. Each major statute is a complex field in its own right. Sidebar 5.1 provides a simple conceptual road map.

It may be useful to think of environmental statutes as addressing four types of subject matter. One basic grouping consists of federal statutes that address some tangible aspect of the environment. There are statutes concerned with protection of air, water, marine mammals, and endangered species. Prominent examples are the CAA and the Clean Water Act (CWA). Other statutes address particular types of pollution. For example, the Federal Insecticide, Fungicide, and Rodenticide Act (FIFRA) directs the EPA to screen pesticides before they are marketed.

A second type of law addresses a broad category of business practices, such as hazardous-waste generation and management. These laws attempt to create standards and incentives for businesses, to encourage safe treatment of waste in the future, and to respond to existing problems caused by inadequate past practices. Here, the chief statutes are the Comprehensive Environmental Response, Compensation, and Liability Act (the "Superfund Law" or CERCLA); the Resource Conservation and Recovery Act (RCRA); and the Pollution Prevention Act of 1990 (PPA).

A third category addresses the problem of information production and distribution. These laws require government agencies to formally study the

> **Sidebar 5.1**
> **Types of Environmental Statutes**
>
> Environmental law is a patchwork of statutes, regulations, and court decisions. The statutory elements of the law can be loosely grouped in four categories:
>
> Laws regulating particular types of pollutants or protecting particular dimensions of the ecosystem. For example, FIFRA directs EPA to screen all pesticides. The CAA and the CWA direct the EPA to regulate discharges into the air and the water. The Safe Drinking Water Act specifically addresses drinking-water quality.
>
> Laws addressing phases of industrial practice, such as waste disposal and management. The chief statutes here are CERCLA, or the "Superfund law"; RCRA, which regulates waste management; and the Pollution Prevention Act.
>
> Information and disclosure requirements that are contained in a number of statutes. NEPA, TSCA, and the various right-to-know laws and regulations focus on the production and dissemination of toxicity and exposure data.
>
> Land-use controls that are contained in some statutes, such as the Endangered Species Act and the wetlands provisions in the Clean Water Act. Land use is also an important theme in constitutional and common law.

environmental consequences of their actions and require polluters and employers to share what they know about the adverse effects of the chemicals they use and produce. These statutes include the National Environmental Policy Act (NEPA); the Toxic Substances Control Act (TSCA); and the Emergency Planning and Community Right to Know Act (EPCRA).

A fourth category of laws restricts the uses one can make of particular parts of the environment. Most of this is what lawyers call land-use law. Traditionally, this law has been made up largely of state law, including local zoning ordinances. However, the evolution of federal land management and increased federal protection of wetlands have expanded the federal law in this area.

Environmental statutes express a range of regulatory strategies and philosophies (See Sidebar 5.2). This diversity is partly caused by the learning process: growing experience with environmental regulation has naturally produced new approaches. At the same time, environmental regulation affects a great many interests in our society and therefore is affected by political processes. Different presidential administrations, for instance, have taken markedly different approaches to regulation.

Sidebar 5.2
Types of Pollution-Control Strategies

Goal-based programmatic laws, sometimes called "command-and-control" regulation, articulate an objective, such as healthful ambient air; direct the EPA to determine the specific discharge limits necessary to reach the goal; and sketch an approach to arriving at the goal. The EPA then writes a specific plan for each industry contributing to current pollution levels, usually with the assistance of the state governments. Permits are issued to each source in compliance and enforced primarily by the states.

Market-based incentive schemes identify a group of sources of a type of pollution, set a limit on overall pollution from all sources, and allow polluters some discretion as to where and how to reduce the overall pollution. This type of regulation may be designed for a single company with multiple sources at one industrial site; for several companies in a limited geographic area; or for a whole industry that produces widespread but fairly homogeneous pollution. The most ambitious scheme of this sort is the CAA's Acid Rain Control Program, which applies to large fossil-fuel-fired electric power plants and aims at reducing national emissions of sulfur dioxide and nitrogen oxides.

Both economic and ecological principles support controls on certain types of products and raw materials. Several statutes (CERCLA, RCRA, and the PPA) together impose limits on and restructure incentives relating to the use of petrochemical products and related technologies. Industries that use chemicals are encouraged to shift to less-polluting processes and products. At the same time, discharges and waste management are scrutinized and limited. Liability for spills and dumping is imposed, both on the industry as a whole and on individuals and firms responsible for improper disposal.

Self-monitoring, reporting of discharges, and labeling of toxic dangers are now widespread requirements in state and federal laws. Emerging standards for performing environmental audits will eventually allow the government to compile more specific and comprehensive pollution inventories and to set regulatory priorities more rationally. At the same time, the growing information base will give firms greater opportunities to be efficient, not just in controlling pollution, but in production, as well.

Liability rules notify firms that they must take care to avoid spills and illegal disposal of hazardous pollutants or they will have to pay for the cleanup. The Superfund Law's broad application of this principle attempts to focus the attention of the entire petrochemical industry on the environmental problems that it produces. This approach is a specific application of the conventional economic principle, expressed in modern tort law, that enterprises should be responsible for their social costs.

The learning curve and political change together have led Congress to adopt several different regulatory strategies. Early approaches were project-oriented. They followed the command and control model of regulation: each law established a goal (such as cleaning up pollution in waterways) and delegated governmental power to the EPA to (1) identify the steps necessary to achieve the goal; and (2) carry out those steps. Often, the control strategies have directed all polluters of a certain type to take identical measures to achieve the goal. This approach has been criticized as inefficient, but it also has many defenders because of its perceived equity and simplicity.

Within the larger command and control framework are four other major types of regulation. Market-based approaches may work best in some settings. These initiatives create economic incentives to reduce pollution. Because pollution itself is the result of market failure, it may take some ingenuity to find ways to let companies profit from reducing pollution and waste. The EPA's most elaborate market-based program was inserted in the CAA as part of the 1990 amendments to the Act. This is the Acid Rain Control Program. The electric industry as a whole must reduce sulfur dioxide and nitrogen oxide emissions from fossil-fuel-fired electric-power plants, but individual companies may buy and sell emissions allowances issued by the EPA. Newer power plants which can efficiently reduce emissions levels may sell their emission credits to owners of less efficient plants.

Another kind of regulation is based upon ecological principles and seeks to minimize the impact of industrial production and human habitats on the ecosystem. Adjusting or reducing the use of raw materials at the beginning of the production process and increasing recycling should prevent environmental damage. The PPA of 1990 reflects this orientation. Laws that force production sectors to control discharges and to pay for waste disposal indirectly encourage dematerialization. The RCRA and the Superfund law are intended to have this effect.

Another regulatory strategy simply requires firms to disclose data about their discharges and the associated health effects. The simple act of reporting a discharge may focus a firm's attention on the problem and lead it to identify and measure its discharges; public disclosure may also be embarrassing and encourage greater efforts to control pollution. Liability requirements constitute another approach. Several statutes require producers, managers, and transporters of hazardous chemicals, wastes, and petroleum products to pay cleanup costs after any spill or unlicensed disposal. This approach builds upon the common law, discussed below.

Agency Actions

Congress typically has either created an administrative agency to carry out its legislative programs or has delegated implementation to an existing agency. The federal EPA is the leading agency in the area of the environment. Most states also have an environmental agency that implements state environmental laws and works with the EPA to enforce and carry out federal law.

Administrative agencies are given power to make legally binding rules, enforce those rules, and decide disputes that arise under them. An agency's authority over a particular matter may be limited or broad, depending on the wording of the statute that authorizes agency action. Agencies promulgate environmental rules and standards after carrying out procedures that generally allow for participation by all interested parties. Usually, the agency issues a proposed rule, receives written comments on it, and then publishes a final version. This final action must include an explanation and justification of the rule and, for most national regulations, demonstrate that its benefits will outweigh its costs.

When the EPA is regulating chemicals that have adverse health effects, it often uses, as part of its measurement of costs and benefits, a method called quantitative risk assessment (QRA). The QRA process gathers, evaluates, and amalgamates data from different studies to calculate the risk of cancer from the expected human exposure to a pollutant. QRAs' estimates are uncertain and can only be expressed properly as a range of risk levels. Also, the QRA process itself is expensive and time consuming because it gathers and synthesizes different kinds of data about ecology, toxicology, and exposure to pollutants. Often, the data are scarce or nonexistent. However, QRA is becoming a common part of regulation because it can be useful when we want to compare the costs and benefits of different regulatory options.

Regulations that establish limits on allowable discharges of pollution are generally implemented through permit systems. These systems are run by state and local environmental authorities under the supervision of the federal EPA's regional offices. Polluters must apply for permits and then comply with the discharge limits stated in the permits. When a permit is violated, it often will be renegotiated, and the discharging firm put on a timetable to come into compliance. This arrangement is called a "consent order." Some environmental statutes require polluters to monitor their own discharges and to regularly report the amounts. An example is the federal CWA. Otherwise, the agency is charged with detecting violations as part of its enforcement mandate.

In addition to regulations, permits, and consent decrees, other official documents may affect an environmental decision. The EPA and other agencies may announce a policy in an independent document or may issue guidelines describing the manner in which they expect to implement a program. When a compliance question comes up in a particular case, the EPA may decide the point and write a letter to the specific firm involved. Later, when the question comes up in a new situation, the EPA's letter may have some precedential value. Decision makers facing technical and compliance issues should find out whether similar cases have been decided.

Court Decisions

Federal, state, and local courts also contribute to environmental law. Courts actively make environmental law by interpreting statutes and state and federal constitutions and by applying the common law, particularly tort law (the law governing civil proceedings to redress wrongs and to compensate for injuries).

The courts' role in reviewing agency decisions is specified in the judicial review provisions of the agency's governing statutes or in the applicable state or federal administrative procedure laws. Judicial review of agency actions leaves a "gloss" on a statute; that is, the interpretation of the court becomes a part of the law unless it is overturned by a higher court or changed later by the same court. Where there are variations of interpretation in different jurisdictions, the Supreme Court may decide the matter, or the EPA may take the matter up again and try to reach a consensus position.

Statutes and regulations may seem quite precise on first reading, but they frequently fall short of answering the specific question posed by a case. This reality should not be surprising for three basic reasons. First, experience shows that anticipating the future in any detail is very hard; life *is* stranger than fiction. Second, finding language that is completely unambiguous is very hard. Words take their meaning from their context, both that of the author and that of the reader. Third, most environmental statutes and regulations are written in an adversarial context and therefore are the product of compromise; one way legislators strike a bargain is to leave the details for the agency and the courts to handle later.

When a court is interpreting a statute or regulation, it is guided by some basic principles. First of all, the intent of the author at the time the statute was enacted should be discovered. This intent is found first in the plain meaning of the words of the statute and in the statute as a whole and its evident purpose. Also, the history of the statute may be used, including documents written by legislative committees before the law was passed. In general, courts give deference to an agency's interpretation of its own statute, particularly where Congress has delegated broad authority to it. However, courts do regularly send the EPA back to the drawing board to reconsider and revise its rules.

Courts may also rule on the constitutionality of a statute. Here, the question is whether the United States Constitution or a state constitution limits Congress' or the state legislature's authority to make the law. One current issue is whether the Commerce Clause of the U.S. Constitution limits an individual state's power to exclude or impose taxes on imports of hazardous waste.

A third area where courts are active is the common law. Until the 20th century, this was the core of the law. State constitutions generally have incorporated the English common law into its own state law. Common law is case-by-case dispute resolution, following doctrinal principles and precedents. Tort law is the common law concerned with accountability for injuries. Tort law develops as courts and juries consider whether a particular defendant or group of defendants should pay money damages to compensate the plaintiffs for injuries. The accumulated court decisions on a particular type of case form the precedent that guides courts in later decisions.

While statutes and agency regulation have replaced the common law as the predominant legal form, tort law is still a significant element in environmental law. If a problem is not already covered by a statute or regulation, the common law may be applied unless the legislature has expressly preempted it. Legislatures rarely do this, however, in part because regulation generally does not compensate injured individuals, and legislatures are reluctant to prevent people from seeking redress by limiting access to the courts. Also, knowledge of potential liability provides a general, though perhaps weak, incentive for producers to take care not to expose people to unnecessary risks (Lyndon, 1995 and sources discussed there). Since the early 1970s, tort law has become an increasingly important and controversial part of the legal system. However, existing data do not support the claim that the number and cost of tort cases has grown relative to other types of litigation (Ostrom and Kauder, 1996; Daniels and Martin, 1995; Rahdert, 1995; Saks, 1992).

The tort of nuisance is the original environmental law. Nuisance doctrine holds that, generally, one cannot use one's land in such a way that it annoys or interferes with one's neighbors. Product liability is a field of tort law that began in the 19th century, but expanded in the second half of the 20th century. Today, in most states, manufacturers and distributors of products that are "unreasonably dangerous" must pay those who are injured by the product. Both nuisance law and product liability form the basis for the emerging field of "toxic torts." Illness caused by pollution or products may be compensated if plaintiffs prove to a court that the defendants caused the injury in violation of common law duties owed to the public.

Environmental and similar types of issues have spawned developments in regulation, common law, and insurance, and the shape of these three and their relationships are in flux. It seems likely, however, that all three will continue to be important influences on behavior associated with environmental impacts.

Procedural Law

Procedural rules identify who may participate in a court case or regulatory proceeding and govern the presentation of evidence and the manner and timing of decisions. Procedures may be specified by statute, by agency regulations, or by court rules. In general, evidentiary rules are more relaxed

5. Characterizing the Regulatory and Judicial Setting 139

before agencies than they are in court. However, all legal proceedings have strict time limits for commencing any action and also establish other parameters with some specificity.

Many environmental statutes outline the procedures that the agency must follow when setting general standards. Usually, the EPA publishes proposed and final agency actions in the *Federal Register*, and must give any interested person the opportunity to submit comments. Similarly, when a site is being chosen for a polluting activity or when a zoning change is being considered, public hearings generally are required. Participants in an administrative proceeding who disagree with the final action usually may seek review by a court. Permitting processes and the negotiation of consent decrees are not necessarily open to the public, though agencies often will allow public comment.

Sometimes the EPA uses a process called "negotiated rulemaking." In this approach, before the agency proposes a rule, it contacts interested private groups and supervises meetings that aim to reach a consensus position; this process is then incorporated into the agency's published proposal. Even when the process does not yield a consensus, the agency still may learn a great deal about the issues; where it does yield one, litigation over the rule may be avoided. Critics of this approach argue that some points of view may not be represented in the negotiation and that the notice-and-comment process could become a mere formality with only limited judicial review available to correct it.

Most federal environmental statutes authorize citizens to sue the EPA when the agency has failed to perform a nondiscretionary statutory duty. These provisions also authorize the courts to award attorneys' fees to prevailing citizen plaintiffs. However, these cases can usually be won only if the statutory mandate that the suit seeks to enforce is very specific, as when an agency is required to promulgate a standard by a particular date. In addition to citizens' suits, any person can generally petition the EPA to commence a rulemaking proceeding. A request for a rulemaking may prod the agency to act, but it may be hard to enforce the request in court.

Laws about Information Production and Access

Uncertainty about the effects of pollution is a dominant theme in environmental law. However, developments in information theory have led to new regulatory strategies. The growth of warranty and labeling provisions in consumer and food and drug law has been followed by an increased reliance on information strategies in the environmental context. Here, the law is a mix of statutory requirements, common-law principles, and "burdens of proof." A quick sketch of the background and basic types of information requirements will help introduce the research tools outlined in the next section.

A central principle of the individualist jurisprudence of the 19th century was that the government would not interfere with people's affairs unless they were at fault and caused harm. The burden of proving harm was placed upon those who sought legal controls or payment of damages. The basic position of the common law and of regulation today is still *ex post facto*, and consequently, uncertainty about the specific adverse effects of pollution often inhibits efforts to control it. Polluters generally have not been required to study pollution or even to disclose any information they may have about its effects. However, with the increasing influence of ecological principles, the law is gradually changing. Although we still allow polluting activities to proceed until a red light goes on, environmental regulation today contains a number of information requirements.

For instance, many firms are required to report to their state agency or to the EPA about certain kinds of discharges into the environment. Firms are also generally required to identify for their employees and for local emergency services (e.g., fire departments and hospitals) any chemicals they handle that may have adverse health effects. In some contexts, firms are required to perform environmental audits of their operations.

Environmental auditing is a relatively new phenomenon, but may soon be commonplace. The International Organization for Standardization (ISO), a long-standing body that works to facilitate international trade, publishes standards for quality-management systems. It has published a series of standards for environmental protection (series ISO 13020) that covers wastes, air quality, water quality, soil quality, occupational safety and industrial hygiene, safety of machinery, domestic safety, noise, vibration and shock, ergonomics, accident and disaster control, fire protection, explosions, excessive pressure, electric shock, radiation, and dangerous goods. It is now developing further environmental standards in five areas: management systems, audits, labeling, environmental-performance evaluation, and life-cycle assessment. As these standards are completed, good business practices and environmental auditing should be facilitated. At the same time, ISO 14,000, as the new environmental standards are called, may affect international trade and national legal responsibilities. Their impact may be contested, and adjustments may be required either in the standards or in the law itself. More can be found about the ISO, its standards, and its operations on its World Wide Web home page at http://www.iso.ch/.

Finally, government agencies are required to publicly and formally assess the environmental impacts of their actions, and federal agencies must perform a cost-benefit analysis of any regulatory actions that may have significant national economic impacts.

The coverage of our information laws is incomplete. However, the database is growing. Each process yields documentation, and much of this is available to anyone who requests it, except for documents that entail privacy, trade secrecy, or law enforcement. In spite of the gaps, we are beginning to develop useful inventories of pollution (See Sidebar 5.3).

Sidebar 5.3
Environmental-Information Statutes

The Freedom of Information Act (FOIA). This federal law and numerous, subsequent state enactments make all documents in the possession of the government available to anyone on request, subject to a number of exemptions designed to protect personal privacy, sensitive law-enforcement matters, national security, trade secrets, and a broad category of other confidential material. Each major federal agency has its own FOIA process and personnel.

The National Environmental Policy Act (NEPA). Federal agencies must assess the environmental impacts of their activities. Each legislative recommendation and all other major federal actions that may significantly affect the human environment must be formally studied and reported upon. Opportunity for public comment must be afforded, and the agency's compliance with these procedures is subject to judicial review.

The Toxic Substances Control Act (TSCA). This act directs the EPA to screen chemicals entering the market. Manufacturers and importers of new chemicals must submit to the EPA information concerning each new chemical's structure, its intended uses, expected quantities of production, estimates of potential human exposure, and any health-effects data that the manufacturer may have. However, no new testing is required. The EPA must screen each chemical to determine whether it presents an unreasonable health risk. To require further tests or to limit production, the agency must demonstrate that the chemical may pose an unreasonable risk to human health.

Right-to-Know laws. Right-to-Know laws require the disclosure of the chemical identity and health effects of discharges into the environment and human exposures in the workplace. Many states enacted right-to-know laws in the late 1970s and early 1980s. State laws covering workers are now partially preempted by the federal OSHA Hazard Communication Standard. This standard requires employers to identify and to warn workers of known chemical hazards. The rules also specify labeling requirements for containers and conduits, require posting of warning notices in the workplace, and mandate chemical-safety training. OSHA's rules have led to the systematic distribution of information about toxic chemicals in industrial use. The federal Emergency Planning and Community Right-to-Know Act (EPCRA), builds upon OSHA's rules, but does not preempt similar state laws. It mandates the formation of state and local emergency-response committees and requires that every company subject to OSHA's rules complete an inventory of chemicals covered by those rules and provide it to the local committee. Industrial facilities that use chemicals on the EPA's list of "extremely hazardous substances" must notify the local committee of the presence of these chemicals, immediately report any unexpected releases, and participate in the committee's planning activities. EPCRA also established the National Toxics Release Inventory (TRI). The statute directs the EPA to compile a national database on routine discharges into the environment of more than 300 chemicals. The inventory is to provide the basis for regulatory planning, and the inventory data must be made accessible to any person on a reasonable-cost-reimbursement basis.

Researching Environmental Law

This section outlines the different ways one can learn more about the law so decision makers can be more sophisticated consumers. Sometimes, researching a question of environmental law is simple and quick (i.e., if your question is simple and the law has already articulated an answer to it). More often, however, research is a confusing odyssey through statutes and regulations that are only partly on point, and through scientific and technical data that are incomplete and uncertain. For those with little experience in this area, it will be difficult to know which situation you are in. You may think you have your answer, but fail to see the background complexities.

Environmental law cannot be separated from environmental and health sciences and engineering. Each legal question refers explicitly or implicitly to a reservoir of knowledge and uncertainty. Experienced lawyers, even those practicing environmental law, do not make decisions without technical advice from scientists and environmental engineers who are familiar with the specific problem at hand. Therefore, take note of this second caveat: If you are facing an environmental decision, it is important to get experienced legal and technical help as soon as you can. Also, remember that, because environmental law is complex and is constantly changing, one should always network to keep up to date. Whether you are in business, a member of an environmental advocacy group, a government official, or a concerned individual, try to develop a speaking relationship with the environmental agency staff who are working on the problem. Do not hesitate to call state or federal help lines and to use other resources for the latest information pertaining to your decision. What follows is an overview of the current resources available to learn about environmental law (See Sidebar 5.4).

Books and Treatises on Environmental Law

The market for books on environmental law can be divided into four main segments:

Sidebar 5.4
Questions Addressed by Environmental Laws

Environmental law addresses a wide variety of issues. Consider the fictional small town of Ames. In its semi-rural neighborhoods, residents are concerned about the occurrence of rashes and respiratory illnesses. Some believe these ailments may be produced by environmental causes and form an association, People Opposed to Pollution (POP), to consider the possible environmental causes of their problems.

One suspect is United Carpet, Inc. (UCI), a manufacturing plant located immediately outside Ames. POP's leaders agree that they need to gather information about UCI's environmental discharges. They must find out whether there are data connecting UCI to their symptoms. These questions seem to be "scientific" in nature, but legal questions are closely related.

Does UCI or anyone else have information on the health or environmental effects of its discharges? If so, must it share this information with POP? If not, must UCI study these effects? Must UCI stop polluting if there is no proof that its discharges are safe, or is it entitled to continue discharging until harm is proven?

To what extent is UCI's operation regulated by law? If it is regulated, how can POP find out if UCI is complying with the law? If it is not regulated, can POP get it regulated? What level of proof of harm might be required to impose controls? Does anyone have a legal obligation to study these questions? If a causal link can be proven, does UCI owe compensation to those who have been injured?

UCI has announced it will expand its operations on a property that is adjacent to its existing plant but within the Ames town line. The company has sought a zoning variance for this expansion, and a public hearing date has been set. POP members may want to oppose this expansion, but they have limited funds and do not want to pay for help, at least until they have a better understanding of the legal issues they face. They may educate themselves by doing some research.

Generally, one can take two different approaches to a legal-research problem. If the question can be formulated in specific terms, one can ask the "system" that question and perhaps get an answer. Or, one can do some research on environmental law generally and also in the particular area of interest. The latter approach will, of course, help refine the questions one asks, and it will also lead to a better understanding of the context. The context may hold the keys to the resolution of an overall problem if there is no clear answer to the

> specific legal question. In the long run, going through the general research could save time.
>
> The research will probably begin by trying to identify the UCI carpeting plant's emissions. Several ways are available to find out about them. For instance, POP could call or write to UCI itself and ask for the information directly. UCI might well respond with a list of all emissions or a less comprehensive list of all the currently regulated pollutants it emits, or it may suggest a meeting to discuss the matter.
>
> If UCI does not respond or its answers are not satisfactory, then POP must determine what it is entitled by law to know. It also needs to know what environmental laws apply to the plant and learn more about UCI's compliance status. In any event, POP should begin to network right away.

- Popular expositions of the law
- Books and journals written expressly for practicing lawyers or for law students
- Books for environmental managers of businesses
- Books on regulatory policy

A fifth category now emerging is books produced by and for community activists who wish to participate in local environmental-law enforcement. See, for example, the *Work Plan for Citizen Participation in Clean Air Act Title V Permitting* (Swanston et al., 1996). This publication and other works are highlighted on the web page of the Minority Environmental Lawyers Association, http://www.concentric.net/~Mstanisl/.

The number of offerings in each of these categories is growing. Indeed, because of the dynamism of environmental regulation, titles that are a few years old will often be outdated. The variety of excellent sources on environmental law is so great that it is not possible to direct the reader to all or even to the best of them. Particular works will be mentioned here only to give a flavor of the types of books and articles available.

The law itself can be found in collections of federal and state statutes, court decisions, and agency regulations. These documents are published in varying formats from jurisdiction to jurisdiction. Weekly and monthly newsletters report on changes in the law and on court decisions.

An enormous volume of work describes, analyzes, and argues about the law. These books and articles are targeted to legal professionals and are likely to contain the most specific information. Works of this sort range from very general summaries that are reissued periodically (see, for example, Findley and Farber, 1996; Eggen, 1995) to less general but still introductory works, such as textbooks designed for use in law-school classes (see, for example, Percival et al., 1996), treatises that summarize and expound upon the whole body of environmental law (see, for example, Rodgers, 1994), and other law-practitioners' books. Some material is published in loose-leaf format and is updated frequently. Numerous magazines and journals focus on environmental issues, and articles on environmental law and regulation often appear in general law reviews.

These materials are available in law libraries, which vary in size from small private collections of law firms to large libraries that are part of law schools. Public libraries should be able to help the researcher locate a law library. While access is generally limited, one can often apply to the librarian for temporary permission to use a law library. Also, some resources from law libraries are now available on the Internet. There are also handbooks for environmental managers and handbooks for environmental activists. These two categories may increasingly build upon and interact with other media, such as the Internet. Finally, many books discuss policy issues. Some are directed primarily to policy makers and academics, and some aim for a larger audience (see, for example, Colborn et al., 1996.)

Standard Computerized Legal Research Services

Two computerized legal-research services, LEXIS® and WESTLAW®, make available the full texts of statutes, court decisions, most agency regulations, and many agency decisions. They also offer access to many legal journals and to other periodicals. In addition, they provide legislative histories and many different kinds of administrative-agency documents that are important to a full understanding of the law. Course materials from practitioners' continuing-legal-education courses are also sometimes available.

Research conducted with WESTLAW® and LEXIS® can be very broad or very focused, depending on the way the query is framed and the size of the database selected. Illustrative databases containing court decisions in WESTLAW® are Cases, U.S. Supreme Court Cases, Multistate Cases, and Individual State Cases; LEXIS® has similar categories. Both systems also provide a wide range of agency decisions. For instance, LEXIS® offers EPA's FIFRA decisions, EPA Consent Decrees, EPA General Counsel Opinions, and many other databases. WESTLAW® has parallel offerings.

Both LEXIS® and WESTLAW® have become essential tools for lawyers and law students. They quickly make available more legal precedent, more data, and more diverse points of view than books can provide. They have provided a different orientation to research, which now may evolve to a new level with the interactive Internet, although it is too soon to tell where this path leads. The two services have different strengths, research formats, and fee structures. Most law offices and law libraries subscribe to at least one of these two services. Both offer a nonsubscriber research service for a fee.

Environmental Law on CD-ROM

Computers may store information in different formats, such as the floppy diskette and the CD-ROM (compact-disc read-only memory). CD-ROMs with law resources offer material that generally parallels the legal materials available in law libraries and on WESTLAW® and LEXIS®. However, the CD-ROM format offers certain advantages in some circumstances. Depending on the user's needs, available technology, and budget, CD-ROMs can offer a large library in a small space. A typical CD-ROM can hold the equivalent of 424 floppy diskettes. The data on a CD are practically indestructible, with a shelf life of approximately 100 years. However, a CD cannot be edited by the user, and the information contained on it may become obsolete after a relatively short time. A mix of the scope and currency of the library and its cost can be tailored to individual needs. In general, if one needs to refer frequently to one segment of the literature, CD-ROM may be a useful approach. Practitioners specializing in a particular field of law may prefer CD-ROMs. In some mass-tort litigation, CD-ROM technology has been used to consolidate the documents received in discovery and to share and standardize legal arguments and procedures.

However, in some settings, using CD-ROMs has disadvantages. For instance, the technology is offered in many different formats and search protocols so that moving from one to another entails fresh investments in learning. Also, technical problems are common when using them in networks. Access to CD-ROM materials on environmental law is likely to be limited to law libraries.

Environmental Law on the Internet

Environmental law is available in a variety of ways on the Internet and the World Wide Web, the Net's most popular protocol. The full range of these resources is so great and is changing so quickly that it is always a good idea to refer to a current guide. The rapid development of the technology means that some sources are already becoming obsolete. The discussion here will focus on the Web, but its central position in the current system may shift; researchers need to be aware of the technical context within which they are working.

One must note carefully the quality of the material one is looking at, because much of the information on the Net is either very thin or outdated. It has been so easy and inexpensive to set up a website that many information resources have been begun with enthusiasm but without the resources to maintain them. Also, it is easy to hang up a shingle on the Web, so that sometimes an "environmental law" search request will yield only the name and address of an attorney.

When you are looking for specific information, you have no way of telling whether you will find it on the Net. Sometimes it will turn out that it would have been easier to go to a library and look it up in a book. However, the amount of available material is growing, and many useful research resources are already on the Net. The curious blend of optimism and opportunism on the Internet expresses excitement over the new opportunities it creates. Its potential is particularly significant here because environmental law and science depend upon integrating dispersed data. The next few subsections of this chapter focus on how to find out about the law, but we will see that the division between environmental data and law is fuzzy.

EPA on the Internet

Like the rest of the federal government, the EPA is in a transitional state with respect to its data systems. Federal information policy has long been plagued by conflicting views about how to develop public databases and data systems. Lack of coordination, combined with the great variety of specialized needs and budgets in different agency offices, has produced a confusing array of information services. Like other agencies, the EPA is working to develop systems that serve a variety of groups, from the general public to its own enforcement offices. As information and technologies develop, the EPA's services will change. Indeed, their information-provision guidelines are as fluid as other EPA policies, perhaps more so. Each information service is, in a real sense, a transitional step to the next one. The EPA seems strongly committed to eventually using the World Wide Web to expand its enforcement capabilities by allowing general access to pollution and compliance information.

The EPA provides the public with a variety of types and levels of legal information at www.epa.gov. The EPA home page offers two ways of getting the information you need. You can choose a user category ("Kids, Students, and Teachers; Concerned Citizens; Researchers; Business and Industry; and State, Local, and Tribal Governments") or a topic category (ranging from "About EPA" to "Systems and Software"). Many different levels of information are available, including documents listed under "Environmental Appeals Based Opinions, Policy and Strategy Documents, and Compliance and Enforcement Documents." There are "plain English" guides to the law, as well as full-text versions of laws and regulations. Under some headings, the EPA offers "Answers to Frequently Asked Questions," which can serve as a primer on subjects new to the researcher. For instance, the Office of Solid Waste and Emergency Response provides nearly ten pages on the EPA's Brownfields Initiative.

The EPA also provides the ENVIROFACTS database on its web server. This a relational database, updated monthly, that integrates information extracted from five other EPA data systems. It contains data available under the Freedom of Information Act; no enforcement or confidential data are included. The EPA recommends ENVIROFACTS for new visitors to its website and provides both text search and online query forms to assist researchers.

Integrated Data for Enforcement Analysis (IDEA) is another EPA offering on the World Wide Web; it is located at http://es.inel.gov/oeca/idea. It was developed by the EPA's Office of Enforcement and Compliance Assurance. IDEA provides access to a variety of data systems, including some of those that ENVIROFACTS builds upon. For example, IDEA's resources include the CERCLA Information System, or CERCLIS, and the Toxic Chemical Release Inventory System (TRIS). CERCLIS provides an inventory of CERCLA (Superfund) sites and integrates data from Superfund removal and remediation processes. The data include general site information, site assessment, removal activities, remedial investigation and studies, and enforcement activities. TRIS contains information about releases of toxic chemicals reported by manufacturers as required by EPCRA. The data include facility identification; chemical-specific information on amounts of chemicals onsite and amounts released or transferred offsite; offsite locations to which waste-containing toxics are transferred; waste-treatment methods and efficiency; and pollution-prevention activities. Through IDEA, users can retrieve data for performing multimedia analyses of regulated facilities. IDEA can be used to produce the compliance history on a specific facility, identify a group of facilities that meet a user's criteria, and produce aggregated data on selected industries. The information can be accessed by selecting topics from a menu or through a keyword search. Some technical knowledge is required to use this resource, but the EPA has an IDEA User Support and Training Team at a toll-free support telephone line, 1-888-EPA-IDEA.

The EPA and state regulatory documents may be reproduced by other sources and made available at a website for a variety of purposes. For instance, a law firm may select and publish a rule or policy to publicize its own work. It is not always easy to tell what is the best or most accurate source of a document, and, in referencing any legal document, one should specifically cite the original document.

Law Libraries on the Internet

Law library resources on the Net are developing rapidly, so that it is not possible to describe them well in this chapter. Two of the more long-standing sites are the Legal Information Institute at Cornell Law School (http://www.law.cornell.edu/comments.html) and the World Wide Web Virtual Library of the Indiana University School of Law at Bloomington (http://www.law.indiana.edu/law/v-lib/lawindex.html). Both contain guidance into the world of environmental law and also provide the text of numerous statutes. Another source is the U.S. House of Representatives Internet Law Library: Environmental, Natural Resource, and Energy Law at http://law.house.gov/101.htm. Other similar resources exist, and the number and variety of this type of service are likely to expand.

Collections of environmental law are also offered by a number of nonacademic organizations. These sources may not be actual libraries, but may provide some similar services or resources. For instance some state bar associations have websites with resources for attorneys. These sites focus on information specific to local concerns, but may also be helpful in other states. Business, civic, and special-interest organizations also provide information about environmental law. Here the access may entail a subscriber fee.

While these library materials are less comprehensive than services like LEXIS® and WESTLAW®, they can be a good place to start researching a problem. However, it is not possible at this point in time to make a query of these libraries and be confident that your answer reflects the full breath and depth of the law and other legal resources.

Networking

There are a number of ways to network on the Internet. The dominant forms currently are mailing lists, Web news groups or discussion groups, and bulletin-board services. The last are private, contained systems that usually must be accessed directly by telephone and modem. Each category has a variety of formats, and the number and content of each group is developing so rapidly that the best approach is to find a guide, either a very recent book or a person who already knows something about your topic. Several books outline the ways to access the different types of connections and list the more active and formal networking nodes. A more free-form approach is to simply state your issue, put it out on the Web at your own site or post a query on an existing site and see who responds. You can also seek out existing sites that invite discussion. Doctors looking for feedback on unusual symptoms,[1] individuals wanting information about particular problems,[2] and researchers gathering data[3] can all make contacts with unprecedented freedom.

---

[1] A doctor in South Carolina reports that a patient has experienced intermittent sensations of "the hottest jalapeno pepper you could imagine" over the ventral surface of her tongue, then a metallic taste appeared as well. The doctor reports there is no evidence of psychopathology and describes the medical tests she has been given, which have revealed nothing; she ends, "Any thoughts would be appreciated."

[2] From Aachen, Germany, an apartment dweller seeks information about pesticides used to exterminate silverfish. He gives the names of the chemical compounds in the pesticide his landlord proposes to apply and asks for e-mail on "possible unhealthy effects.... Thank you very much in advance, Ralf."

[3] A medical researcher at the University of California at San Diego has set up the Antibody Resource Page, that seeks to provide a wide variety of medical pharmaceutical and commercial information.

Innovations in Environmental Law

Given the complexity of environmental law and the pace at which both technology and the environmental sciences are evolving, it is impossible to predict the forms future laws will take. Many current developments are likely to continue and may extend indefinitely. Or they may branch into different patterns of change that blend the old and the new. However, some trends seem robust enough to rely upon, and other signs of change tempt one to speculate. Disclaiming any certainty, I first suggest in this section current trends that we may expect to continue and expand. I then identify some ideas that seem novel today, but are worth greater attention.

The New Environmental Economics

Ecological perspectives are becoming influential as the scientific database expands. Our growing knowledge helps to identify the costs of pollution more fully, including secondary and latent effects. This learning process should bring about new technologies and new products that have less impact on the environment. Systems analysis and ecological economics suggest that the law should shape incentives so that long-term and large-scale resource management is facilitated. This finding has implications for the law's use of cost-benefit analysis and for legal "burdens of proof." The old assumption that the costs of regulation outweigh its benefits is being reversed as we increasingly recognize that the effects of pollution are systemic and long-lasting.

Experience also shows that regulation can boost incentives to economize. For example, when the plastics industry objected to the EPA's regulation of vinyl chloride emissions, the costs of regulation were anticipated to be enormous, but compliance ended up saving the industry money. When the first Toxics Release Inventory results were about to be released, some large manufacturers announced they would voluntarily reduce toxic emissions, one by as much as 90 percent. In the current debate over the Superfund liability scheme, opponents of liability generally cite the costs of cleaning up hazardous waste sites but ignore the incalculable long-term economic benefits of instituting clear and strong disincentives to dumping wastes. Ecological approaches to decision making articulate these benefits more clearly than traditional approaches do and place the burden of uncertainty on the proponents of pollution rather than on environmentalists, regulators, or the public.

In environmental regulation, the law is already taking a more "holistic" approach. The old method of regulating, identifying, and treating pollution chemical by chemical has been supplemented by more comprehensive definitions of pollution (e.g., the law addresses hazardous "waste streams" under RCRA and requires treatment or containment of chemical-waste soups under CERCLA). Today, instead of just trying to find out how each "chemical" may cause a given type of cancer, environmental science is also reaching for the ability to recognize immune system, neurological or reproductive effects of chemicals and generic groups of chemicals, such as environmental estrogens. At the same time, research in biology and medicine is increasing our understanding of particular health effects and metabolic mechanisms. At some point, scientists may actually be able to identify specific effects caused by specific exposures, thus limiting the universe of pollutants of concern and allowing greater regulatory precision.

Information Strategies

Concern with the problems of uncertainty may be giving way to a greater focus on the usefulness and flexibility of information in environmental regulation and in common law. As we rethink environmental law in information terms, legal requirements to produce and share knowledge about the environmental impacts of pollution are likely to expand.

Because of the prospect of regulation and the possibility of liability, firms are increasingly interested in identifying their own pollution. The environmental-auditing procedures emerging as part of ISO 14000 are, in essence, information-production and information-management tools. As auditing methods are developed and refined, they will change the culture of business. Leading companies will adopt auditing practices, and others will follow. Eventually, auditing should become a common practice, a "custom in the trade." Regulators may try to expedite this development and, particularly in fields where hazardous chemicals are used, may impose auditing requirements.

Of course, auditing raises new legal questions. For instance, should a firm that performs a thorough audit and uncovers violations be given amnesty in exchange for a genuine commitment to correct them? The EPA and a number of states have clashed on how best to handle this situation.

A related question is whether audits that are submitted to regulators should be held in confidence or be made available to the public. In the long run, making audit results public, or at least partially available, could serve several broad social goals. It would enhance our understanding of the economy's environmental dimensions. Ecological economists suggest that current measures, such as the gross national product or gross domestic product, are inadequate and should be supplemented by other data, including the environmental costs that could be assessed with a national mass-balance accounting system.

In addition, the potential for increased sharing of environmentally sound techniques and production processes would be facilitated by a database that showed who is using different chemicals and materials. Here, however, proprietary concerns pose a serious obstacle to information sharing. Some suggestions as to how we might overcome this difficulty are discussed below.

Finally, releasing cumulative audit data could support and expedite research. The best evidence of human-health effects from pollution is epidemiology, and most of this evidence has focused on worker health. Opponents of regulation have often argued that worker health effects cannot be extrapolated to the general population because the exposures are so different in the two categories. Although making these connections is difficult, if more data were available on worker and nonworker exposures and on general environmental impacts, we would understand our overall environmental risks better than we do now.

Participation

In the early stages of each new technology, hopes run very high. Today, the claims that are made for information technologies, particularly the Internet and the World Wide Web, sometimes seem extravagant. Yet, if there is any field in which universal interconnection may make a difference in understanding and performance, the environment is such an area. Optimism is warranted here precisely because environmental science and law both depend heavily on integrating dispersed data. Environmental knowledge consists of statistical compilations and syntheses of different kinds of data, and the component data must come from widely varied sources, including manufacturers, engineers, workers, neighbors, amateur naturalists, practicing physicians, academic researchers, and regulators. The capacity to bring these sources together is now becoming a reality. In late 1996, representatives to the Global Information Society's Environmental and Natural Resources Management project agreed on a standard for locating environmental information in libraries, in data centers, or on the Internet. The service standard is designed to make information easy to find and, if widely practiced, will facilitate environmental research and analysis. Expert and nonexpert networking should increase and may yield powerful results.

The new information technologies will also facilitate participation in legal processes affecting the environment. Wider participation in decisions may be expected, although expertise will still be a limiting factor. Expertise may be more available for sharing, however. In any event, basic information, such as that provided by the geographic information system (see Chapter 6), will encourage participation in local and regional decisions, such as facility siting.

Informal processes may increase, and as participation grows, mechanisms of alternative dispute resolution (ADR), such as informal settlement conferences or arbitration, should become more appealing. However, greater procedural flexibility may not be an unmitigated improvement. Whether ADR is the solution to the shortcomings in formal legal systems is not clear.

Of course, information, like everything else, has its drawbacks. Too much information and information of poor quality will inhibit decision making. Quality control (and honesty) are at the heart of current debates on the law and management of the new communication technologies. However, in the environmental area, decision making needs to be genuinely democratic to have long-term validity. On the whole, more participation makes better law and policy. Given this dynamic, it should be worth the aggravation of verifying and sifting through incomplete or inadequate scientific and legal arguments.

Environmental Law Tomorrow: Trends and Suggestions

Today, computerized law libraries allow a researcher to go directly to the answer to almost any question if the matter has already been addressed by lawmakers or commentators. Analysis and synthesis must still be accomplished by the researcher, however, and new questions are always arising. The sheer amount of material available on computerized research services seems to cut both ways in terms of its effects on the legal process. Because more material is available, it may become an overwhelming task to integrate everything that is related to a particular problem. Legal thinking could become more fragmented as a result. On the other hand, analysis and commentary itself is more available, so that one can build more easily on the legal analysis of others. In any event, the computerization of law necessarily entails some shift in perspective and form. In the environmental area, it seems to allow the law to specialize its formats and requirements to deal differently with the many facets of environmental quality and control.

Greater specificity may be a liability in a context in which coalitions are fluid and shift with each issue. The networking capability of the Internet may foster a new kind of environmental politics. Each decision could become more seriously contested, and environmental proceedings may become even more fractious than they are today. The current paradigm for decision making is balancing costs and benefits. It is an information-intensive method that entails making new basic value judgments with each exercise in decision making. If new production technologies could be encouraged to minimize environmental impacts, through containment or other means, it might be possible to develop "bright-line" rules in environmental law, perhaps with cost-benefit balancing for residual risks. This approach would make the procedural dimension of the law less costly and would allow greater investor certainty; of course, depending on the context, the criticisms that have been leveled at the current command-and-control system of regulation may also apply.

Within the evolving framework of legal and regulatory principles, many possible specific laws or strategies may be adopted. Some excellent prototypes for new regulatory strategies have been suggested, and so many seeds have already been planted that it would take a separate book to describe and evaluate them. One such suggestion is that Congress should make the EPA into an independent commission, like the Federal Communications Commission, so it could make environmental law with greater shelter from political winds. An alternative proposal is to establish an independent environmental-information commission, perhaps building on the model of the Bureau of Standards. To describe and evaluate even these two proposals would take an entire chapter. What follows, therefore, is simply a list of a few suggestions that should be considered, among many other candidates worthy of study.

We could, for instance, enact laws that would structure and support incentives to produce green technology and information, such as an expansion of EPA's reward scheme for innovations that have environmental value. We could move beyond this initiative and establish a patent or registration system that would grant a period of exclusive control over environmental innovations so investors could be more confident that they will recoup their investment. Economists are divided over the efficiency of the patent system, yet recent intellectual property laws have crafted systems to support innovation in specific areas, such as the development of superconductors. A patent or registration system might facilitate the sale and exchange of useful data and technologies that would otherwise languish because firms have no means of making a profit from them and, indeed, their release might assist rivals. An overall background of strict regulation is essential to create opportunities for such improvements.

Because lack of knowledge about environmental effects is a key limitation on controlling pollution and deciding which technologies to foster, the law should increase its focus on learning. The market creates strong disincentives to research and disclose the negative effects of pollution; in many cases damning information has been withheld from the public or has been distorted. The time may have come to consider a program that would charge polluting industries for comprehensive and independent study of their health and environmental impacts. The current Superfund law provides that the federal Centers for Disease Control and Prevention may charge the costs of studies of the health impacts of some Superfund sites to the parties responsible for the contamination at each site. This is a limited program, and waste cleanup is not the best point at which to levy such a charge, but the concept is a good one, and we should examine the different forms such a system could take.

We could also harness the information that already exists but is not disclosed or fully used because of legal and market disincentives. Legal schemes like California's Proposition 65 replace resistance to regulation with support for it by more easily holding polluting firms liable for injuries if they are not meeting specific regulatory standards.

Recent empirical research challenges the assertion that the tort system imposes an undue burden on society. At the same time, theoretical work supports the idea that some kind of liability mechanism is a valuable component in an overall scheme to guide technological change. The decentralized, flexible, and responsive nature of a liability system makes it particularly useful for coping with emerging environmental and toxics problems. Injured individuals and their immediate support network (neighbors, doctors, etc.) are often the first ones to notice a problem.

The central theme in these proposals is a change in our posture toward knowledge itself. If we recognize knowledge as a product of investment and coordination, then the burden of proof of harm should be on the party best situated to identify the risk of harm, convene the appropriate points of view, and fund the study. The market failure of environmental pollution stems in part from the improper pricing of products that are not fully understood before they become the subject of massive investment and distribution. Ecology and medicine both have influenced the law to be more conservative (i.e., more protective) of the environment. This is an appropriate stance in the global community we now inhabit.

Key Resources

Environment Law by Rodgers William H., JR. 1994; is a leading treatise on the law.

Environment Reporter is published in a looseleaf format by the Bureau of National Affairs in Washington and has been in print since 1970; an enhanced CD-ROM product is also available.

Environmental Law Reporter is published in a looseleaf format by the Environmental Law Institute in Washington and has been in print since 1971; an enhanced CD-ROM product is also available.

Law of Environmental Protection is a three-volume work in looseleaf format with updates, authored by the staff of the Environmental Law Institute; edited by S.M. Novick, D.W. Stever, and M.G. Mellon; and published in Deerfield, IL., by Clark Boardman Callaghan starting in 1987.

Law of Chemical Regulation and Hazardous Wastes is a three-volume work in looseleaf format with updates, authored by D.W. Stever and published in New York by C. Boardman Co. starting in 1986.

Toxics Law Reporter: A Weekly Review of Toxic Torts, Hazardous Waste, and Insurance Litigation is published in a looseleaf format by the Bureau of National Affairs in Washington and has been available since 1986.

Treatise on Environmental Law by F.P. Grad is a five-volume work in looseleaf format with updates, published in New York by M. Bender starting in 1973.

An Introduction to Ecological Economics by R. Costanza, J. Cumberland, H. Daly, R. Goodland, and R. Norgaard (Boca Raton, PL: St. Lucie Press, 1997) is a good starting point for understanding this cross-disciplinary field.

LEXIS® provides Environmental Resources among its research services and capabilities; subscription information is available at 1-800-227-4908.

WESTLAW® provides Environmental Resources among its research services and capabilities; information about their products and services is available at 1-800-336-6365; inquiries may be addressed to Marketing Support, Do-06, 620 Opperman Dr., P.O. Box 64833, St. Paul, MN 55164-9752.

References

Colborn, T., Dumanoski, D., and Myers, J.P. 1996. *Our Stolen Future: Are We Threatening Our Fertility, Intelligence, and Survival?* New York: Penguin Group.

Daniels, S. and Martin, J. 1995. *Civil Juries and the Politics of Reform.* Evanston, IL: Northwestern University Press and the American Bar Foundation.

Eggen, J.M. 1995. *Toxic Torts in a Nutshell*. St. Paul, MN: West Publishing Co.

Findley, R.W. and Farber, D.A. 1996. *Environmental Law in a Nutshell*. St. Paul, MN: West Publishing Co.

Locke, J. 1690. Of property, Chap. V in *Concerning Civil Government, Second Essay: An Essay Concerning the True Original Extent And End of Civil Government*, gopher://gopher.vt.edu:10010/02/116/1, visited Nov. 17, 1997.

Lyndon, M.L. 1995. Tort law and technology. *Yale Journal on Regulation* 12:137–176.

Ostrom, B.J. and Kauder, N.B. 1996. *Examining the Work of State Courts, 1994*. Williamsburg, VA: National Center for State Courts.

Percival, R.V., Miller, A.S., Schroeder, C.H., and Leape, J.P. 1996. *Environmental Regulation: Law, Science, and Policy*, 2nd ed. Boston: Little, Brown and Co. See also Updates to Environmental Regulation Law, Science, and Policy (Second Edition) at www.law.umab.edu/courses/environment/, visited Nov. 17, 1997.

Rahdert, M.C. 1995. *Covering Accident Costs: Insurance Liability, and Tort Reform*. Philadelphia: Temple University Press.

Rodgers, W.H., Jr. 1994. *Environmental Law*. St. Paul, MN: West Publishing.

Saks, M.J. 1992. Do we really know anything about the behavior of the tort litigation system, and why not? *University of Pennsylvania Law Review* 140:1147–1292.

Swanston, S.F., Stanislaus, M.V., Crawford, N.N., Jewell, D.A., Lewis, R., and Pasquerella, J. 1996. *Work Plan for Citizen Participation in Clean Air Act Title V Permitting*. New York: Minority Environmental Lawyers Association; New York: Center for Constitutional Rights; and Washington, D.C.: Environmental Exchange.

Decision-Maker Response

DEAN HILL RIVKIN

Mary Lyndon's comprehensive and insightful paper introduces the complex legal landscape that confronts environmental decision makers at all levels, from the grass roots to top government and corporate officials. The web of statutes, regulations, and court decisions, intertwined with a host of less formal guideposts, is neatly unraveled in this piece. The strength of Lyndon's paper is not only in its explication of existing legal sources, but also in its identification of the interstices of the law.

She observes, for example, that "[s]tatutes and regulations may seem quite precise on first reading, but in fact they frequently fall short of answering the specific questions posed by a case." For lawyers, the malleability and contingency of the language of the law is almost a truism. For many nonlegal decision makers, if the face value of the words in a law or regulation do not give answers, they are hard-pressed to know where to turn.

This is where the role of context comes in. Legal environmental experts understand that many key words in a statute are, as Lyndon points out, compromises in the legislative and administrative processes. To gain an understanding of what a word or phrase means, one must often resort to dictionaries, legislative history, agency archives, or prior court decisions. In these materials must rest the grounded predictions about the meaning of a legal provision. A facial reading can often mislead decision makers. As Lyndon keenly recognizes:

More often, however, research is a confusing odyssey through statutes and regulations that are only partly on point and through scientific and technical data that are incomplete and uncertain. For those with little experience in this area, it will be difficult to know which situation you are in. You may think you have your answer but fail to see the background complexities.

The real strength of Lyndon's piece lies in her analyses in the sections on "Participation" and "Environmental Law Tomorrow." Here, she stresses the importance of expert and nonexpert participation in creating "environmental knowledge." She acknowledges the omnipresence of "informal processes" and their key role in shaping outcomes in environmental disputes. She also describes the role that law might play in fostering "learning" and dispersing knowledge about promising environmental technologies and methods. She cautions that knowledge about the environment should be a necessary prerequisite before products are unleashed into the marketplace with a blind faith that they will not cause environmental or health harm. Do today's environmental laws ensure that future products—the cigarettes or asbestos of tomorrow—will not inflict harm on future generations as they have on past ones?

The communication of legal decision-aiding tools is a complicated subject. Most nonlawyer decision makers rely heavily on lawyers for guidance about a particular action or problem or to determine whether there is a realm of legal/nonlegal judgment that would allow decision makers to resolve the environmental puzzles they are facing. This is a very difficult but central question in understanding the types of environmental challenges that will face decision makers in the future. An example from my service as a member of the Southern Appalachian Mountain Initiative (SAMI) will illustrate the conundrums of using law to assist in structuring resolutions to imponderable environmental questions.

SAMI is a nonprofit organization whose membership includes the environmental regulatory agencies of eight states (Alabama, Georgia, Kentucky, North Carolina, South Carolina, Tennessee, Virginia, and West Virginia), federal agencies, industry, academia, environmental organizations, and other stakeholders across the region. SAMI's mission is as follows:

Through a cooperative effort, identify reasonable measures to remedy existing—and to prevent future—adverse effects from human-induced air pollution on the air quality related values of the Southern Appalachians, primarily those of Class I parks and wilderness areas, weighing the environmental and socioeconomic implications of any recommendations.

SAMI focuses on air-quality issues in the Southern Appalachian Mountains and their effects on resources, including visibility, water, soils, plants, and animals. Specifically, SAMI is interested in visibility degradation, acidic deposition, aquatic and terrestrial systems, and ozone impacts to terrestrial systems. SAMI is unique because it is a voluntary regional initiative, unlike those mandated by the Clean Air Act (CAA), such as the Northeast Ozone Transport Commission and the Grand Canyon Visibility Transport Commission. Some view SAMI as a prototype for decision making on transboundary environmental issues.

Initially, SAMI faced the test of calculating the emission reductions that were contemplated in the 1990 CAA Amendments. Assuming full compliance, often a dubious assumption, these reductions would yield baseline levels of permissible emissions. The tricky part came next: how to calculate further emission-reduction opportunities, taking into account the socioeconomic impacts (almost always a factor in modern environmental decision making) of these reductions. This daunting task has taxed even the most sophisticated decision makers. Simply knowing the boundaries of the legal and regulatory framework was insufficient to help SAMI further its mission. A range of other disciplines (engineering, environmental sciences, and economics to name a few) were necessary to make progress. Even with expert help, SAMI floundered. Why? Because modern environmental disputes involve heavy doses of politics and power, which provide much of the

5. Characterizing the Regulatory and Judicial Setting 159

context for the legal and regulatory frameworks that Lyndon so ably describes.

How are seemingly objective decision-making tools used in this type of setting? First and foremost, these tools, when wielded by narrow-interest groups, can confound decision making, not aid it. Cost-benefit analysis, for example, is a common tool for obtaining guidance about alternative decisions in environmental conflicts. Most experts would acknowledge that cost-benefit analysis is infused with subjectivity and value-laden underpinnings. In SAMI, cost-benefit analysis was transformed into an integrated-assessment framework, a computerized data system that presumably would answer questions about the environmental and socioeconomic impact of SAMI's recommendations.

The problem that occurred is that each participant brought different values to the integrated-assessment framework. A seemingly useful regulatory tool became a nightmare of conflicting assumptions. The winners in the lengthy discourse over integrated assessment were those with the resources and staying power that other individuals and organizations (mostly in the environmental community) could not muster.

I learned a great deal from observing an industry lawyer who represented the utility industry on air-quality issues in the SAMI process. He was an expert at "doing meetings." He knew when to make a concrete proposal and when to filibuster. He carried his proposals in a laptop computer, which he often used to knock out a first draft of language that was contested during a particular session. He was also able to draw on experiences he had had in similar projects, experiences that few others in the room had gone through. In the SAMI meeting, this deployment of strategic knowledge thwarted the consensus-building process, just as a lawsuit would have done. I often think that we in legal education should teach a course called "Strategic Behavior for Lawyers in Meetings." In the environmental field, this is where much of the action is.

Another example from SAMI about the subtleties of power in decoding the legislative and regulatory settings involves allocating the burden of proof in environmental decision making. This burden of proof is often a critical factor in determining whether a particular change should occur under then-current laws or regulations. If the burden of proof is too high on those (most often the environmental interests) who wish to change the status quo, no progress (a loaded word) will occur.

In SAMI, a lawyer representing one of the SAMI states not well known for its environmental leadership suggested that all of SAMI's recommended emission reductions must meet the test set out by the United States Supreme Court in *Daubert vs Merrill-Dow Pharmaceuticals, Inc.*, 509 U.S. 579 (1993), which was a mass tort case, not an environmental one. The *Daubert* case held that, to be admissible in court, expert testimony first had to be evaluated by the trial judge based on several criteria, including: (1) Have the data and methods underlying the testimony been tested? (2) Has it been subjected to peer review? (3) What is the potential rate of error? (4) Is the technique widely accepted? The *Daubert* decision expressly acknowledged that this process was not suitable for social-science testimony as opposed to scientific and technical testimony.

In the context of SAMI, where industry groups have access to a range of experts and the ability to carry out targeted studies and produce specific data, the prospect that a proposal will "pass" the *Daubert* test is relatively small. But to compare the judicial setting with the quasiadministrative setting of SAMI and to view all issues as strictly scientific, rather than social and economic, undermines the consensus process established by SAMI. Virtually no meaningful proposal for emission reductions, except what came to be called the "no-brainers" (e.g., turning the lights out at night), can pass muster under the *Daubert* test. Today, SAMI formally has not adopted this decision-making standard, but much of SAMI discourse is filled with issues of burden of proof, uncertainty, reliability, and scientific validity. Even the best formal decision-making tools are inadequate in this type of modern environmental setting.

My pessimistic assessment of the "usefulness" of the "tools" that Lyndon's paper reviews is rooted in the reality of modern-day environmental disputing. Every so often, the "law" can be deployed to solve environmental problems. But in an increasing number of environmental disputes, the law must be combined with a broad range of interdisciplinary tools to shed light on resolutions for complex, polycentric problems. Very often, resolution means compromise. But for this process to work, the production, use, and evaluation of knowledge (which is now largely in the possession of experts) must be democratized. If the process is not opened up in this way, politics and power will prevail, and ordinary people and the environment will lose.

6
Integration of Geographic Information

Jeffrey P. Osleeb and Sami Kahn

Decision makers are often faced with difficult environmental decisions, such as where to locate a landfill, whether to build an incinerator, and how to assess environmental risks. They usually make these decisions on the basis of overwhelming amounts of information using data that are difficult to interpret and often conflicting (Anderson and Greenberg, 1982). Because environmental data are collected by so many sources and methods, conclusions are often at odds with each other, depending upon the derivation of the data and how the data are presented.

Moreover, environmental decisions often require the use of geographic or spatial data, which are data that can be placed on a map. For example, the results of a water-sample test may be presented in nonspatial terms, such as parts per million or pH. However, these results are not useful for comparison with other sites unless the location of the source of the sample is a part of the data set. When that information is present, the water-sample results from site A may be compared with the results from sites B and C. Without that geographic information, such comparisons could not be made.

As an indication of its usefulness, ever-growing amounts of geographic data are being gathered and disseminated for the analysis of environmental problems. The Office of Management and Budget estimated that in FY 1994 the federal government expended more than $4 billion on spatial data activities (Commission on Geosciences, Environment, and Resources, 1997).

Given that information for making decisions comes in different forms and from varied sources, the trick to integrating information is to find a suitable format to tie the information together. The technologies used today to correlate such information range from the relatively simple spreadsheet to complex information systems, such as database-management systems, that allow linking and arranging the numbers within a data set. Geographic information systems (GIS), the most sophisticated of the existing technologies, go beyond that level of performance and allow linkages among multiple spatial and nonspatial data sets.

The particular challenge addressed by this chapter is that, while much environmental data has a spatial component, those data are often presented in a tabular form, making the data difficult to interpret. More importantly, the spatial data must be linked with other spatial and nonspatial data sets to maximize the use of the data in decision making. This chapter focuses on methods that integrate data, and in particular, geographic data. Geographic data are data for which there is a street address, a latitude and longitude, or some other means of placing the information on a map. Examples of geographic data include the locations of hazardous-waste facilities within a state; the distribution of populations of different socioeconomic levels around those facilities; the proximity of schools to those facilities; and the distribution of health impacts that might be produced in the surrounding populations by those facilities. The analysis of such geographic data requires the use of technologies that can integrate such spatial information.

Spatial-Information-Integrating Technologies

Spatial-information-integrating technologies are decision-aiding tools that can be used to organize, analyze, integrate, and present geographic data in a more comprehensible form. These technologies can provide a graphic representation of the geography of an environmental situation. They are comprised of digital maps and other models (such as routing systems, buffering methods, and location algorithms) that perform spatial analysis. These technologies permit the user to access and combine environmental data with demographics, facility information, health data, and infrastructure characteristics to answer a variety of questions concerning the well-being of the region.

A number of approaches employ techniques of spatial analysis. Factorial ecology (Berry and Rees, 1969; Murdie, 1969) uses factor analysis to combine layers of information. Trend surface mapping (Haggett, 1965) looks for patterns in a three-dimensional surface. Spatial autocorrelation analysis (Cliff and Ord, 1973) reflects upon the quality of information contained in spatial data. GIS has been used in different ways by varied disciplines to analyze and to integrate information. Each of these techniques permits, to varying degrees, the structuring and layering of different types of information, thereby providing the capability of integrating information for answering both simple and complex questions. Such questions may require only the description of a specific facility at a given location or may require the combination of several spatial approaches that integrate a wide variety of data and phenomena. Examples of issues that might be addressed with these approaches are:

- The location of noxious facilities
- The status of community health
- The equity reflected in the distribution of risk
- The environmental load

The spatial techniques and applications considered in this chapter are:

- Geographic information systems
- Spatial decision-support systems
- Geographic plume analysis

A summary of these integrative approaches appears in Table 6.1.

TABLE 6.1. Summary of integrative approaches to spatially explicit environmental data.

Approach	Uses	Strengths	Weaknesses
Geographic information systems (GIS)	To integrate and map spatially explicit data; techniques for incorporating nonspatial data are available	• Require the user to organize the data and document the data sources • Provide information in an organized map format • Can incorporate diverse spatial scales • Can be linked to spatially explicit computer simulation models	• Require sophisticated computers and other technologies (scanners, digitizers, etc.) • Require specialized training that must be updated to keep abreast of current approaches • Time and labor intensive
Spatial decision-support systems (SDSS)	To merge a geographic information system with mathematical models	• Present solutions to decision problems in a variety of modes • Have limited data requirements • Can increase citizen involvement	• Costly because unique models must be developed • Require highly trained individuals matched to each environmental problem • Require sophisticated computers
Spatially explicit computer models	To model particular situations to analyze potential decision impacts and effects	• Can be tailored to specific needs • Can produce results for different scenarios for decision opportunities	• Require sophisticated computers • Require technical training • Require unique models to be developed for each situation • Include uncertainties, and therefore results are subject to statistical error • Often cannot model the full complexity of the situation • Require a great deal of data • Time consuming and cumbersome • Must be manually integrated with GIS

Geographic Information Systems

A geographic information system (GIS) is a management-support system that permits the decision maker to view and analyze spatial information at speeds and in ways that were never possible in the past. A GIS combines data-capture technologies (such as scanners, digitizers, and global-positioning systems) and spreadsheet and database-management software with mapping, graphics, and statistical routines. Together, they permit the presentation and analysis of spatial data in a highly sophisticated manner. GIS permits the analyst to look at all the spatial and nonspatial information that has been collected about a particular location by merely pointing to the location on a computerized map or typing in an address. In a similar manner, information can also be obtained about locations within a designated radius or distance from a given location. In addition, facilities of a similar nature at different locations can be identified. Finally, information from maps (even those originally produced at different scales) can be overlaid and related.

GIS uses computerized data of two types: the base map and the attribute data. A base map is just that—a graphic representation of the geographic layout. It may show streets, census-tract boundaries, streams and bodies of water, topographic contours, or all of these simultaneously. The federal government has invested large sums of money to collect information for base maps and to make that information available at a very low cost. An example of such a base map is seen in the TIGER (Topologically Integrated Geographic Encoding and Referencing) files, which give all street maps and census-tract boundaries for the United States (ESRI, 1990). Private companies have refined these government-produced base maps by adding information, enhancing accuracy, and including additional geographic features.

These maps can be used in conjunction with attribute data that describe features like locations of hazardous-waste facilities, the types and amounts of materials stored, the frequency of inspection, and the demographics of the surrounding area. The sources of information for these attribute data might include the U.S. Bureau of Census, the U.S. Environmental Protection Agency (EPA), telephone books, local-government records, residents, or businesses. These data may be tagged with some locational information, such as a street address, a ZIP code, or a census tract. Similarly, data produced by a socioeconomic model and containing a location can be used as attribute data in a GIS. In short, any geographic data, be it quantitative or qualitative, can be used in a GIS.

A GIS analysis results in spatial information that is organized in a clear, graphic manner. Such analyses can yield patterns that confirm hypotheses, such as those associating negative health effects with a source of pollution. These results can then be used by decision makers, who must ultimately evaluate their validity and usefulness.

The greatest strengths of GIS are:

- Presentation of spatial information in a visual manner
- Accumulation of information from various sources and the representation of all that information in the same geographic scale
- Allowing one to point to a location on a map and obtain information about that location
- Ability to perform spatial analysis on a site to determine its impact on other locations

This last function cannot be performed in any other manner. GIS represents a major technological breakthrough for undertaking environmental analysis because it is flexible in its ability to add and analyze new information.

However, GIS does have some limitations. It can be a relatively expensive tool because of its technical requirements (e.g., skilled technicians, high-level computers, and the collection and maintenance of large amounts of data). The expense increases drastically when the available data are not computer readable. An organization cannot just purchase the necessary hardware and software, collect data, and then let an untrained person run the GIS and expect to get meaningful analysis. As with any technology, GIS can only be used effectively if it is properly integrated into the entire decision-making process; it cannot merely be tacked on as an afterthought. Therefore, an organization should expect to train GIS analysts in all phases of GIS technology, including cartography, database management and spatial statistics. In addition, managers and decision makers should also be trained in the technology so that both the requirements of the technology and the appropriate applications for which the technology may be used are well understood. Using GIS successfully requires not only investments in hardware, software, and data, but also the hiring of properly trained GIS personnel and the retraining of current personnel to use this technology properly and effectively.

Ultimately, GIS is only as good as the data put into it. As with other methods, the analysis undertaken with GIS generally requires current information. Similarly, the results of GIS can be misinterpreted or misused. As one can lie with statistics, one can lie with maps (Monmonier, 1996).

The environmental applications of GIS are numerous and varied. Kim et al. (1995) used GIS to build an environmental information system for efficient water-quality management. Their system included a water-quality database, a database-management system, and a water-quality model to estimate pollutant loadings. To simulate the effects of both point and nonpoint pollution sources into rivers and lakes, nine digital attribute layers were used, including roads and hydrography. The authors found that a GIS-based system was highly advantageous in modeling pollutant loadings and identifying cost-effective mitigation strategies. Similarly, Kim et al. (1993) used GIS to model urban nonpoint-source pollution into Lake Superior. The base map included street networks, while attribute layers included city limits, hydrography, land use, and urban storm-sewer networks. They expect that short-term use of the model will include the identification of critical sewer sub-basins with significant amounts of nonpoint pollution loads from each of the surrounding communities, while the long-term use will be to aid communities in the siting and implementation of control practices.

Emani et al. (1993) used GIS to assess socioeconomic vulnerability of a coastal community to extreme storm events and sea-level rise. They recognized that, unlike the slow changes that may occur from global climate change, extreme storm events leave little time for response and should be anticipated. They used data on land use, coastline and estuaries, transportation, and socioeconomic indicators to produce indices of vulnerability for the test community.

Moreno and Siegel (1988) used GIS to conduct corridor selection for a proposed highway project in Arizona. The GIS was used to determine the suitability of various highway locations through consideration of various environmental and highway-engineering factors. Using the graphic and statistical results of the GIS, the research team was able to develop measures that reduced potential impacts associated with the highway. This approach is applicable to the selection of other linear facilities, such as power-transmission lines and pipelines.

Finally, Stein et al. (1995) used GIS to process spatial data to assess the risks of environmental contamination in The Netherlands. They argue that the three stages in which GIS is crucial are in the application of geostatistics, the choice of appropriate models, and decision making.

GIS has become a multibillion dollar industry during the past decade and is now widely used by decision makers at all levels of government, the private sector (including developers), attorneys, real-estate companies, insurance companies, and utilities. Recently, information that can be used with GIS has become readily available on the Internet, a growing source of GIS information.

Spatial Decision-Support Systems

A spatial decision-support system (SDSS) is a specialized application of GIS that merges that technology with powerful mathematical models. SDSS allows the decision maker to consider a series of "what if" questions (Ralston, 1991; Arentze et al, 1996; Peterson, 1993; Carver, 1991). It permits the analysis of an existing environmental problem relative to some optimal situation (e.g., minimal cost), a level of pollution, or some maximized net benefit. The decision maker is able to address "semistructured" problems that typically require the selection of a set of solutions from a set of alternatives (Densham and Goodchild, 1989). In addition, SDSS permits the decision maker to track such measures as the cost of various solutions in solving environmental problems while determining the efficiency of each solution. This analytical capability is extremely helpful in determining not only which solution is preferred, but also whether a problem should be solved at all. With this technique, the cost of solving the problem can be evaluated and compared to the cost of leaving the problem unresolved.

SDSS requires the development of a model, gathering the necessary information, and running the model with the information (Armstrong and Densham, 1990; Carver et al., 1995). The results of the model are then compared with the existing situation. SDSS is a very powerful tool that provides a normative (optimal) solution to a problem, permitting the decision maker to evaluate a situation or a number of situations. One of the strengths of this decision-making tool is that it has limited data requirements because the approach provides a highly structured model for gathering data; at the same time, information on many cases is not required, as is the case with a statistical model. Therefore, the data needed for SDSS are relatively inexpensive and, generally, easily gathered. However, a disadvantage of this model is that front-end costs can be high because a unique model must be developed by highly trained personnel for each environmental problem. Once developed, though, the model can be used to address many similar situations (by using different data) and to assess various scenarios. Smotritsky et al. (1993) found that SDSS would increase citizen participation because citizens could propose alternative scenarios that could then be evaluated against the current situation and other proposed solutions. The authors found that this approach could ease the decision-making process when determining corridors through which highways, pipelines, and similar facilities were to be built.

SDSS is used by all levels of government and by large organizations within the private sector, such as oil companies, railroads, distributors, and manufacturers. Unlike GIS however, SDSS is used almost exclusively by large organizations because of the high costs that arise from the need for a highly trained staff to develop the model and the uniqueness of each model. Generic software for SDSS that will greatly reduce the cost of model development is expected in the near future.

An example of the use of an SDSS is an evaluation of the location of the facilities within a regional health-care system. The optimal location of the facilities provides maximal accessibility to the population within a specified budget. SDSS allows various scenarios to be run, each reflecting different budgets and accessibilities to facilities. This approach was used by Osleeb and McLafferty (1992) in assessing the optimal solution to the problem of eradicating dracunculiasis (guinea worm disease) in West Africa. There, a model was developed to determine the best combination of numbers and locations of water wells and schools. A multiple-attribute tradeoff curve identified potential solutions for eradicating the disease by presenting the optimal combinations of water wells and schools for given budgetary expenditures.

SDSS was also used by Ratick and White (1988) to locate sites for noxious facilities. In their model, various locations and sizes of facilities were compared in order to minimize overall public opposition to the site. The authors found that a small number of cost-efficient large facilities concentrates risk in a few areas, while a greater number of smaller, less noxious facilities, shared the risk among the entire population and produced lower opposition because each individual is more likely to be incurring their fair share of the burden imposed by the facility.

Finally, Baiamonte (1996) developed an SDSS that analyzes the health risks and equity considerations associated with siting new hazardous facilities, given the already existing distribution of environmental burdens. In this model, it was assumed that risks should be spread equally among a population rather than concentrated near a few. This SDSS was applied by the author to the Greenpoint/Williamsburg section of Brooklyn, New York, a study that will be discussed in detail later in this chapter.

Geographic Plume Analysis

Another specialized application of geographic information systems is geographic plume analysis (GPA), an analytical tool that complements air-dispersion models that require and produce large amounts of information (Maitin and Klaber, 1993). Through spatially explicit modeling, GPA now allows decision makers to overlay the results of air-dispersion models with census information to estimate the demographic impacts of releases of toxic substances. GPA consists of two major components: a chemical-dispersion model that is integrated with a GIS. Other situations in which simulation models have been linked to a GIS may be found in Dale et al. (1993) and in Emmi and Horton (1995).

The dispersion model typically uses attributes of the chemical and the atmosphere to predict how airborne particulates will be deposited at different distances and directions from the source. This dispersion model creates a "plume footprint" that can then be overlaid with a GIS that might include demographic and socioeconomic characteristics of an area, as well as the features of the built environment. Thus, the potential impact of a chemical release on a given population can be predicted.

Dispersion models are simulation models (Gilbert and Conte, 1995). Simulation models start with a model of a complex system. Various alternatives are evaluated and calibrated against existing information. Dispersion models typically incorporate information about stack height, weather conditions (such as prevailing winds, precipitation, relative humidity, and temperature), pollutant type and amount, plume conditions (e.g., temperature, density, water content, and buoyancy), and the built and natural environment. This information can then be used to predict the direction the particulates travel; the distances they go; their rate of fallout and deposition; and/or their concentrations as they are being transported and after they have been deposited. The Areal Locations of Hazardous Atmospheres (ALOHA) model is one widely used tool for estimating the movement and dispersion of gases. This model, developed by the National Oceanic and Atmospheric Administration and the EPA, provides estimates of pollutant concentrations downwind from the source of a release. It takes into consideration many physical characteristics of the release site, atmospheric conditions, and circumstances of the release. Through the use of ALOHA, a plume footprint can be created for a particular release or spill. Then this information is combined with a site-specific GIS database to determine the effects of the release on the surrounding environment.

The strength of a GPA is its ability to predict impacts on a population produced by specific concentrations of particulates at ground level at various distances from a point source or from several sources. Simulation models, including GPAs, have a number of drawbacks. Because they include uncertainty, the results are subject to statistical error. Also, they tend to reflect very complicated systems that are beyond the capability of standard modeling techniques of operations research or statistical analysis, which tends to makes model building and interpretation difficult (Wagner, 1975). A specific weakness of dispersion models is that they are difficult to use and require a great deal of data that are not easily obtained; all of the data must be gathered for the atmospheric conditions, the characteristics of the release, and the nature of the surrounding terrain *at the time of the release*. In addition, these models are time-consuming and cumbersome to use and require highly skilled personnel. Finally, dispersion models are not yet integrated with GIS; they must be integrated manually. Because of these constraints, dispersion models and GPA are primarily used by large government agencies, specialized consulting firms, and academicians.

Chakraborty and Armstrong (1995) developed a GPA to assess whether different racial and income groups were disproportionately affected by the release of chlorine from highway truck accidents in Des Moines, Iowa. Employing the ALOHA model, the authors found that the areas most susceptible to the release of the chlorine had higher proportions of minorities and low-income households than the city as a whole. The authors found that this finding was consistent with other research demonstrating environmental inequity.

GPA has been employed by Lao and Sharma (1995) to determine the extent of a hazardous spill and the geographic limits of the population that might be placed at risk from such an event. With this information, they used the system to establish plans for emergency response and evacuation.

Osleeb et al. (1996) used GPA to evaluate the load from TRI facilities in the Greenpoint/Williamsburg section of Brooklyn, New York. The authors used the Industrial Sources Complex model (ISC3) model developed by the EPA (1995) to develop a plume analysis for a number of carcinogenic and noncarcinogenic toxins. This plume footprint was used to calculate an aggregate load for each of the 159 census-block groups within the community stemming from all TRI stacks. The authors found that the high loads were concentrated in a few block groups rather than being evenly dispersed.

Additionally, the high loads were not deposited in relation to distance from the stacks. The GPA had integrated information on stack height, prevailing winds, pollutant type and amount, mixing conditions in the atmosphere, and seasonality to perform this analysis. This use of GPA was part of the investigation described in detail in the following case study.

Case Study: Greenpoint/Williamsburg in Brooklyn

Background

The Greenpoint/Williamsburg Environmental Benefits Program (G/WEBP) is a community-based project that incorporates the use of GIS, SDSS, and GPA.

Greenpoint/Williamsburg (population 154,000) is a section of Brooklyn, New York (see Figure 6.1) that is a well-known, multiethnic residential

FIGURE 6.1. The Greenpoint/Williamsburg section of Brooklyn in relation to New York City.

FIGURE 6.2. The sewage treatment plant operated by the City of New York that was once a major source of pollution.

community. It is also recognized as a community where numerous private and public noxious facilities are located. Since the 1850s, much of the industry in this community has been concentrated in what has been called the five black arts: printing, pottery, petroleum and gas refining, glassmaking, and ironmaking (Baiamonte, 1996). While the activity of these industries has decreased since the end of World War II, many environmental burdens remain. For example, more than 17 million gallons of refined oil products have leaked into the water table of the community. Today, the neighborhood includes a sewage treatment plant (Figure 6.2) and an incinerator (Figure 6.3) that are both run by the City of New York; the only low-level-radioactive-waste repository in New York City; approximately 20 EPA TRI sites (Figure 6.4); more than 200 hazardous-material processors; a major expressway; and a large number of chemical and petroleum bulk-storage tanks (Figure 6.5).

In 1991, the community learned, as they had suspected for a long time, that the sewage treatment plant had not been properly maintained and operated. These inadequacies caused considerable air pollution and particularly foul smells in the community. As a result, the community won a consent decree that mandated that the City of New York make improvements to both of the facilities that it operates in the area and to pay substantial financial damages to the community. These fines were then used to establish the G/WEBP. An integral part of this program was the

FIGURE 6.3. The city-owned incinerator that was once another source of air pollution.

FIGURE 6.4. Leviton Industries, one of many TRI facilities located in Greenpoint/Williamsburg.

FIGURE 6.5. Sunoco's petroleum holding tanks. Many tanks like these have leaked oil into groundwater.

development of a GIS that is currently used by the community to monitor environmental conditions within the community.

Approximately 40 percent of the land use in Greenpoint/Williamsburg is industrial, and 30 percent is residential. This mixed-land-use zoning results in stark contrasts between adjacent industrial and residential sites. (See Figures 6.6 and 6.7; Figure 6.8 is a map showing the location in the community where each picture was taken.) With all of this environmental activity nearby and the large amounts of information available, the GIS has proven to be extremely useful to this community to help to locate facilities, to monitor and update facility activities, and to assess impacts on human safety. The GIS has also been helpful in producing grant proposals for the collection of additional information and for education and remediation purposes.

Metadata Development

In 1990, the Spatial Analysis and Remote Sensing (SPARS) Laboratory of the Department of Geography at Hunter College was retained by the New York City Department of Environmental Protection (NYCDEP), Office of Environmental Quality, and the Greenpoint/Williamsburg Citizens Advisory Committee to document facilities that stored and produced toxic sub-

FIGURE 6.6. Mixed land-use zoning in Greenpoint/Williamsburg. Here, a park is located adjacent to a highly industrial TRI facility.

FIGURE 6.7. One of the many beautiful streets of Greenpoint/Williamsburg. Stacks are visible only blocks away from meticulously maintained homes.

FIGURE 6.8. A map showing the locations of the pictures of Greenpoint/Williamsburg.

stances. In addition, a secondary task of this project consisted of the development of a data inventory and the collection of all available facility-based and environmental data for Greenpoint/Williamsburg. A data dictionary was developed in the form of metadata to identify:

- The availability of data
- The agency source of the data
- The format of the data
- The cycle of data collection
- The nature of the data in their present form (i.e., whether they could be mapped or not)

Finding applicable data for the inventory and assessing the quality of these data were very time-consuming tasks. A variety of city, state, and federal agencies were canvassed for relevant data. Those agencies were scattered throughout the city and the state, and the data they had were often stored at off-site locations. In many cases, the data were not in a format that was machine readable. The agencies surveyed included: NYCDEP, New York State Department of Environmental Conservation (NYSDEC), Brooklyn Fire Department, NYC Department of Transportation, NYC Department of Finance, NYC Department of City Planning, and the EPA. Representatives from each of these agencies were interviewed for information regarding data sources, locations, formats, etc., and the data were inventoried.

Development of a Pilot GIS

Upon completion of the metadata compilation, the Greenpoint/Williamsburg Citizens Advisory Committee engaged the SPARS Laboratory to develop a pilot GIS to explore the possible use of such a tool by the community (see Ahearn and Osleeb, 1993). The primary task was to collect the relevant data that had been identified and to link a series of data layers describing the Greenpoint/Williamsburg community to two geographies (base maps):

- The Department of City Planning's block/lot COGIS map files, which defines the 15,000 property boundaries of every property in the community
- The U.S. Bureau of the Census TIGER line files, which contain street and address information, census-block boundaries, and census-tract boundaries

Additional relevant data were then identified and gathered for the project, and these data were also attached to the applicable base map for display and analytical purposes by importing the databases into the GIS and linking them either by block/lot geography (COGIS) or through address matching to the TIGER files on a block or census-tract level. The data layers tied to the TIGER line files included demographic data, data on industrial facilities, health data, environmental data, and information on environmentally sensitive facilities (such as schools, hospitals, nursing homes, and daycare facilities). Land-use data, city-owned parks, schools, and other facilities were linked to the COGIS base map along with data concerning industrial firms, including companies covered under the Superfund Amendments and Reauthorization ACT of 1986 (SARA) Right to Know Law and those requiring permits. The information about facilities that was entered into the database included hazardous-material storage, air emissions, discharges to water and land, permits, inspections, complaints, violations, and other applicable data. These data were provided by the NYC Department of

Environmental Protection (NYC DEP) and the NYS Department of Environmental Conservation (NYS DEC). As part of the pilot study, the SPARS Laboratory developed a set of test queries to demonstrate the potential of the GIS and the applications of these databases. In addition, analytical models were reviewed that could be used to assess risk to community residents and workers.

Table 6.2 provides a summary of the data used and their sources.

The attributes are arranged in various data layers that can be superimposed onto the base maps as needed. The GIS thus has the capability to answer a range of questions, from very simple questions about a particular facility at a given location to very complex questions that require models to relate various phenomena. The questions may be divided into three levels of inquiry (see Table 6.3): (1) Simple queries ask about a given location and use only one data layer. (An example would be, what facility is at a given address, and what is stored at that facility?) (2) More-involved queries

TABLE 6.2. Data layers in the Greenpoint/Williamsburg GIS.

Geographical base map layers	
Source	Data
COGIS, NYC Department of City Planning	Lot/block geography, double-line street map
LION, NYC Department of City Planning	Block/street geography, single-line street map with topology
TIGER, U.S. Bureau of the Census	Census tract, block group, block, ZIP code

Attribute Layers	
Source	Data
U.S. Bureau of the Census	Census data, including demographics, income, and ethnicity
NYC Department of City Planning	Schools, police stations, fire stations, parks, public housing
U.S. Environmental Protection Agency (EPA)	Toxic Release Inventory
NYS Department of Environmental Conservation (NYS DEC)	Underground bulk petroleum storage
NYC Department of Environmental Protection (NYC DEP)	Right-to-know reporters (hazardous-material processors); chemical bulk storage; major oil-storage facilities; petroleum bulk storage; TRI (facilities that release toxic chemicals to the air, water, or land); complaints about air, water, and noise pollution
NYC Department of Finance	Tax information
NYC Department of Health	Cancer rates by census tract; childhood lead-poisoning cases
NYC Department of Transportation	Truck routes
NYC Transit Authority	Bus and subway routes
Greenpoint/Williamsburg Community Board	Solid-waste transfer stations

TABLE 6.3. Sample queries that can be answered by the Greenpoint/Williamsburg GIS.

Query Level	Examples
Using one data layer	What facility is at 100 Main Street, and what is stored there? Find all facilities with violations in 1996. When was the facility last inspected?
Exploiting the relationships among multiple layers	Show all the facilities within a quarter mile of my house, which is located at 243 Greenpoint Ave. What are the reported cancer cases within a half mile of all TRI facilities? Where are all childhood lead-poisoning cases in homes over 40 years old?
Evaluating a new or proposed condition	What would be the effect on community environmental load of operating an incinerator within one mile of the community? Are large numbers of people near the site of a new hazardous-material user potentially at risk?

use multiple data layers and probe the relationships between or among those data. (For example, what schools are within a quarter of a mile of a facility?) (3) Complex queries might seek to evaluate a new or proposed condition and not only combine multiple layers but also require a model to be used to explore future conditions. (An example would be, what might be the effect on community health of operating an incinerator within a given proximity to the community?)

The result of these queries are maps and tables that show the specified entities (such as schools, fire houses, and facilities) and any relationship that may have been requested (such as the TRI or right-to-know facilities located within a particular distance). Figure 6.9 shows the result of a query that asked to show all TRI and right-to-know facilities.

The GIS can also be used to retrieve information in real time by just pointing (on the computer screen) to a facility of interest and clicking a button. For example, you can point to a particular TRI facility and find out its address, what it produces, and when it was last inspected.

One of the most important aspects of the G/WEBP is that the GIS is located within the community. Through careful planning with the Community Board, it was decided that an Environmental Watchperson Office would be set up in the community. Manuals explaining the proper use of the GIS were created, and the watchperson and the office's staff were trained by the GIS developers at Hunter College. (These manuals can serve as self-administered tutorials.)

The watchperson acts as liaison among the community, government, and the private sector and is able to take any findings to the Department of

6. Integration of Geographic Information 179

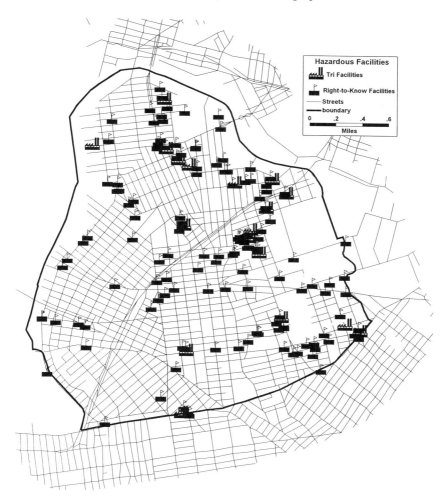

FIGURE 6.9. The results of a query to the GIS database requesting the locations of all TRI and Right-to-Know facilities.

Environmental Protection or other government agencies to promote greater vigilance on the part of the agencies. The office is open to the public and can provide information from the GIS for many different users. For example, developers and planning board members can ascertain information on hazardous facilities to develop zoning based on populations at risk, while community members can use the information to encourage proper prioritization of facility inspections. The possibilities are limitless.

The pilot GIS in Greenpoint/Williamsburg has provided the community with much-needed information about its surroundings. The GIS has also been combined with an SDSS to evaluate environmental equity within the community and to analyze a series of "what if" scenarios for future planning.

Using SDSS to Locate Environmentally Hazardous Facilities in Greenpoint/Williamsburg

In conjunction with the G/WEBP, an SDSS was developed to assess the location of proposed environmentally hazardous facilities in Greenpoint/Williamsburg. As noted earlier, SDSS combines a GIS with mathematical models to test various scenarios. In this case, because Greenpoint/Williamsburg was already extremely burdened by environmental hazards, it was imperative to consider the distribution of existing facilities.

Deciding where to locate hazardous facilities involves many complex and often conflicting factors. For this model, it was decided that risk and equity were the most important criteria. Therefore, the model assesses the interplay between equity and risk. Here, equity refers to the concept that environmentally hazardous facilities should be widely distributed, in smaller less hazardous facilities, so that any burdens imposed are shared equally among all populations as opposed to concentrated in one or two very large facilities that burden a select population. Consideration of risk, on the other hand, seeks to minimize the population that is exposed to these hazards. The SDSS that was developed for Greenpoint/Williamsburg produces compromise solutions to these conflicting goals.

The measure of equity was based on an index that represents the integrated impacts on environmental quality at a specific location. This index combines information on noise, odor, air pollution, and risk of industrial accident, and was determined for each census-block group in the community. The equity component seeks to minimize variations in the value of this index for different areas when a new facility is located. In other words, the equity component seeks to avoid making bad places worse.

The risk component of the model seeks not only to minimize the overall population exposed to the hazardous materials, but also to minimize the exposure of vulnerable populations. That is, it seeks to minimize the exposure for the elderly and for children under five, who are particularly sensitive to the adverse effects of pollutants. Interestingly, as shown in Figure 6.10, these populations are concentrated in certain parts of the community. If the model only sought to minimize the exposure of these groups, the equity would clearly be further unbalanced, with hazardous facilities being increasingly concentrated in other, already burdened areas (Figure 6.11). The model was able to resolve these problems.

Several scenarios were run with the SDSS to produce a tradeoff curve between the equity and risk associated with the siting of any new hazardous facility in the Greenpoint/Williamsburg community. Such a

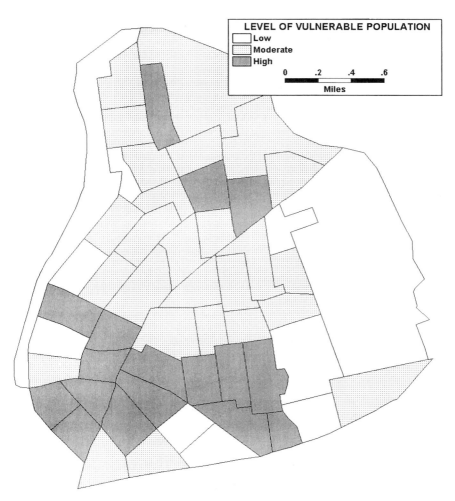

FIGURE 6.10. Locations of populations that are particularly vulnerable to air pollution (the elderly and children under five). Note that they are concentrated in the southern region.

curve (as shown in Figure 6.12) quantifies the risk and equity factors associated with various locations. The decision maker can then choose a solution.

Running the SDSS with equity only, it was found that in Greenpoint/Williamsburg, contrary to the findings of Ratick and White (1988) and others, a smaller number of large facilities would produce a more equitable distribution of facilities than a large number of small facilities. These results can be attributed to the model's consideration of the pre-existing hazardous sites. The model determined that, in areas like Greenpoint/Williamsburg that are heavily burdened in specific areas with hazardous facilities, locating large numbers of small facilities only perpetuates the already existing inequities.

Environmental equity (justice), another area of concern, considers whether the location of undesirable facilities disproportionately burdens low-income and minority populations (Bowen et al., 1995). One of the uses of an SDSS, such as the one created for Greenpoint/Williamsburg, would be to evaluate the distribution of environmental burdens along socioeconomic lines to facilitate decision making.

Further questions that could be addressed using similar SDSS models might include:

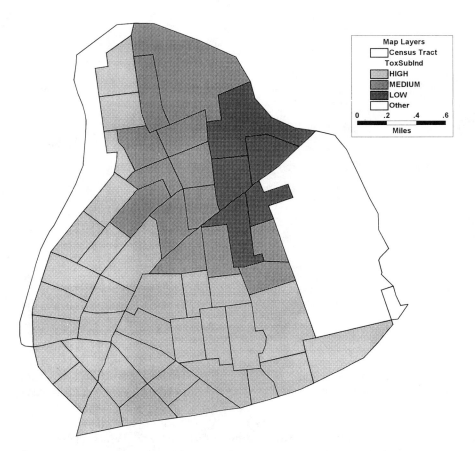

FIGURE 6.11. A map of the Integrated Environmental Quality Index. A high rating indicates lower environmental burdens. The SDSS model will seek to avoid the further burdening of already low-quality areas.

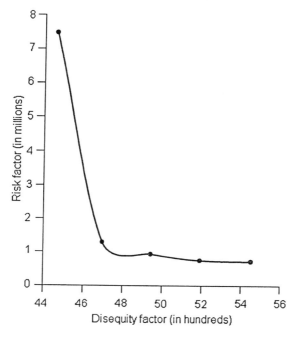

FIGURE 6.12. Multi-objective tradeoff curve, plotting risk versus disequity with data for a given scenario from Baiamonte (1996, p. 47).

- Where to locate schools on the basis of the location of hazardous-waste sites while considering financial constraints
- How to prioritize inspections of noxious facilities given limited resources
- Whether minorities and economically disadvantaged populations are being disproportionately affected by a present or future environmental hazard

Clearly, SDSS is an extremely powerful tool that allows decision makers to analyze various environmental scenarios and assess the tradeoffs that occur. This type of analysis is critical in areas such as Greenpoint/Williamsburg that are subject to great environmental burdens. But it is applicable to any area, even one that is considered environmentally healthy, to assess the potential effects of a change in the environment.

Geographic Plume Analysis for Greenpoint/Williamsburg

Assessing the baseline aggregate environmental load (Osleeb et al., 1996) was the third phase of the G/WEBP. This effort involved estimating the environmental load placed on each census-block group in the community from noise, odor, air releases, and the hazards associated with stored materials. To determine the impact of air releases, a GPA was undertaken.

The GPA used the EPA's Industrial Source Complex (ISC3) dispersion model, as well as the GIS developed for the G/WEBP. For this project, the most appropriate unit of analysis was determined to be the census-block group, of which there are 159 in Greenpoint/Williamsburg. This areal unit was chosen because a smaller unit (such as a census block) would be unrealistic for presenting the generalized results of the dispersion model, and a larger unit (such as a census tract) would be too encompassing and would show the impacts as being very extensive, making the results meaningless. Therefore, census information that had been previously furnished at the census-block level in the GIS was augmented with census-block-group data.

The ISC3 model is very appropriate for the urban setting of Greenpoint/ Williamsburg. In addition to average weather conditions by day, month, or season (seasonal weather conditions were used) and temperature gradients, the model takes into account such urban factors as stack-tip downwash, plume rise, and the influence of nearby buildings on the lateral dispersion of the plume. In addition, each chemical being released is treated separately on the basis of particulate size. Deposition of each type of particulate is calculated for an x-y coordinate that can be transformed into a universal transverse macerator projection, a commonly used map projection, and can then be assigned to the centroid of the appropriate corresponding census-block group. (For a complete discussion of map projections, see Campbell, 1991.) The deposition resulting by each pollutant from each pollution source is then aggregated for each block group to determine the load.

The results of the geographic plume analysis for the TRI facilities located in the community are shown in Figure 6.13. While airborne toxicants definitely concentrate adjacent to TRI reporters with air releases, much of the load still occurs in block groups that are a significant distance away from the TRI sites. This result indicates that the pollution clearly affects populations that are well beyond the point source of the pollution; and in many cases, block groups with no pollution source had high deposition rates, a nonintuitive and important finding.

Conclusion

Techniques for integrating and graphically displaying spatial information have evolved during the past 25 years. While limited in the past by the difficulties associated with mainframe computers, their uses have been greatly enhanced by the recent developments of the workstation and the new, powerful personal computers. GIS software that is easily adaptable to a wide range of problems has been developed, and it is becoming easier to use these systems as a result of the development of user-friendly programming languages. Additionally, these changes have facilitated the coupling of GIS with spatial mathematical models to produce SDSS and GPA. Some

FIGURE 6.13. The results of a geographic plume analysis, showing the environmental particulate load on various block groups from TRI facilities.

progress has also been made in three-dimensional representations (Mitasova et al., 1993) and in the incorporation of highly accurate orthodigital photography, which is now being used to supplement the more common digitized map.

These technologies have enjoyed wide acceptance, as demonstrated by the tremendous growth of the industry. As an example, the first Geographic Information Systems/Land Information Systems Conference in 1987 attracted 500 attendees. In 1997, the annual conference attracted more than 5,000 people. The industry's sales have grown to $8 billion annually. The tools that this industry is producing are very useful to environmental decision makers, and with advances in technology, they can be expected to be used widely in the future to integrate spatially referenced environmental information.

Key Resources

Ahearn, S. and Osleeb, J.P. 1993. Greenpoint/Williamsburg Environmental Benefits Program: Development of a Pilot Geographic Information System. In: *Proceedings of GIS/LIS*, Minneapolis, MN pp. 1–12.

DeMers, M.N. 1997. *Fundamentals of Geographic Information Systems*. New York: John Wiley and Sons.

Densham, P.J. and Goodchild, M.F. 1989. Spatial Decision support systems: A research agenda. In: *GIS/LIS '89: Proceedings, Annual Conference and Exposition, Orlando, Florida, November 26-30, 1989*. Bethesda, MD: pp. 707–716. American Society for Photogrammetry and Remote Sensing and American Congress on Surveying and Mapping Washington, D.C.: Association of American Geographers and Urban and Regional Information Systems Association, and Englewood, CO: AM/FM International.

Glickman, T.S. and Gough, M. Eds. 1991. *Readings in Risk*. Washington, D.C.: Resources for the Future.

Huxhold, W.E. 1991. *An Introduction to Urban Geographic Information Systems*. New York: Oxford University Press.

Maitin, I.J. and Klaber, K.Z. 1993. Geographic Information Systems as a Tool for Integrated Air Dispersion Modeling. *Proceedings of GIS/LIS*, Minneapolis, MN. (2):466–474.

National Research Council. 1993. *Toward a Coordinated Spatial Data Structure for the Nation*. Washington, D.C.: National Academy Press.

Websites on the Internet at www.ucgis.org and www.fgdc.gov

References

Ahearn, S. and Osleeb, J.P. 1993. Greenpoint/Williamsburg environmental benefits program: Development of a pilot geographic information system. In: *GIS/LIS Proceedings: 2–4 November 1993, Minneapolis Convention Center, Minneapolis, MN*. Bethesda, MD: American Society for Photogrammetry and Remote Sensing. Pp. 1–12.

Anderson, R.F. and Greenberg, M.R. 1982. Hazardous waste facility siting: A role for planners. *American Planning Association Journal* 48:204–218.

Arentze, A.T., Borgers, A.W.J., and Timmermans, H.J.P. 1996. Design of a view-based DSS for location planning. *International Journal of Geographical Systems* 10(2):219–236.

Armstrong, M.P. and Densham, P.J. 1990. Database Organization Strategies for Spatial Support Systems. *International Journal of Geographical Information Systems* 4(1):3–20.

Baiamonte, A. 1996. *An Equity Model for Locating Environmentally Hazardous Facilities in Densely Populated Urban Areas*. Thesis, Department of Geography, Hunter College, City University of New York.

Berry, B.J.L. and Rees, P.H. 1969. Factorial Ecology of Calcutta. *American Journal of Sociology* Vol. 74.

Bowen, W.M., Salling, M.J., and Haynes, K.E. 1995. Toward environmental justice: Spatial equity in Ohio and Cleveland. *Annals of the Association of American Geographers* 85(4):641–663.

Campbell, J. 1991. *Map Use and Analysis*. Dubuque, IA: Wm. C. Brown.

Carver, S.J. 1991. Integrating multi-criteria evaluations with geographical information systems. *International Journal of Geographical Systems* 5(3):321–339.

Carver, S.J., Heywood, I., Cornelius, S., and Sear, D. 1995. Evaluating field-based GIS for environmental characterization, modeling, and decision support. *International Journal of Geographical Systems* 9(4):475–486.

Chakraborty, J. and Armstrong, M.P. 1995. Using geographic plume analysis to assess community vulnerability to hazardous accidents. *Computers, Environment, and Urban Systems* 19:1–17.

Cliff, A. and Ord, J. 1973. *Spatial Autocorrelation*. London: Pion.

Commission on Geosciences, Environment, and Resources. 1997. *The Future of Spatial Data and Society: Summary of a Workshop*. Washington, D.C.: National Research Council, National Academy Press.

Dale, V.H., Southworth, F., O'Neill, R.V., Rosen, A., and Frohn, R. 1993. Simulating spatial patterns of land-use change in Rondonia, Brazil, *Lectures on Mathematics in the Life Sciences* 23:29–55.

Densham, P.J. and Goodchild, M.F. 1989. Spatial Decision support systems: A research agenda. In: *GIS/LIS '89: Proceedings, Annual Conference and Exposition, Orlando, Florida, November 26–30, 1989*, Bethesda, MD: American Society for Photogrammetry and Remote Sensing and American Congress on Surveying and Mapping Washington, D.C.: Association of American Geographers and Urban and Regional Information Systems Association and Englewood, CO: AM/FM International. Pp. 707–716.

Emani, S. et al. 1993. Assessing vulnerability to extreme storm events and sea-level rise using geographic information systems (GIS). In: *GIS/LIS Proceedings: 2–4 November 1993, Minneapolis Convention Center, Minneapolis, MN*. Bethesda, MD: American Society for Photogrammetry and Remote Sensing. Pp. 201–205.

Emmi, P.C. and Horton, C.A. 1995. A Monte Carlo simulation of error propagation in a GIS-based assessment of seismic risk. *International Journal of Geographical Systems* 9(4):447–461.

Environmental Systems Research Institute (ESRI). 1990. *Understanding GIS: The ARC/INFO Method*. Redlands, CA: Environmental Systems Research Institute.

Gilbert, N. and Conte, R. (Eds.). 1995. *Artificial Societies: The Computer Simulation of Social Life*. London: University College London Press.

Haggett, P. 1965. *Locational Analysis in Human Geography*. London: Edward Arnold.

Kim, K., Hart, D., and Prey, J. 1993. Geographic data base generation for urban nonpoint source pollution modeling in the Lake Superior region. In: *GIS/LIS Proceedings: 2–4 November 1993, Minneapolis Convention Center, Minneapolis, MN*. Bethesda, MD: American Society for Photogrammetry and Remote Sensing. Pp. 351.

Kim, K., Kim, J., Choi, J., and Sung, M. 1995. Applications of GIS to water quality management. In: *GIS/LIS Annual Conference and Exposition: Proceedings,*

November 14–16, 1995, Nashville Convention Center, Nashville, TN. Bethesda, MD: American Society for Photogrammetry and Remote Sensing. Pp. 554–562.

Lao, Y. and Sharma, H.P. 1995. A GIS approach to managing gaseous hazardous spills on highways. In: *GIS/LIS Annual Conference and Exposition: Proceedings, November 14–16, 1995, Nashville Convention Center, Nashville, TN*. Bethesda, MD: American Society for Photogrammetry and Remote Sensing. Pp. 581–590.

Maitin, I.J. and Klaber, K.Z. 1993. Geographic information systems as a tool for integrated air dispersion modeling. In: *GIS/LIS Proceedings: 2–4 November 1993, Minneapolis Convention Center, Minneapolis, MN*. Bethesda, MD: American Society for Photogrammetry and Remote Sensing. Pp. 466–474.

McMaster, R.B. 1988. Modeling community vulnerability to hazardous materials using geographic information systems. In: *Proceedings: Third International Symposium on Spatial Data Handling*, August 17–19, 1988, Sydney, Australia, Columbus, OH: International Geographical Union, Commission on Geographical Data Sensing andProcessing, and Department of Geography of Ohio State University. Pp. 143–156.

Mitasova, H. et al. 1993. Multidimensional interpolation, analysis and visualization for environmental modeling. In: *GIS/LIS Proceedings: 2–4 November 1993, Minneapolis Convention Center, Minneapolis, MN*. Bethesda, MD: American Society for Photogrammetry and Remote Sensing. Pp. 550–556.

Monmonier, M. 1996. *How to Lie with Maps*. Chicago: The University of Chicago Press.

Moreno, D. and Siegel, M. 1988. A GIS approach for corridor siting and environmental impact analysis. In: *GIS/LIS '88 Proceedings: Accessing the World, Third Annual International Conference, Exhibits, and Workshops, San Antonio, Marriott Rivercenter Hotel, San Antonio, Texas, November 30-December 2, 1988*. Falls Church, VA: American Society for Photogrammetry and Remote Sensing. Washington, D.C.: Association of American Geographers and McLean, VA: Urban and Regional Information Systems Association. Pp. 507–514.

Murdie, R.A. 1969. Factorial Ecology of Metropolitan Toronto, 1951–1961. Research paper No. 116, Department of Geography, University of Chicago.

Osleeb, J.P. and McLafferty, S. 1992. A weighted covering model to aid in Dracunculiasis eradication. *Papers in Regional Science* 71(3):243–257.

Osleeb, J.P. 1995. The development of a comprehensive citywide geographic information system for the City of New York, remarks to the New York City Council, Committee on Technology, May 22.

Osleeb, J.P. et al. 1996. *Greenpoint/Williamsburg Baseline Aggregate Environmental Load Study*. New York: NYC Department of Environmental Protection.

Peterson, K. 1993. Spatial decision support systems for real estate investment analysis. *International Journal of Geographical Systems* 7(4):379–392.

Ratick, S.J. and White, A.L. 1988. A risk-sharing model for locating noxious facilities. *Environment and Planning* B15:165–179.

Ralston, B.A. 1992. Implementing a spatial decision support system or take my computer, please, *Abstracts, Association of American Geographers Annual Meeting, San Diego, 1989*. Washington, D.C.: Association of American Geographers.

Stein, A., Staritsky, I., Bouma, J., and Van Groenigen, J.W. 1995. Interactive GIS for environmental risk assessment. *International Journal of Geographical Systems* 9(5):509–525.

Smotritsky, Y., Osleeb, J., and Mellot, J. 1993. *Route Location Algorithm for Transportation and Utility Corridors*. Washington, D.C.: National Science Foundation.

Wagner, H.M. 1975. *Principles of Management Science*, 2nd ed. Englewood Cliffs, NJ: Prentice Hall.

Decision-Maker Response

SURYA S. PRASAD

The chapter is based on a case study that required the integration of spatial information and is intended to inform investigators and researchers about the fundamental nature and applications of available tools in the environmental decision-making process. However, the complexity of environmental input parameters involved in their applications make their presentation somewhat vague.

The authors begin by noting that the overwhelming amount of environmental information requires the integration of that information before analysis can proceed. The Introduction lays a good foundation and delivers an impressive array of observations and concepts. It adds a new dimension to the needs for methods for data integration. The sections following the Introduction systematically describe methods for linking spatial data with other spatial and nonspatial data sets, focusing on the needs and methods to integrate geographic data.

Today's decision makers need powerful tools to integrate data and help in interpretations for decision making. Supporting information is necessary to integrate analysis data derived from trivial operations, including environmental sampling and analysis. The chapter provides a complete review of the major spatial information integrating technologies and identifies resources for solving complex integrated environmental assessments. The evaluations resulting from the use of these analytical tools assist in meeting the challenges encountered in protecting public health and the environment.

The emphasis of the chapter is on spatial data, but there are many cases in which nonspatial information needs to be integrated for environmental decision making. In those cases, the data of the information sources must have compatible units of measurement and spatial and temporal resolution. In some situations, extrapolation between data points may be necessary. In all cases, the manipulations of the information must be clearly set forth for the analyst.

The chapter provides a good overview of geographic information systems, providing information on such aspects as what features are used in developing base maps and addressing the use of mathematical models in support of management decisions. Several existing models are described, and the merits of each are presented in a logical manner. The authors used current literature to support the need for existing data elements and in aiding information management. The requirements of spatial decision-support systems are defined with an enumeration of their strengths and weaknesses.

A large number of spatial simulation models could have been discussed in this paper. The diversity of these models in terms of the processes they simulate and the management questions that are of concern are immense. For examples, spatial simulation models are used for such diverse purposes as deciding where to site new transportation routes, where and when to schedule military training missions so as to protect endangered species, and how to remove pollutants from a site. Decision makers need to be aware of the diversity of models that are available. However, these models are constantly evolving, and new ones are being developed. Therefore, the best way to learn about these models may be to access the NCEDR web site that serves as a companion to this book and points to other web sites that describe examples of spatial simulation models.

It may not be feasible or necessary in one chapter to communicate how decision-aiding tools within a category are used and how results are derived from that use. In most instances, an explanation of the factual data used, assumptions made, and the methods and analyses conducted may be all that is needed for supporting an interpretation and the resulting decisions. Overall, the chapter presents overwhelming information designed for a technical audience that is able to interpret the findings. This chapter will not be of great value to a nontechnical audience. However, with the assistance of skilled professionals, administrators, supervisors, and regulators may benefit from reading the chapter. In the case studies, the results seem to support the model scenarios for the technologies employed in the decision-making process. However, additional scenarios and case studies are necessary to support validation.

The chapter describes the use of tools for manipulating factual data. Such manipulation may be perceived as altering the nature to influence the outcome of an interpretation. The authors take pains to identify types of situations that specifically warrant data integration in forecasting, assessing, and conducting postdecision scenarios.

Among the factors that constrain communication of tools and their results is the application of modern concepts. This point is important because of the rapid evolution of concepts and models in the field of information integration.

7
Forecasting for Environmental Decision Making

J. SCOTT ARMSTRONG

"The Ford engineering staff, although mindful that automobile engines provide exhaust gases, feels that these waste vapors are dissipated in the atmosphere quickly and do not present an air pollution problem." Official spokesperson for the Ford Motor Company in 1953 in response to a letter from the Los Angeles county supervisor

Cerf and Navasky, 1984, p. 38.

Those making environmental decisions must not only characterize the present, they must also forecast the future. They must do so for at least two reasons. First, if a no-action alternative is pursued, they must consider whether current trends will be favorable or unfavorable in the future. Second, if an intervention is pursued instead, they must evaluate both its probable success given future trends and its impacts on the human and natural environment. Forecasting, by which I mean *explicit processes* for determining what is likely to happen in the future, can help address each of these areas.

Certain characteristics affect the selection and use of forecasting methods. First, the concerns of environmental forecasting are often long term, which means that large changes are likely. Second, environmental trends sometimes interact with one another and lead to new concerns. And third, interventions can also lead to unintended changes.

This chapter discusses forecasting methods that are relevant to environmental decision making, suggests when they are useful, describes evidence on the efficacy of each method, and provides references so readers can get details about the methods. A key consideration is whether or not the forecasting methods are designed to assess the outcomes of interventions. The chapter then examines issues related to presenting forecasts effectively. Finally, it describes an audit procedure for determining whether the most appropriate forecasting tools are being used.

A Framework for Forecasting

Figure 7.1 shows possible forecasting methods and how they relate to one another. The figure is designed to represent all approaches to forecasting. The methods are organized according to the types of knowledge. As one moves down the chart, the integration of statistical and judgmental methods increases.

Judgmental methods are split into those involving one's own behavior in given situations (intentions or role playing) and expert opinions. The intentions method asks people how they would act in a given situation. Role playing examines how people act in a situation where their actions are influenced by a role. Experts can be asked to make predictions about how *others* will act in given situations. They can also identify analogous situations, and forecasts can be based on extrapolations from those situations.

Intentions and expert opinions can be quantified by relating their "predictions" to various causal factors with, for example, regression analysis. Expectations about one's own behavior are referred to as conjoint analysis (e.g., given alternatives having a bundle of features that have been varied according to an experimental plan). Expert opinions about the behavior of others (which can also be based on an experimental design, but are often based on actual data) are referred to as judgmental bootstrapping.

The statistical side of the methodology tree has univariate and multivariate branches. The univariate branch leads to extrapolation methods. By drawing upon expert opinions, one can develop rule-based forecasting. This procedure uses domain knowledge to select and weight extrapolation methods.

Expert systems use the rules of experts. These rules might be based on protocol sessions in which experts are asked to describe how they make forecasts while they are actually in the process of forecasting. Alternatively, experts' rules could be formalized by drawing upon estimates produced by judgmental-bootstrapping models. Quite commonly, developers of expert systems also draw upon empirical studies of relationships, with some of those studies involving econometric models. Another possible source of information is embodied in relationships estimated by conjoint analysis.

The multivariate branch is split into data-based and theory-based branches. In the theory-based approach, the analyst formulates a model and then refines the parameter estimates based on information gleaned from experts and data. These constructs are referred to as econometric models. Data-based approaches try to infer relationships from the data. I refer to them as multivariate models.

In all, then, 11 approaches to forecasting are proposed. More attention will be given here to those for which there is stronger evidence. For

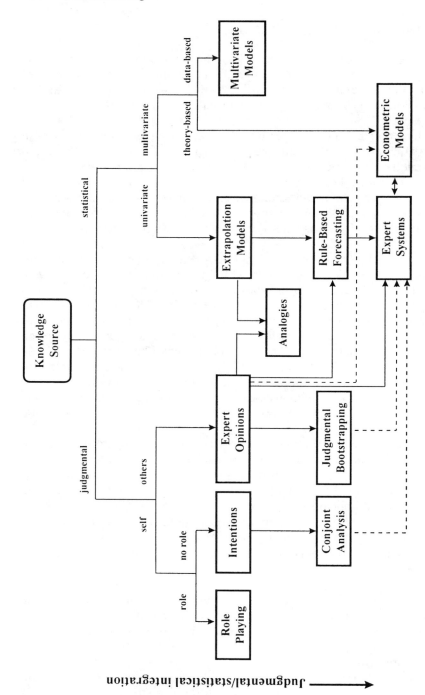

FIGURE 7.1. Characteristics of forecasting methods and their relationships; (dotted lines represent possible relationships.)

example, despite an immense amount of research effort, the evidence that multivariate methods provide any benefits to forecasting is weak. The case for expert systems is stronger, but, I am not aware of studies that assess the use of expert systems for producing environmental forecasts.

Forecasts Without Interventions

Measurement procedures for environmental decision making are improving, and more relevant data are being collected. As a result, environmental decision makers have increasing amounts of information about trends in the environment and related socioeconomic, political, and legal conditions. Some of these trends, such as increases in pollution, may seem unfavorable; whereas other trends, such as advances in technology, may seem favorable.

Whether or not a trend is favorable depends upon one's point of view. For example, some observers fear the effects of global warming, while others believe that its effects would be beneficial. In structuring the forecasting problem, then, it is important to forecast the *trends and their effects on all affected populations, human and nonhuman.* For example, what are the predicted effects of increasing amounts of pollution on residents, land owners, people living downwind or downstream, product manufacturers, consumers, waste-disposal firms, and wildlife?

Forecasts with Interventions

The primary reasons for making explicit forecasts are to determine whether to intervene, and if so, how. Forecasting can help decision makers to assess alternative interventions.

Interventions can affect many aspects of the system. For example, to reduce pollution in Santiago, Chile, in the early 1990s, the government restricted the use of automobiles in the central business district by allowing entry to only those cars whose license plates ended in even numbers on one day and those ending with odd numbers on the next day. Such a plan might have an impact on those who sell automobiles (perhaps refurbishing old cars so people can have cars with both even- and odd-numbered license plates), commuters (perhaps becoming less productive because they spend more time on public transit), and so forth. Forecasters need to consider the effects on each group and how they will react to an intervention.

A forecast of the effects on only part of the system might be worse than no forecast at all because it might lead to unwise decisions. For example, the concern over disposal problems with plastic packaging put so much public pressure on McDonald's that they switched to paper packaging. Some analysts have concluded that paper packaging not only is less convenient for workers and customers, but also creates more pollution when one considers the entire cycle from producing paper packaging, using it, and disposing of it.

Use of Forecasts

Plans are often confused with forecasts. Plans are sets of actions to help deal with the future. Forecasting (or predicting) is concerned with determining what the future *will* be. A plan is an input to the forecasting model. If the forecasts are undesirable, then one might change the plan, which, in turn, could change the forecast. The point to remember is that good plans depend on good forecasts.

In practice, forecasts are sometimes used to motivate people. More properly, people should be motivated by plans (e.g., "meet this plan, and we will pay you a bonus of 25 percent").

Decisions have often been made before any formal forecasting has been done. In such cases, the forecast serves little purpose other than to annoy people if it conflicts with their decision or to please them if it supports their decision. For the forecast to be used effectively, it should be prepared before decisions are made.

Not only the expected outcome, but also other likely outcomes (such as the best and worst outcomes) should be forecast. If the worst outcome poses too much risk, forecasts should be made for alternate interventions.

Methods for Environmental Forecasting

Environmental forecasting often involves decisions that have long-term consequences. In addition, these forecasts are likely to be subject to severe biases, depending on one's perspective. To address these issues, I suggest four principles:

- Use relevant information
- Ensure that the information and procedures are objective
- Structure the inputs to the forecasting procedure
- Use methods that are no more complex than necessary

Relevant information Few people would argue that obtaining more information is not useful, but in some situations, more information does not improve forecast accuracy. If the additional information is irrelevant, it is likely to reduce forecast accuracy. For example, in many professions, the height of people is not important to job performance; however, height plays a role in job interviews when people make predictions about a candidate's ability to do a job. The danger of using irrelevant information is especially serious when using judgmental procedures. For example, knowing that a celebrity has taken a strong stand on an environmental issue might detract from one's ability to make a good judgment on the issue; more than likely, the person's celebrity status has nothing to do with his or her making an informed judgment. Also, more information is not likely to improve forecasting accuracy if it is used in an unstructured manner (Armstrong, 1985, pp. 100–102).

Objective information and procedures Before data collection and analysis begin, decision makers should agree on what information is relevant and how it should be analyzed. These decisions should be made by people who are not biased to any particular viewpoint, at least in that they will not personally gain. It would seem sensible here to draw upon those with expertise in forecasting methods.

To avoid collecting data that might be biased, one should seek alternative sources of data. Brenner et al. (1996) concluded that subjects who received biased data were less accurate in their predictions but more confident, even when it was painfully obvious that the data were biased and incomplete.

Structured inputs Forecasting can be improved by structuring the problem to make efficient use of available information. One of the most useful tools for structuring forecasting problems is decomposition. The safest way to decompose problems is to break them into additive elements. For example, population is often decomposed to forecast births, deaths, immigration, and emigration; these separate forecasts are then added to form a composite forecast of population. Alternatively, multiplicative decomposition is sometimes useful. For example, certain air pollutants might interact, so it would not be sufficient to merely add their effects. But multiplicative decomposition can be risky because the errors in each element get multiplied by those in all the other elements. MacGregor (1999) summarized research on this topic and concluded that multiplicative decomposition improves accuracy when uncertainty is high and extreme numbers (large or small) are involved. For example, it would be difficult to make judgmental estimates on the number of pounds of harmful automobile pollution produced each year in the United States. Typically, one can decompose problems to avoid forecasting extreme numbers. However, multiplicative decomposition is sometimes detrimental when numbers are not large and uncertainty is not great.

Simple methods One of the more interesting findings from empirical research on forecasting is that relatively simple methods provide forecasts that are as accurate as those from more complex methods. It follows, then, that forecasting methods should be no more complex than necessary. This conclusion seems to cover a wide range of conditions. Simple methods reduce costs and they aid understanding among forecasters and decision makers. They also reduce the likelihood of mistakes. Mistakes do occur, however, even for relatively simple methods, such as exponential smoothing. Gardner (1984) cited 23 books and articles containing errors in model formulations for exponential smoothing. The more complex a process is, the more likely it is that a mistake might creep in and remain undetected.

The use of simple methods should be welcome to decision makers. Yokum and Armstrong's (1995) survey of forecasters concluded that ease of understanding, implementation, and use were almost as important as accuracy when selecting forecasting methods.

One way to avoid imposing one's own bias is to seek forecasts that have been provided by independent third parties. This procedure might also save money. Before using a published forecast, you should have details about the source of the forecast along with the methods and data that were used. Unfortunately, these details are often omitted.

Our major concern in this chapter is with the most useful methods for environmental forecasting. Most approaches from Figure 7.1 are relevant. However, data-based multivariate forecasting models are inappropriate because they ignore the substantial body of expertise that is typically available on environmental issues. In addition, they are expensive, difficult to communicate, complex, subject to the imposition of biases or mistakes that are hard to detect, and often misunderstood, even by leading proponents. Finally, despite enormous efforts in their development, little testing has been done of their predictive validity, and what has been done suggests that this approach is not promising.

Below are descriptions of the methods that seem most relevant given the guiding principles for environmental forecasting. Table 7.1 lists these methods along with brief descriptions of their uses, advantages, and disadvantages.

Judgmental Methods

Judgmental forecasting involves methods that process information by experts, rather than by quantitative methods. The experts might have access to data, and their approach might be structured, but the final forecasts are the result of some process that goes on in their heads.

Before discussing tools that aid judgmental forecasting, it is important to mention one tool that is widely used and well accepted, but which typically harms accuracy and leads to an unwarranted gain in confidence. The culprit is the traditional (unstructured) group meeting. Besides the biases inherent in unstructured meetings (such as the influence of the boss), the group's information is likely to be poorly used (Armstrong, 1985, pp. 120-121).

Judgmental forecasts are susceptible to various biases. To reduce biases, one should select unbiased experts (i.e., those who have nothing to gain from a forecast that is either too high or too low). In addition, care should be given to how the forecasting problem is formulated. Questions should be structured to use the judges' knowledge most effectively, pretested to ensure that the experts understand them, and worded in different ways to see if that affects the forecasts. Such procedures are particularly important when forecasting sensitive issues, such as the effects of global warming.

TABLE 7.1. Methods for environmental forecasting.

Tools	Description	Advantages	Disadvantages
Judgment			
Role-playing	Simulates the interaction of conflicting groups in a decision process	Taps the experience and knowledge of decision makers in a realistic interaction	Group dynamics may influence players in ways that are unintended; selected players may not be knowledgeable
Intentions	Determines what actions decision makers would take in certain circumstances	Provides the perspective of those who will actually make decisions	Subject to sampling and questioner bias; changes with time as other factors come into play
Expert opinion (Delphi)	Experts forecast how others will behave	Information is derived from knowledgeable sources; inexpensive to perform; useful even when data are poor or lacking	Overly influenced by the current situation
Analogies	Examines how similar situations have turned out	Based on real-world experiences	May have poor correspondence to current situation
Conjoint analysis	To gauge citizens' reactions to aspects of an intervention	Citizens often have a good sense of how they will respond	Expensive
Judgmental bootstrapping	To gauge citizens' reactions to aspects of an intervention	Experts often have useful information; inexpensive	Experts may lack relevant knowledge
Extrapolation			
Exponential smoothing	Extends historical values into the future	Reliable; reproducible; simple methods produce accurate results; limits introduction of bias	Inaccurate, given discontinuities or unstable trends; ignores domain knowledge; especially risky when trends are contrary to expectations
Econometric Methods			
Single-equation, theory-based models	Forecast based on causal relationships	Results are firmly grounded in domain knowledge and theory; especially useful when large changes might occur; alternative interventions can be compared; can aid in the construction of confidence intervals	Complex procedures not easily understood by decision makers; may lack data on causal variables; may overlook key variables; expensive

TABLE 7.1. *Continued*

Tools	Description	Advantages	Disadvantages
Integrated Forecasts			
Rule-based forecasting	Assign differential weights to extrapolative forecasts	Based on published research and expert advice about forecasting methods; improves accuracy; offers protection against large errors; aids objectivity; incorporates managers' knowledge	Added complexity and cost
Expert systems	Apply rules determined by experts and by empirical studies	Formalizes available knowledge about a situation	Expensive; little information about forecast validity
Combined Forecasts			
Equal weights	Combine forecasts from several methods, giving the results of each method the same influence on the final result	Improves accuracy; offers protection against large errors or mistakes; aids objectivity; useful given uncertainty about which method is best	Ignores domain knowledge; not appropriate when better method is known

The use of structured procedures can greatly improve the accuracy of judgmental forecasts. Structure is easy to apply and involves only modest costs. I discuss four structured judgmental procedures that should be of interest for environmental forecasting: (1) role-playing, which uses subjects to act out relevant interactions to determine what they would do when affected by an intervention; (2) intention surveys, which use statements by key participants in the system about what they expect to do given certain trends or interventions; (3) Delphi, which uses expert judgment to forecast trends or the effects of intervention; and (4) analogies, where experts try to generalize from similar situations. Brief attention is given to conjoint analysis and to judgmental bootstrapping.

Role-Playing

Role-playing involves asking subjects to adopt the viewpoints of groups in a negotiation situation and having them act out the interactions. When the interactions of conflicting groups are important to the outcome, role-playing provides a way to simulate this interaction. If new and important interventions would lead to behaviors that are dependent upon the interactions among decision makers, then role-playing is likely to be more relevant than intentions. With intentions, decision makers would have to predict what they would do initially, how they would modify their decisions in reaction to the decisions made by others, how others would respond to this reaction, and so on. This chain of events is often too complex for the respondent, so it makes sense to act it out.

To use role-playing to forecast the outcome of an intervention, such as a tax on air pollution, one would write short descriptions of the problem and of the roles of key decision makers. Different materials can be prepared to test alternative interventions. These guidelines should be followed:

- Use props to make the situation realistic.
- Select subjects who can act the role (interestingly, the selection of subjects does not seem to be a critical aspect for the accuracy of role-playing).
- Subjects should receive their roles before they receive any information about the situation, and they should not step out of their roles.
- Subjects should act as they would act if they were actually in such a role.
- Subjects should improvise as needed.

Forecasts would be based on the outcomes of the role-playing sessions. Ideally, possible outcomes can be identified in advance. However, if the range of possible outcomes is uncertain, one should leave the materials open-ended and ask research assistants to code the outcomes of the role-playing sessions. If the session does not lead to an outcome, one can ask the players to predict what would have happened had it continued to a conclusion.

Prediction intervals can be constructed by assessing the proportion of times that a certain outcome occurs in a set of role-playing sessions. The standard error of this estimate can then be obtained by using the formula for the standard error of a proportion, with the number of role-playing sessions as the sample size. Prediction intervals would be expected to be larger than this estimate because of possible response biases.

While role-playing has been used as a predictive device in the military and in the legal profession for many years, research on its value as a predictive technique is limited. Armstrong (1987) and Armstrong and Hutcherson (1989) report on studies that compared unaided opinions with role-playing for eight situations. These included the conflict between Mexico and the United States, which led to the United Stated acquiring Texas; the marketing of Upjohn's drug, Panalba, after a commission concluded that it was causing unnecessary deaths; the presidential-election conventions held by the party that was out of power; an attempt by Philco to gain the agreement of supermarket owners to allow them sell appliances in supermarkets; negotiations between the National Football League players and the owners in 1982; an attempt by artists in The Netherlands to have the government buy their artwork if they could not find anyone to purchase it; a negotiation over the royalties for an academic journal; and whether bombing North Vietnam would be a good strategy for the United States in the 1970s. Role-playing was superior to opinions on seven of the tests; it tied on one (that involving political conventions). Averaging across the eight situations, which involved 226 role-playing sessions, role-playing was correct for 63.6 percent of the cases versus only 18.2 percent for unaided opinions. A listing of these experimental comparisons is provided in Table 7.2.

Although none of the eight validation situations involved environmental decisions, they all involved conflicts between groups. Thus, I would expect that role-playing would be useful for forecasts involving environmental conflicts, such as whether to charge farmers more for cattle-grazing rights on government lands or what restrictions would be effective to control certain types of fishing. Based on research to date, role-playing would seem to be more accurate than other methods (except experimentation) for forecasts involving environmental conflicts.

Role-playing, however, is inexpensive relative to experimentation. Many of the studies were conducted using role-playing sessions that lasted less than an hour. In five of these situations (indicated by asterisks in Table 7.2), such "low-fidelity" role-playing sessions were used.

The key aspect of role-playing is that it simulates interactions. It is not enough just to tell people about the roles and ask them to consider the interactions. When subjects were given information about the roles of the parties involved and were asked to consider this, it did not improve the accuracy of their forecasts (Armstrong, 1987).

Role-playing would be relevant to trash-disposal fee problems. Various regulations could be presented to individuals who play the roles of household members. They would also be informed about the decisions of their neighbors. The government might respond to some of the consequences (e.g., illegal dumping or increased trash compacting by households) with

TABLE 7.2. Role-playing versus opinions.

Situation	Conflict among	Percent correct (Sample size)	
		Opinions	Role play
US-Mexico	Countries	1 (1)	57 (96)
*Panalba (prescription drug)	Stockholder and consumer	34 (63)	79 (57)
US Political Convention	Candidates	67 (12)	67 (12)
*Philco appliances	Manufacturer and retailer	3 (37)	75 (12)
*NFL Football	Players and owners	27 (15)	60 (10)
*Artists in Holland	Artists and government	3 (31)	29 (14)
*Journal royalties	Publisher and editors	12 (25)	42 (24)
North Vietnam bombing	Countries	0 (1)	100 (1)
	Averages	18.2	63.6

*Based on low-fidelity role-playing.

new regulations, which, in turn, could be assessed. Town meetings could be simulated. Actions by trash collectors could also be predicted. The cost of such forecasts would be low compared with the cost of actually conducting and monitoring a trial of the various proposals.

Intention Surveys

Intention studies are surveys of individuals about what actions they plan to take in a given situation or, if lacking a plan, what they expect to do. Such surveys are useful for predicting the outcomes of interventions. When a situation depends on the decisions of many people (such as with the trash collection for a community), surveys are much more expensive than Delphi. However, they provide the perspective of those who will actually be making decisions. For example, consider the situation when the prohibition of Freon TM as a coolant was first proposed. Surveys might have been made of manufacturers of refrigerators and coolants to see how they would respond. In addition, one could have presented this situation to consumers and asked them how they would respond.

Tools for surveys have been improving since the 1936 *Literary Digest* poll predicted that Landon would easily defeat Roosevelt for president. Squire (1988), in a re-analysis of that event, concluded that the forecast was incorrect primarily because of nonresponse bias and secondarily because of sampling error. (People often assume that sampling error was the major cause.) Procedures for controlling sampling error are now well-known. Nonresponse error, where people fail to respond at all to the survey instrument, can be controlled by a variety of procedures, such as making extrapolations across waves (Armstrong and Overton, 1977). Perhaps the primary source of error is that caused by the nature of the response. Numerous improvements have been made to control for response error. Given these improvements, it is not surprising that the total error for political-election forecasts decreased substantially in the United States from 1950 through 1978 (Perry, 1979).

Despite improvements in dealing with response bias, the problems for environmental forecasting are substantial. Citizens may have difficulty in predicting how an event or a change might affect them and in deciding how they will feel about the event. Lowenstein and Frederick (1997) discuss these issues and conclude that little evidence exists on the ability of people to predict how environmental changes will affect them. They did present evidence about how people would react to rain-forest destruction, restricted sport fishing because of pollution, and recovery of certain endangered species. They concluded that people greatly overestimate the effects of such changes on their life satisfaction.

As with other methods, objectivity is a key concern. When surveys are conducted by biased organizations, such as by political candidates, errors are often substantial. Shamir (1986) classified 29 Israeli political surveys according to the independence of the pollster. The results showed that the more independent the pollster, the more accurate the survey.

When interventions would create large changes and where the behavior of decision makers is dependent upon decisions by others, respondents may find it difficult to predict how they would behave. Surveys are of less value in such cases.

Given all the ways that intentions or expectations may be wrong, it should not be surprising to find that sampling error alone provides a poor way to estimate prediction intervals. In a study of 56 trial polls in the 1992 presidential election, Lau (1994) concluded that the sample size of the poll was not closely related to the relative forecast errors for a set of surveys. When Buchannan (1986) examined errors for 155 political elections in nine countries from 1949 to 1985, they were twice as large as those expected from sampling error alone. This finding occurred with voting for political candidates, a behavior that was familiar to the respondents. For environmental concerns, where the future behavior may be less familiar to the respondents, one might expect that response and nonresponse biases would constitute large sources of error. These errors should be reflected in the assessment of prediction intervals.

Consider the trash-disposal fee problem again. If people have had no experience with such a system, it may be difficult for them to anticipate how they would behave. Their behavior depends to some extent on the behavior of their neighbors. Furthermore, if people did not want to comply with this new procedure and instead planned to illegally dispose of trash, would they be willing to admit it in a survey? Could you imagine a respondent saying "Well, if it is going to cost that much to dispose of this waste legally, I will probably dispose of it illegally." Some procedures can help mitigate these problems. One way to do this is to use the random-response technique to help ensure confidentiality. For example, in a telephone survey, you could ask, "If a tax of 80 cents per bag of trash was implemented, would you dispose of any of your trash illegally? To answer, first flip a coin. If the coin turns up heads or if you expect that you would dispose of some trash illegally, then answer 'yes'." The amount of expected illegal disposal for the sample can then be teased out statistically. For example, if the average response for the respondents was 50 percent, then there would be no illegal disposal expected; if it was 100 percent, then the assumption is that everyone would dispose of some trash illegally. Projective questions could also be used, such as, "Would your neighbors dispose of waste illegally?" Sudman and Bradburn (1982, Chap. 3) provide a discussion along with a 12-item checklist of how to ask threatening questions about behavior.

Delphi

Delphi involves the use of experts to make independent anonymous forecasts. Delphi goes beyond expert surveys in that it is conducted for two or more rounds. After the first round of forecasts, each expert receives a quantitative summary of the group's forecasts. In addition, anonymous explanations of their choices might be provided by the experts. Typically, two rounds are sufficient; however, if the cost associated with error is high, conducting three or four rounds may be worthwhile. Delphi is usually conducted by mail, and honoraria are paid to the participating experts. Stewart (1987) discusses the advantages and limitations of Delphi.

Delphi can be used to forecast trends, such as "What do experts expect to happen to the levels of New York City air pollution during the next 20 years?" It can also be used to forecast the effects of interventions: "What would be the impact of a $1/gallon federal tax on gasoline?"

Experts need some level of domain expertise to make forecasts of change. Surprisingly, expertise beyond a modest level seems to have little relationship to accuracy (Armstrong, 1985, pp. 91–96). As a result, there is little need to pay large honoraria to members of Delphi panels. Perhaps the primary criterion for the selection of experts for a Delphi panel is that they be unbiased.

Delphi requires only a few experts. The number of experts should be at least five but seldom more than 20 (Hogarth, 1978; Libby and Blashfield, 1978; Ashton and Ashton, 1985). As a result, Delphi studies can be relatively inexpensive to conduct. This approach may be much less expensive than surveys that obtain information of individuals' intentions or expectations. For example, in predicting the outcomes of voter referendums on land use and property tax, Lemert (1986) found that, for the same level of accuracy, asking a few politicians for their predictions was more cost-effective than conducting a large-sample voter-intention survey.

When information is coming from a variety of sources, such as a number of Delphi respondents, the question comes up whether each source's information should be given the same weight. Rather than weighting by expertise, the preferred procedure is to weight each panel member's forecast equally, as long as each possesses at least some expertise. Based on studies to date, the required level of expertise is surprisingly low (Armstrong, 1985, pp. 91–96). Simple averages are commonly used and are often sufficient. McNees (1992) found little difference between the accuracy of means and medians in a study of economists' forecasts. However, trimmed means (throwing out the highest and lowest estimates) are likely to be more accurate in cases involving high uncertainty. The median, the ultimate trimmed mean, may be the safest way to summarize forecasts (Larreche and Moinpour, 1983) if one has more than, say, 10 experts.

Delphi is relevant when data are lacking, the quality of the data are poor, or experts disagree with one another. As a result, Delphi is applicable when new interventions are proposed or where a trend has recently undergone a shock. Nevertheless, judgments tend to be too conservative in the face of rapid change. In particular, judgment underestimates exponential growth (Wagenaar and Sagaria, 1975) and exponential growth is common in environmental problems. For example, Wagenaar and Timmers (1979) presented a computer-screen simulation of the growth of duckweed on a pond, an exponential process. Subjects asked to forecast when the pond would be covered greatly underestimated the time it would take. In another study (Wagenaar and Timmers, 1978), subjects were given information about pollution problems; when information was provided to subjects at more frequent time intervals, their predictions became less accurate. In a third study, Timmers and Wagenaar (1977) found that better judgmental predictions were made when the variable reflected a decrease with time (e.g., instead of predicting population per square mile, predict square miles per person).

Because Delphi is based on (1) acting on prior research about the use of more than one expert; (2) using unbiased experts; (3) using structured questions; and (4) summarizing in an objective way, one would expect it to be more accurate than unaided judgment. It is. The few studies conducted on the validity of Delphi support its contribution to accuracy. Armstrong (1985, pp. 116–120); Stewart (1987); and Rowe et al. (1991) summarize these studies. Delphi is much more accurate than unaided judgmental forecasts, especially when the unaided forecasts are made by only one or two people or where they are made in traditional group meetings.

Consider again the problem of trash collection. An impartial group of experts might be asked to predict what would happen if a fee were applied to trash containers. If the experts have direct experience with such systems in other localities or if they know the research literature on this topic, Delphi would seem to offer a reasonable way to forecast the effects of this policy.

One disadvantage of Delphi is that experts tend to be optimistic and overconfident; when they think about a problem, their confidence goes up much more rapidly than their accuracy. A tool that helps overcome this problem is the devil's advocate procedure, where someone is assigned for a *short time* to raise arguments about why the forecast or its interpretation might be wrong. The devil's advocate procedure led to more accurate forecasts in a study by Cosier (1978). Merely developing arguments against the validity of a forecast should produce a better assessment of confidence in a forecast (Koriat et al., 1980; Hoch 1985).

The variance among experts' forecasts offers a rough approximation of uncertainty (Ashton, 1985). For example, in McNees's (1992) examination of economic forecasts from 22 economists over 11 years, the actual values fell outside the range of their individual forecasts about 43 percent of the time. Little evidence exists on this topic, and it is not clear how to translate such information into a prediction interval.

For a more direct approach to an uncertainty estimate, one can ask each expert to provide 95 percent confidence intervals. However, experts are usually not well calibrated, and in some cases, about half of the estimates fall outside the 95 percent confidence intervals (Fischhoff and MacGregor, 1982; O'Connor and Lawrence, 1989). Experts are well calibrated when they receive good feedback about the accuracy of their forecasts. This issue is discussed by Plous (1993, Chap. 19); he compares the excellent calibration for weather forecasters, who receive frequent, well-summarized feedback, to the poor calibration of physicians, who receive only occasional and poorly summarized feedback.

Analogies

To forecast the outcome of interventions, it is common for experts to search for cases where similar interventions have been conducted at different times or in different geographic areas and then to generalize from them. For example, some people generalize that socialist systems' poor environmental record is evidence that government regulation harms the environment. Such an assumption is counterintuitive to other people, who point out that socialist and free-market systems differ in many ways. The key point here is that the use of analogies is fraught with dangers.

Stewart and Leschine (1986) discuss analogies with respect to risk assessments. In making a decision about an oil refinery to be established at Eastport, Maine, the analysts rejected the use of worldwide estimates of tanker spills and instead relied on a comparison with one British port, Milford Haven. Although this decision-making group believed that this was a better comparison, one can reasonably attack the use of a single site as being risky because bias could easily enter into the selection of a single analogous case. To prevent such problems, it helps to select a large number of analogous situations. In the case of oil spills, it might be possible to rate all ports for similarity (without knowledge of their oil spill rates), then select a large sample of the most similar.

To picture how analogies might be properly used in an environmental-decision process, consider the following problem. A community is considering alternative procedures for trash collection. Analogies might be useful if various trash-collection procedures had been tried in other communities and researchers had reported on the effects of these trials. Although each locality likes to think of itself as unique, a useful starting point would be to assume that people in a community would react to a given plan the same way others, on average, had reacted to similar plans in similar communities.

It is possible to structure the use of analogies by analyzing data from a sample of analogous situations. Fullerton and Kinnaman (1996) summarize some of these studies and report on the imposition of an $0.80 fee per 32-gallon can or bag of garbage in Charlottesville, Virginia. Their study examined the effects of this change on all key interest groups. The plan reduced the volume of garbage, but weight reductions were modest (only 14 percent). Illegal dumping also increased, so the true weight reduction was estimated at 10 percent. Considering administrative costs and the effects of illegal trash disposal, the program resulted in a net loss for the community. Such experience could guide a forecast of the effects of imposing such a fee in similar communities.

Conjoint Analysis

Conjoint analysis can be used to predict what strategy would be accepted. For example, one could propose different possible plans that would have various effects. The effects could be varied according to an experimental design. Once a model is developed, predictions can be made for changes in the design.

Judgmental Bootstrapping

Experts could be asked to predict the reactions to various possible interventions. A model could then be developed by regressing these predictions on the various elements of the intervention.

Extrapolation

Extrapolation involves making statistical projections using only the historical values for a time series; it is an appropriate tool to use when the causal factors will continue to operate as they have in the past. Furthermore, if one has little understanding of the causal factors, it might be best to use extrapolation.

Extrapolation has some useful characteristics. For one thing, it is fairly reliable. If agreement can be reached on the definition and length of the time series and on the statistical procedure, the same forecast will be achieved irrespective of who makes the forecast. Extrapolation can also be relatively simple and inexpensive. Although many complex procedures have been developed for extrapolation, such as the well-known Box-Jenkins methods, they have not produced gains in accuracy (Armstrong, 1984; Makridakis et al., 1993).

The opportunities for the introduction of biases in extrapolation are limited. Perhaps the major potential source of bias is that extrapolative forecasts can differ substantially depending on the time period examined. This bias can be reduced by selecting long time series and by comparing forecasts when different starting and ending points are used. Another source of bias associated with extrapolative forecasts involves the selection of the extrapolation method. To combat this bias, one should use simple, easily understood methods and preferably more than one method.

Extrapolation suffers when a time series is subjected to a shock or discontinuity. Few extrapolation methods account for discontinuities (Collopy and Armstrong, 1992b). Instead, when discontinuities occur, extrapolation will lead to large forecast errors. For example, nuclear power plant construction experienced a strong upward trend from 1960 through the mid-1970s and then a strong downward trend after that (Brown et al., 1994, p. 53). An extrapolation of nuclear power plant construction made in the early 1970s would have produced large errors.

There are many approaches to extrapolation. Most of them share the assumption that recent trends will continue. They vary primarily in how they weight the historical time periods. Exponential smoothing is widely used for this purpose. It is a moving average where the heaviest weight is placed on the most recent observation. Exponential smoothing is useful when one might expect a continuation of the forces that have operated in the past. It is less relevant for interventions, because these actions can change the direction or magnitude of the causal forces.

Exponential smoothing has several desirable qualities. First, it is simple and easy to understand. Second, it is inexpensive. And third, as noted, it puts more weight on the most recent data. However, this last benefit poses a limitation; that is, it is relevant only if the data have no seasonal effects (or have been seasonally adjusted) and if the most recent observations are free of unusual events.

Different forms of exponential smoothing have been proposed, such as Brown's model to estimate the current smoothed average, $\overline{Y}_t$ using the latest value, Y_t, and the previous smoothed average $\overline{Y}_{t-1}$.

$$\overline{Y}_t = \alpha Y_t + (1 - \alpha)\overline{Y}_{t-1}$$

A smoothing factor (alpha) of 0.4 would put 40 percent of the weight for the level onto the latest period, Y_t, and 60 percent onto all preceding periods, $\overline{Y}_{t-1}$. This value means that 24 percent of the average (0.4 times 0.6) is applied to the period immediately preceding, with weights declining exponentially as older observations are treated. A similar procedure is used for estimating trends although the smoothing parameters will differ. (Monthly or quarterly data may first need to be adjusted for seasonal effects.) The need for seasonal adjustments is obvious in many cases, such as forecasts of electric-power demand. For a detailed discussion of exponential smoothing, see Gardner (1985).

Once the quantitative extrapolations have been made, it is risky to adjust the forecast judgmentally. Nevertheless, if those making the adjustments are unbiased and have good domain knowledge, and if the adjustments are made by a group of experts following structured procedures, then the adjustments are likely to improve accuracy. Even better is to use judgmental information as inputs to a quantitative model (Armstrong and Collopy, 1998).

Prediction intervals are easy to construct for exponential smoothing. The intervals should not be based on the fit to the data but, rather, on *ex ante* forecasts. Even so, these estimates are likely to underestimate uncertainty because they assume that the effects of causal factors will be the same in the future as they have been in the past.

Rule-Based Forecasting

When one has domain knowledge and large changes are involved, rule-based forecasting can be used. Rule-based forecasting is a validated, fully disclosed, and understandable set of conditional actions to make forecasts by assigning differential weights to extrapolation forecasts. In Collopy and Armstrong (1992a), domain knowledge and forecasting expertise led to a rule base with 99 rules conditioned on 18 features of time series. The features involved characteristics of the time series (such as the presence of a significant long-term trend), the amount of variability about the trend line, the presence of an unusual last observation, and so on. Some of these features are determined by judgment, although Adya et al. (1998) obtained good results by statistically determining all but four features: causal forces, irrelevant early data, cycles, and suspicious patterns. The rules assigned weights to four extrapolation methods to produce a combined forecast. These differential weights were shown to be more accurate than equal weights (Collopy and Armstrong 1992a; Adya et al., 1998). The key points are that managers' knowledge should be applied to forecasting, and it should be done in a structured way.

Extrapolations typically ignore managers' domain knowledge. Rule-based forecasting integrates this knowledge by asking managers to describe their expectations about the future trend in a series. These expectations represent the overall effects of the various causal forces that are acting on a series. To do this, managers would be asked to classify a series as growth, decay, supporting, opposing, regressing, or unknown. The forces are listed in Table 7.3 along with some examples, and the procedure is described in Armstrong and Collopy (1993). For example, a manager's expectation that automobiles will produce less pollution per gallon of fuel (a decay series) should be reflected in the forecast. Causal forces provide a simple and inexpensive way to use domain expertise when making statistical extrapolations.

TABLE 7.3. Relationship of causal forces to trends.

Type of causal forces	Causal forces direction when		Examples
	Trend has been up	Trend has been down	
Growth	Up	Up	Gross national product; electricity consumption
Decay	Down	Down	Resource prices
Supporting	Up	Down	Short-term land prices?
Opposing	Down	Up	Wildlife
Regressing	(Toward a mean)		Demographic (% male births)

One rule for the use of causal forces is to avoid extrapolating a trend if it would be contrary to the expected trend. The expected trend is based on a specification of the causal forces affecting the series. Consider the situation in 1980, when Julian Simon made the following challenge: "Pick any natural resource and any future date. I'll bet the [real] price will not rise." He based this on long-term trends and argued that there had been no major changes in the long-term causal factors. Paul Ehrlich, an ecologist from Stanford University, accepted the challenge; he selected 10 years and five metals (copper, chromium, nickel, tin, and tungsten) whose prices had been rising in recent years. However, the causal forces for the prices of resources are "decay" because of improved procedures for prospecting; more efficient extraction procedures; lower energy costs; reduced transportation costs; development of substitutes, more efficient recycling methods; and more open trade among nations. The exhaustion of resources might lead to increased prices; however, this seldom has a strong effect because new sources are found. For example, Ascher (1978, pp. 139–141) showed that forecasts of the ultimate available petroleum reserves *increased* from the late 1940s to the mid-1970s. Such changes seem common for resources because of improvements in exploration technology. Thus, in my judgment, the overall long-term causal force is decay, so metals prices would be expected to decrease. They did, and Simon won the bet; his predictions were correct for all five metals (Tierney, 1990).

Rule-based forecasting is especially useful when domain knowledge indicates that recent trends may not persist. In the case of metals forecasting, Ehrlich assumed that recent price trends would continue. I implemented this assumption in Figure 7.2 by using Holt's exponential smoothing to extrapolate recent trends for one of his five metals, chromium, and obtained a forecast of sharply rising prices.[1] In contrast, although the rule base initially forecasts an increase in prices (because it allows that short-term trends might continue), over the 10-year horizon, the forecast becomes dominated by the long-term trend, which is downward and consistent with the causal forces (see Figure 7.2). This same pattern was found for each of the five metals for forecasts made in 1980.

Much work is currently being done on the integration of judgment and statistical forecasting. Accuracy is almost always improved if the integration uses unbiased and structured inputs and if the judgmental inputs are made independently of the statistical forecasts (Armstrong and Collopy, 1998).

---

[1] The forecasts were prepared by Monica Adya, using a version of rule-based forecasting that is described in Adya, Armstrong, Collopy, and Kennedy (1998). The data were obtained from *Metals Week*.

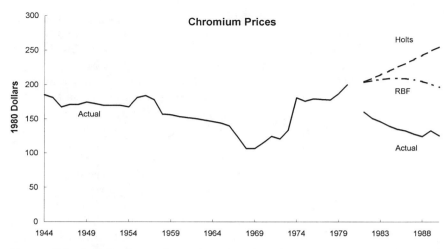

FIGURE 7.2. Actual commodity prices for chromium from 1944 to 1990 and prices forecasted for the period 1981 to 1990 by extending the trend with Holt's exponential-smoothing technique and by rule-based forecasting.

Expert Systems

Expert systems seem ideal for cases involving environmental forecasting. One can draw upon the expertise of the best experts. If econometric models have been developed, such as the above-mentioned Turner et al. (1992) study, the resulting information about relationships could be incorporated into the expert system. Further refinements could be made by quantifying experts' rules by judgmental bootstrapping. Information about citizen responses could be incorporated by using conjoint studies. Thus, expert systems allow for the systematic and explicit integration of all extant knowledge about a situation. Expert systems are being used for a variety of problems. Unfortunately, information about the predictive validity of expert systems is limited, but positive.

Econometric Models

Econometric models use information about causal relationships to make forecasts. The causal relationships should be specified by using domain knowledge (i.e. information that a manager has about the problem). Well-established theories should also be used; thus, we know that income should, in most cases, be positively related to demand for an item, and price should be negatively related. Given a description of the product and market, we can also use prior research to determine the approximate magnitude of the relationship. So if a community makes it more expensive to pollute, one would expect less pollution if the plan is properly designed, and perhaps more graft if the plan is poorly designed. Theory or domain knowledge can be used to identify key variables, specify the direction and form of the relationships, and set limits on the values that coefficients may take.

The value of well-established theories should normally take precedence over domain knowledge. For example, Winston (1993) examined 30 published studies where economists, using theory, made predictions about the effects of deregulation. The predictions in these studies conflicted with the opinions of the people affected by the deregulation. The economists predicted that deregulation would, in general, be good for consumers, whereas those who would be affected by the change predicted the opposite. As it turned out, the economists' predictions were almost always correct.

While extrapolations assume that everything continues as in the past, econometric models assume only that the relationships will remain constant. Given an estimate of the relationship of the causal variables to the dependent variable, one must forecast changes in the causal variables in order to calculate a forecast for the variable of interest. For example, to forecast changes in the level of automobile pollution, one might need to forecast the number of miles driven, average vehicle weight, average speed, engine efficiency, fuel type, and effectiveness of emissions-control equipment. If a causal variable changes direction or if it changes at a much different rate than it has in the past, the econometric model will reflect this in its forecasts. An econometric model can also estimate the effects of potential changes such as a new type of engine, a new regulation on automobiles, or a large change in the tax on gasoline.

Methods that do not use domain knowledge or theory, such as step-wise regression, should not be used for forecasting (Armstrong, 1985, pp. 223–225). Not only might the analysts ignore useful information, but also they may be misled by spurious correlations. Also, I see little hope for neural nets, despite their current popularity. Chatfield (1995) discusses the limitations of neural nets.

Econometric methods are most useful where (1) large changes are expected; (2) *a priori* information about relationships is strong; (3) good data are available; and (4) causal factors are easier to forecast than the variable of interest. These conditions are often encountered in forecasting. For example, over the long-term, the effects of air pollution are likely to be substantial. Studies about the causes of pollution provide a reasonable level of knowledge about the causal relationships. Some of the causal variables, such as population and production of various goods, can be forecast more easily than one could directly forecast air pollution. In addition, relationships can sometimes be estimated by laboratory or field experiments, as illustrated by Turner et al. (1992). Finally, the quality of the data is improving. Not surprisingly, then, many researchers use econometric methods to forecast various types of environmental impacts.

One of the major advantages of econometric methods, in comparison with other forecasting methods, is that alternate interventions can be compared with one another. In effect, one is comparing results in an objective way with an attempt to hold all other influences constant. Turner et al. (1992) used this approach in their 50-year forecasts of acid rain to examine the effects that different policies might have.

The technology for econometric methods has become much more complicated since the least-absolute-value method was introduced in 1757, followed by the least-squares method in 1805. But highly complex procedures are not easily understood by decision makers. Worse, little validation research has been conducted on complex procedures. What has been done suggests that complexity seldom leads to improved accuracy. Dielman (1986), using simulated data, concluded that the least-absolute-value method still works well, especially for data that suffer from outliers. In general, theory-based single-equation ordinary least-squares methods have forecast well when compared with alternative procedures.

Simple econometric models aid understanding and reduce the potential for errors. The benefits seem to translate into practice: In a field study involving the forecasting of state-government revenues, the use of simple econometric models was associated with improvements in accuracy, while the use of complex econometric models was associated with reduced accuracy (Bretschneider et al., 1989).

Mechanical adjustments are often necessary to adjust for errors in the current status (i.e., to adjust the starting value). For example, one useful procedure is to add half of the latest error to the forecast. Once an econometric forecast has been prepared, generally speaking, it should not be adjusted judgmentally (Armstrong and Collopy 1998). However, if there has been a major recent event that has not yet been reflected in the data, structured judgmental procedures might be used to adjust the level.

Econometric methods can aid in the construction of confidence intervals. Such intervals are expected to be underestimated if, as is almost always the case, they make no provision for the uncertainty involved with predicting causal variables or for the possibility that relationships might change. Thus, the use of the traditional standard error of a model as the foundation for estimating prediction intervals should be supplemented by other approaches. This practice is illustrated by Turner et al. (1992), who compared forecasts from different models and also compared forecasts given different assumptions about the forecasts of the causal variables. In addition, they examined limitations of the models, such as the effects of excluded variables. Excluded variables seemed to be a serious limitation, although their effects were not quantified. They then tested their model in different geographic regions. Finally, they tested the model by making long-term backcasts to prehistoric times and compared the results with independently obtained lake-chemistry estimates.

If many communities have tried different plans for trash disposal, an econometric model might be estimated to predict the outcomes of various plans. The econometric model could help to control for differences among communities and also for factors that change with time. Such a model would aid in determining the effects of alternative trash fees.

Selecting and Combining Forecasts

Assuming that data exist for using each of the above forecasting methods, which method should be used? If factors that caused changes in the past continue to operate in the same way in the future, the choice of a method is not so important; each method would be expected to have reasonably good accuracy. But given the large changes expected in many environmental problems, the selection of a forecasting method is likely to be important.

Judgment is helpful for estimating levels, while extrapolation and econometric models are better at forecasting changes. Extrapolation is good at forecasting changes when the causal factors continue to operate as in the past, whereas econometric models can compensate for substantial changes in the causal forces. As a result, econometric forecasts are generally more accurate than extrapolation or judgment when large changes are involved (Armstrong, 1985, pp. 391–420). Fildes's (1985) review adds further support and also suggests that econometric models provide small improvements in accuracy for short-range forecasts. Note that these studies were, for the most part, conducted in situations that did not involve environmental forecasting. However, Ascher (1978, p. 119), in examining 10-year forecasts of electricity consumption, concluded that extrapolation and econometric methods were more accurate than judgment. Also, Rausser and Oliveira (1976), in a study of wilderness-area use, found that econometric methods were more accurate than extrapolations and that a combination of econometric forecasts was even more accurate. In general, assuming adequate data and a good understanding of causal relationships, econometric methods would be the preferred forecasting method because they use much relevant information in a structured way.

Rather than trying to chose the single best method, the problem is better framed by asking which methods would help to improve accuracy. Baker et al. (1980) illustrated this use of multiple forecasting methods. They forecasted the impact that offshore nuclear power plants would have on beach visitation by using expert surveys, analogies (visits to beaches near land-based nuclear plants), and intentions of potential beach visitors.

Combined forecasts are those where one uses different methods to make forecasts for the same situation and then combines the forecasts. Combined forecasts are especially useful where much uncertainty exists about which method is likely to produce the most accurate forecasts. Combined forecasts typically improve accuracy because each forecasting method makes some contribution.

Much research suggests that combined forecasts are generally more accurate than forecasts prepared with a single method. Furthermore, they are sometimes more accurate than the best component. Combining also offers protection against mistakes because their effects are muted by the other forecasts. Finally, combining forecasts from different methods and data will add to objectivity and to the appearance of objectivity.

Combining of forecasts should be done mechanically to help assure users that the procedure is objective. That is, a rule should be used, and it should be fully described. An example would be "equal weights," which states that one adds each of the forecasts and calculates an average. This objectivity in the weighting process is expected to improve accuracy, and equal weights is robust across situations (Clemen 1989). For example, Bretschneider et al. (1989), in a field study, found that U.S. states that used mechanical combinations of forecasts had more accurate revenue forecasts than those using subjective combinations.

Uncertainty

The assessment of uncertainty in forecasting should not include tests of statistical significance because they do not relate well to issues of importance in forecasting and because they are so often misinterpreted (McCloskey and Ziliak, 1996). Instead, one can provide prediction intervals. The prediction interval represents the proportion of times that the actual forecasts are expected to fall within a specified range. Thus, 95 percent prediction intervals should be expected to contain 95 percent of the true values.

Estimates of prediction intervals might be obtained by comparing forecasts from different methods. While agreement inspires more confidence and disagreement less, the translation of these differences to prediction intervals must be done subjectively. Some improvements might be achieved if the prediction intervals are estimated independently by a number of approaches, and the estimates are then combined mechanically.

To develop prediction intervals, it is generally best to make forecasts by simulating the situation facing the forecaster and then calculating *ex ante* forecast errors that can be used to construct prediction intervals for each forecast horizon. The resulting limits can be smoothed over the forecast horizon.

Using the Forecasts

Interestingly, researchers, educators, forecasters, and decision makers all use similar criteria for judging which forecasting models are most useful (Yokum and Armstrong, 1995). Accuracy is generally rated as the most important criterion. These experts also agree that ease of understanding and ease of use are nearly as important as accuracy. These agreements on criteria suggest that it might be possible to reach agreement of what *forecasting methods* perform best for a given situation.

Forecasts are used by decision makers, politicians, special-interest groups, manufacturers, lawyers, and the media. Given their different needs, they may desire different forecasts. So agreement on *forecasts* is a difficult matter, especially if no prior agreement has been reached with the decision makers about the proper forecasting methods and if the forecast is surprising to some.

Unfortunately, adjustments to forecasts are often made by biased experts. In a survey of members of the International Institute of Forecasters, respondents (n = 269) were given the following statement: "Too often, company forecasts are modified because of political considerations." On a scale from 1 = "disagree strongly" to 7 = "agree strongly," the average response was 5.37.[2] Fildes and Hastings (1994), in an intensive study of forecasting in a large multidivision firm, found 64 percent of their respondents agreeing that forecasts are frequently modified for political reasons.

Subjective adjustments may expose the forecaster to charges of bias. Glantz (1982) describes how a subjective adjustment of a weather forecast led to the prediction of an extreme drought in Yakima in 1977. Farmers took appropriate action for such a drought, but, as it turned out, the drought did not occur. Serious losses resulted from the farmers' actions. As a result, the farmers sued the government, and the subjective adjustment was challenged as evidence of malpractice.

Often, the most useful forecasts tell us something new, that is, they challenge existing expectations. They are valuable because they allow us to take corrective action. However, such forecasts are frequently ignored or resisted by organizations, as found in Griffith and Wellman's (1979) study of the need for hospital beds. Fortunately, there are tools to aid users of forecasts in such situations. These tools involve presentation techniques and scenarios.

Presentation Techniques

If decision makers are biased to favor certain forecasts, it would seem natural for them to suspect that the forecasting tools are inadequate when forecasts are not to their liking. Thus, in presentations to decision makers, do not start off with the forecasts. The initial emphasis should be on the

---

[2] This survey was conducted in 1989 by Thomas Yokum and me. The responses to this question were analyzed depending on whether the respondents identified himself or herself as primarily a decision maker, practitioner, educator, or researcher. While the practitioners stated the strongest agreement, there were no statistically significant differences among these groups.

forecasting *methods* in order to gain agreement on which methods are appropriate.

Forecasts involve uncertainty, so prediction intervals should be presented along with forecasts. Planners can then prepare contingencies depending on the range of possible futures. Unfortunately, it is common to report only expected values and not prediction intervals. Rush and Page (1979) examined 372 long-term metals forecasts from 1910 to 1964. Explicit references to uncertainly were not the rule. Furthermore, their use decreased from 22 percent of the forecasts published up to 1939 to only eight percent afterward. In a survey of "marketing/forecasting managers" at 134 U.S. firms (Dalrymple, 1987), fewer than 10 percent said that they usually used prediction intervals, and almost half said that they never used them.

Although the capability exists to provide entertaining graphics along with forecasts, these may not improve the message. Wagenaar et al. (1985) compared the delivery of weather forecasts by radio and TV. They concluded that recall was not improved by TV except when written summary statements were also provided on TV.

People tend to understand and remember examples. Thus, it makes sense to reinforce your forecasts with vivid examples. In a study about firefighters' preference for risk, Anderson (1983) found that the use of concrete examples was more effective than was the presentation of relevant statistical results. An experiment by Read (1983) suggests that politicians may be more influenced by a single historical event (e.g., "Here is what happened at Three-mile Island") than by a generalization from a broad range of situations.

Scenarios

Forecasts that call for changes are often resisted by decision makers. For example, Baker (1979) found that hurricane warnings are frequently ignored. People do not like to receive information about the potential destruction of their homes, and they tend to revise the forecast to make it less threatening.

The use of scenarios can help decision makers deal with forecasts that have unpleasant consequences. A scenario is a story about what happened in the future. The choice of tenses in the preceding sentence was intentional. The use of the past tense helps to add realism and to gain the decision maker's commitment to a course of action for a given forecast.

Scenarios are likely to lead people to overestimate the likelihood of an event. Certainly, the forecaster should point out that the function of the scenario is not to forecast but to decide how to use forecasts. The fact that the event will seem more likely should aid people to take unfavorable forecasts more seriously.

Research by psychologists suggests that effective scenarios:

- Use concrete examples
- Include events that are representative of what one might expect
- Link the events by showing causal relationships
- Ask the decision makers to project themselves into the situation
- Ask the decision makers to describe how they acted in this scenario

Additional details on the scenario procedure are provided in Armstrong (1985, pp. 40–45).

Scenarios allow decision makers to report (predict) how they would act given certain future circumstances. People's responses about their intentions may affect their subsequent behavior (Greenwald et al., 1987). Such predictions of behavior have been shown to affect people's behavior when similar situations have been encountered later (Gregory et al., 1992). The generation of scenarios might also be beneficial because people tend to predict that they will act in socially responsible and rational ways. Thus, by asking citizens to describe how they would react to a new trash disposal plan, one might affect the respondents' behavior in such a way as to increase the plan's chances for success if they liked the plan.

Innovations

Innovative methods for forecasting are currently available. The problem is not that tools are lacking, but rather that the methods are not being used. Two innovations are worthy of note: One is software that can incorporate the latest forecasting methods. Another is auditing procedures that can identify areas where forecasting methods are not being applied or are being applied improperly.

Software

Software can incorporate the latest findings about forecasting. Thus, once the software is selected, one is guided by this research and would have to take specific action not to use it. Unfortunately, the providers of forecasting software have been slow to incorporate new findings.

Software can also help to reduce error in the application of the method. This assumes that the software is correct, I have always been surprised to note that users often fail to notice serious errors in software.

Auditing Procedures

Given the potential for manipulation of the forecasts and the high likelihood that unintended consequences may arise from interventions, forecasting procedures should be audited by one or more independent groups of people with expertise in forecasting methods. Environmental-forecasting audits could be made public to add to their impact and to encourage properly conducted forecasts.

Stewart and Glantz (1985) conducted an audit of the National Defense University's forecast of global climates by comparing the forecasting process with findings from research on judgmental forecasting. This audit revealed a number of instances where the procedures of this expensive project were not in agreement with research findings about the best procedures.

The success of the audit can be enhanced by the use of a structured guide. Table 7.4 provides a framework for auditing forecasts. It covers the data, forecasting methods, uncertainty, and considerations for using the forecasts.

Full disclosure of the forecasting procedure is important to allow for replication. A failure to provide such disclosure should be cause for concern. For example, when Stewart and Glantz (1985) attempted to get details on the above-mentioned study of global climate, they were informed

TABLE 7.4. Checklist for environmental forecasting.

Chosen Forecasting Method:	Procedure Follows Best Practice		
	No	?	Yes
Data			
• Objective procedures used to select the relevant data?	—	—	—
• Long time-series used?	—	—	—
• Test sensitivity or data selection (e.g., vary start and finish periods)	—	—	—
• Full disclosure provided?	—	—	—
Forecasting Methods			
• Separate documents prepared for forecasts and plans?	—	—	—
• Forecast independent of key interest groups?	—	—	—
• Forecast used objective methods?	—	—	—
• Structured techniques used to obtain judgments?	—	—	—
• More than one method used to obtain forecasts (e.g., Delphi)?	—	—	—
• Forecasts combined mechanically?	—	—	—
• Forecasts free of unstructured judgment?	—	—	—
• Forecasts prepared for alternate interventions?	—	—	—
• Econometric forecasts not revised subjectively?	—	—	—
Uncertainty			
• Provide prediction intervals	—	—	—
• Do not use tests of statistical significance	—	—	—
• Objective procedures used to assess uncertainty?	—	—	—
• Alternative procedures used to assess uncertainty?	—	—	—
• Arguments against each forecast listed?	—	—	—
Using Forecasts			
• Gain agreement on *methods* prior to discussing forecasts?	—	—	—
• Users understand the forecasting methods?	—	—	—
• Forecast presented in a scenario?	—	—	—

that the raw data had been destroyed, and their request about details of the procedure went unanswered.

It is likely that those who use better methods are also aware of the importance of full disclosure in scientific studies, so they will also be better at reporting on the methods they use. Perhaps for this reason, Weimann (1990) found that the more methodological deficiencies that were reported in a political poll, the more accurate the poll was.

Conclusions

This chapter has described tools to improve forecasting of trends and of the effects of interventions. Among these methods, role-playing and rule-based forecasting have seldom been used for environmental forecasting. Role-playing is appropriate when forecasting the outcome of a situation involving conflict among various parties. Rule-based forecasting is relevant when the forecasters have time-series data and relevant domain knowledge, and it is especially useful when recent trends conflict with expectations.

A key theme running throughout these procedures is objectivity. The attainment of objectivity is critical in the use of judgment. The route to objectivity is through structure, and it is enhanced by using quantitative methods, using structured judgment as an input to these quantitative methods, providing full disclosure, and employing auditing procedures, such as review panels.

Acknowledgments. Helpful comments were received on earlier drafts from John Carstens, Fred Collopy, Virginia Dale, Fred O'Hara, Howard Kunreuther, Robb Turner, and Dara Yang.

Key Resources

For a general description of quantitative forecasting methods and how to apply them, see Makridakis, S. Wheelwright, S.C., and Hyndman, R.J. 1998. *Forecasting: Methods and Applications.* New York: John Wiley and Sons.

For a review of judgmental methods, see Wright, G. and Goodwin, P. 1998. *Forecasting with Judgment.* Chichester, England: John Wiley and Sons.

For a review of the evidence on which methods (judgmental and quantitative) are appropriate for a given situation, see Armstrong, J.S. 1985 (2nd ed.). *Long-Range Forecasting: from Crystal Ball to Computer.* New York: John Wiley and Sons.

If the evidence draws upon the intentions or opinions of a large number of people, the following book provides the best summary of the evidence on survey research: Dillman, D. 1978. *Mail and Telephone Surveys.* New York: John Wiley.

For a general overview of forecasting principles, see Armstrong, J.S. (Ed.), forthcoming 1999 *Principles of Forecasting: A Handbook for Researchers and Practitioners*. Norwell, MA: Kluwer.

Research from many fields has led to substantial improvements in forecasting methods, especially since 1960. These findings are translated into principles for use by researchers and practitioners in the "Forecasting Principles Project," located at the website http://www-marketing.wharton.upenn.edu/forecast. This website will report continuing developments in forecasting.

References

Adya, M. Armstrong, J.S., Collopy, F., and Kennedy, M. 1998. Automatic identification of time series features for rule-based forecasting. (Working paper), Cleveland, OH: The Weatherhead School, Case Western Reserve University.

Anderson, C.A. 1983. Abstract and concrete data in the perseverance of social theories: when weak data lead to unshakeable beliefs. *Journal of Experimental Social Psychology* 19:93–108.

Armstrong, J.S. and Overton, T. 1977. Estimating nonresponse bias in mail surveys. *Journal of Marketing Research* 14:396–402.

Armstrong, J.S. 1984. Forecasting by extrapolation: Conclusions from 25 years of research. *Interfaces* 14(Nov./Dec.):52–66.

Armstrong, J.S. 1985. *Long-range Forecasting*, 2nd ed. New York: John Wiley and Sons.

Armstrong, J.S. 1987. Forecasting methods for conflict situations. In: G. Wright and P. Ayton (Eds.). *Judgmental Forecasting*. Chichester, England: John Wiley and Sons. Pp. 157–176.

Armstrong, J.S. and Collopy, F. 1993. Causal forces: Structuring knowledge for time series extrapolation. *Journal of Forecasting*. 12:103–115.

Armstrong, J.S. and Collopy, F. 1998. Integration of statistical methods and judgment for time series forecasting: Principles from empirical research. In: G. Wright and P. Goodwin. *Forecasting with Judgment*. Chichester, England: John Wiley and Sons. Pp. 269–293.

Armstrong, J.S. and Hutcherson, P. 1989. Predicting the outcome of marketing negotiations: Role-playing versus unaided opinions. *International Journal of Research in Marketing* 6:227–239.

Ascher, W. 1978. *Forecasting: An Appraisal for Policy Makers and Planners*. Baltimore: Johns Hopkins University Press.

Ashton, A.H. 1985. Does consensus imply accuracy in accounting studies of decision making? *Accounting Review* 60:173–185.

Ashton, A.H. and Ashton, R.H. 1985. Aggregating subjective forecasts: Some empirical results. *Management Science* 31:1499–1508.

Baker, E.J. 1979. Predicting responses to hurricane warnings: A reanalysis of data from four studies. *Mass Emergencies* 4:9–24.

Baker, E.J. et al. 1980. Impact of offshore nuclear power plants: Forecasting visits to nearby beaches. *Environment and Behavior* 12:367–407.

Blattberg, R.C. and Hoch, S.J. 1990. Database models and managerial intuition: 50 percent model + 50 percent manager. *Management Science* 36:887–899.

Brenner, L.A., Koehler, D.J., and Tversky, A. 1996. On the evaluation of one-sided evidence. *Journal of Behavioral Decision Making* 9:59–70.

Bretschneider, S.I. et al. 1989. Political and organizational influences on the accuracy of forecasting state government revenues. *International Journal of Forecasting* 5:307–319.
Brown, L.R., Kane, H., and Roodman, D.M. 1994. *Vital Signs 1994.* New York: Norton.
Buchannan, W. 1986. Election predictions: An empirical assessment. *Public Opinion Quarterly* 50:222–227.
Cerf, C. and Navasky, V. 1984. *The Experts Speak.* New York: Pantheon.
Chatfield, C. 1995. Positive or negative? *International Journal of Forecasting* 11:501–502.
Clemen, R. 1989. Combining forecasts: A review and annotated bibliography. *International Journal of Forecasting* 5:559–583.
Collopy, F. and Armstrong, J.S. 1992a. Rule-based forecasting: Development and validation of an expert systems approach to combining time series extrapolations. *Management Science* 38:1394–1414.
Collopy, F. and Armstrong, J.S. 1992b. Expert opinions about extrapolation and the mystery of the overlooked discontinuities. *International Journal of Forecasting* 8:575–582.
Cosier, R.A. 1978. The effects of three potential aids for making strategic decisions on prediction accuracy. *Organizational Behavior and Human Performance* 22:295–306.
Dalrymple, D.J. 1987. Sales forecasting practices: Results from a United States survey. *International Journal of Forecasting* 3:379–391.
Dielman, T.E. 1986. A comparison of forecasts from least absolute value and least squares regression. *Journal of Forecasting* 5:189–195.
Fildes, R. 1985. The state of the art: Econometric models. *Journal of the Operational Research Society* 36:549–586.
Fildes, R. and Hastings, R. 1994. The organization and improvement of market forecasting. *Journal of the Operational Research Society* 45:1–16.
Fischhoff, B. and MacGregor, D. 1982. Subjective confidence in forecasts. *Journal of Forecasting* 1:155–172.
Fullerton, D. and Kinnaman, T.C. 1996. Household responses to pricing garbage by the bag. *American Economic Review* 86:971–984.
Gardner, E.S. 1984. The strange case of the lagging forecasts. *Interfaces* 14:47–50.
Gardner, E.S. 1985. Exponential smoothing: The state of the art (with commentary). *Journal of Forecasting* 4:1–38.
Glantz, M.H. 1982. Consequences and responsibilities in drought forecasting: The case of Yakima, 1977. *Water Resources Research* 18:3–13.
Greenwald, A.G. et al. 1987. Increasing voting behavior by asking people if they expect to vote. *Journal of Applied Psychology* 72:315–318.
Gregory, W.L., Cialdini, R.B., and Carpenter, K. 1982. Self-relevant scenarios as mediators of likelihood estimates and compliance: Does imagining make it so? *Journal of Personality and Social Psychology* 43:88–99.
Griffith, J.R. and Wellman, B.T. 1979. Forecasting bed needs and recommending facilities plans for community hospitals: A review of past performance. *Medical Care* 17:293–303.
Hoch, S.J. 1985. Counterfactual reasoning and accuracy in predicting personal events. *Journal of Experimental Psychology: Learning, Memory, and Cognition* 11:719–731.

Hogarth, R.M. 1978. A note on aggregating opinions. *Organizational Behavior and Human Performance* 21:40–46.

Koriat, A., Lichtenstein, S., and Fischhoff, B. 1980. Reasons for confidence. *Journal of Experimental Psychology: Human Learning and Memory* 6:107–118.

Larreche, J. and Moinpour, R. 1983. Managerial judgment in marketing: The concept of expertise. *Journal of Marketing Research* 20:110–121.

Lau, R.R. 1994. An analysis of the accuracy of "trial heat" polls during the 1992 presidential election. *Public Opinion Quarterly* 58:2–20.

Lemert, J.B. 1986. Picking the winners: Politician vs. voter predictions of two controversial ballot measures. *Public Opinion Quarterly* 50:208–221.

Libby, R. and Blashfield, R.K. 1978. Performance of a composite as a function of the number of judges. *Organizational Behavior and Human Performance* 21:121–129.

Lowenstein, G. and Frederick, S. 1997. Predicting reactions to environmental change. In: M.H. Bazerman et al. (Eds.). *Environment, Ethics, and Behavior: The Psychology of Environmental Valuation and Degradation*. San Francisco: New Lexington Press. Pp. 52–72.

MacGregor, D.G. 1999. Decomposition for judgmental forecasting and estimation. In: J.S. Armstrong (Ed.). *Principles of Forecasting: A Handbook for Researchers and Practioners*. Norwell, MA: Kluwer Academic Publishers.

Makridakis, S. et al. 1982. The accuracy of extrapolation (time series) methods: Results of a forecasting competition. *Journal of Forecasting* 1:111–153.

Makridakis, S. et al. 1993. The M2-Competition: A real-time judgmentally based forecasting study. *International Journal of Forecasting* 9:5–22.

McCloskey, D.N. and Ziliak, S.T. 1996. The standard error of regressions. *Journal of Economic Literature* 34:97–114.

McNees, S.K. 1992. The uses and abuses of consensus forecasts. *Journal of Forecasting* 11:703–710.

O'Connor, M. and Lawrence, M. 1989. An examination of the accuracy of judgmental confidence intervals in time series forecasting. *Journal of Forecasting* 8:141–155.

Perry, P. 1979. Certain problems with election survey methodology. *Public Opinion Quarterly* 43:312–325.

Plous, S. 1993. *The Psychology of Judgment and Decision Making*. New York: McGraw-Hill.

Rausser, G.C. and Oliveira, R.A. 1976. An econometric analysis of wilderness areause. *Journal of the American Statistical Association* 71:276–285.

Read, S.J. 1983. Once is enough: Causal reasoning from a single instance. *Journal of Personality and Social Psychology* 45:323–334.

Rowe, G., Wright, G., and Bolger, F. 1991. The Delphi technique: A re-evaluation of research and theory. *Technological Forecasting and Social Change* 39(3):235–251.

Rush, H. and Page, W. 1979. Long-term metals forecasting: The track record: 1910–1964. *Futures* 11:321–337.

Shamir, J. 1986. Pre-election polls in Israel: Structural constraints on accuracy. *Public Opinion Quarterly* 50:62–75.

Squire, P. 1988. Why the 1936 *Literary Digest* poll failed. *Public Opinion Quarterly* 52:125–133.

Stewart, T.R. 1987. The Delphi technique and judgmental forecasting. *Climatic Change* 11:97–113.

Stewart, T.R. and Glantz, M.H. 1985. Expert judgment and climate forecasting: A methodological critique of "Climate Change to the Year 2000." *Climatic Change* 7:159–183.

Stewart, T.R. and Leschine, T.M. 1986. Judgment and analysis in oil spill risk assessment. *Risk Analysis* 6(3):305–315.

Sudman, S. and Bradburn, N.M. 1982. *Asking Questions: A Practical Guide to Questionnaire Design.* Jossey-Bass, San Francisco.

Tierney, J. 1990. Betting the planet. *New York Times Magazine*, p. 52, December 2.

Timmers, H. and Wagenaar, W. 1977. Inverse statistics and the misperception of exponential growth. *Perception and Psychophysics* 21:558–562.

Turner, R.S. et al. 1992. Sensitivity to change for low-ANC eastern U.S. lakes and streams and brook trout populations under alternate sulfate deposition scenarios. *Environmental Pollution* 77:269–277.

Wagenaar, W.A. and Sagaria, S.D. 1975. Misperception of exponential growth. *Perception and Psychophysics* 18:416–422.

Wagenaar, W.A., Schreuder, R., and van der Heijden, A.H.C. 1985. Do TV pictures help people to remember the weather forecast? *Ergonomics* 28:765–772.

Wagenaar, W.A. and Timmers, H. 1978. Extrapolation of exponential time series is not enhanced by having more data points. *Perception and Psychophysics* 24:182–184.

Wagenaar, W.A. and Timmers, H. 1979. The pond-and-duckweed problem: Three experiments in the misperception of exponential growth. *Acta Psychologica* 43:239–251.

Weimann, G. 1990. The obsession to forecast: Pre-election polls in the Israeli press. *Public Opinion Quarterly* 54:396–408.

Winston, C. 1993. Economic deregulation: Days of reckoning for microeconomists. *Journal of Economic Literature* 31:1263–1289.

Yokum, J.T. and Armstrong, J.S. 1995. Beyond accuracy: Comparison of criteria used to select forecasting methods. *International Journal of Forecasting* 11:591–597.

Decision-Maker Response

JULIA A. TREVARTHEN

Forecasting is integral to planning and environmental decision making. If offers a glimpse of the future that is based upon more than rhetoric and adds value to the environmental dicision-making process.[1] It is a tool for practitioners to use in developing, clarifying, and refining alternative recommendations for action. Decision makers consider forecasts along with other data and analysis in choosing between alternative recommendations. At their most useful, forecasts help clarify the potential outcomes and impacts of a particular decision on all interest groups and affected systems.

Armstrong's discussion of forecasting methods, their strengths and limitations, and of the issues to consider when choosing among them, is complete and concise. The methods discussed are those commonly used in practice. The discussion and presentation are particularly useful for practitioners who are not full-time forecasters, but who still need to understand the appropriate use of forecasting methods.

The Role of Forecasting in Decision Making

Forecasting can plan an integral role in all types of public and private decision-making processes. Many practitioners assist decision makers in making different types of decisions at different scales simultaneously. They need tools that are accurate, effective, simple, timely, and understandable. Forecasting can be such a tool. Whether the decision to be made deals with large-scale ecosystem restoration or permitting a single lot-development project, forecasting can help to illuminate the effects or the choices.

Planning and environmental decisions are typically made in an iterative fashion. Forecasting can play an important part in each iteration. On the public side, governments make plans for future land use, capital investment, service provision, and community character. Resource agencies make plans to restore, preserve, and manage natural systems. Transportation agencies make plans to build roads or provide transit. Forecasts help predict the effects of planning and management decisions. Some state growth-

---

[1] Although the tools discussed throughout this book are framed in the context of environmental decision making, they can be equally effective in a variety of public-policy, planning, and growth-management decision-making processes.

management systems legislatively mandate that certain forecasts be used in planning for the future. For example, in Florida's growth-management system, forecasting methods for assessing the impact of development on the natural and built environments are codified in legislation and administrative rules.

On the private side, developers forecast potential economic return and market viability in deciding what, where, and when to develop. Once a development application is submitted, public review and permitting agencies assess and forecast potential impacts, both positive and negative, to develop recommendations for the final decision makers. Decision makers review the forecasted impacts and proposed mitigation strategies as part of the decision process. Once a decision is made, developers reassess their own forecasts based on the conditions of approval to decide whether to appeal the decision or move forward with the project.

The useful life of a forecast does not end with a decision and its subsequent implementation. The forecast can be updated and revised in light of experience, changing circumstances, and additional data. Both public and private planning and decision-making processes are repeated as plans are periodically revised and updated.

The Limitations of Forecasting

Forecasts are used (and misused) to incite people to action. A forecast that predicts a crisis, such as rapid population change or inadequate water supply, whether legitimate or not, can provide the initial impetus to plan. However, a forecast that has no context or analysis accompanying it is merely a number. Do not rely on a particular forecast as the singular basis for choosing or rejecting a course of action. Forecasting is only one tool of many that should be routinely used and considered in making decisions.

Most environmental and public-policy decisions are complex by definition. They are characterized by complicated and dynamic systems that may be as severely impacted by the cumulative effect of numerous individual decisions as they are by sweeping changes in public policy. Few decision makers have jurisdiction over an entire system. Faced with a choice between equally compelling yet competing interests, relying on that which can be quantified for the "answer" can be seductive.

Choosing a Forecasting Method—Some Practical Considerations

First, know your limitations and get some advice. If forecasting is not your specialty, consult an expert. Resist any temptation to assume that the most complex (or most expensive) method will yield the most accurate results.

Growth management and environmental issues are complex enough without introducing additional obfuscation. Seriously consider combining forecasting methods. Beyond improving accuracy, combining forecasts gives practitioners the ability to present multiple reasons why a particular intervention may or may not be successful. Most importantly, pick the right methods for the problem. Consider using a forecast audit group that includes knowledgeable representatives from all affected interest groups to begin building consensus-based recommendations for interventions from the earliest opportunity.

Using impartial experts in a Delphi survey can be helpful in some purely practical ways. First, involving experts on a particular issue can broaden the range of possible alternatives to consider. Second, experts can provide a degree of objectivity and distance that cannot always be guaranteed when those who have a stake in the outcomes are formulating the choices. And third (and only somewhat facetiously), do not underestimate the cachet that using out-of-town experts can bring to the discussion.

Intention surveys are valuable in that they provide the perspective of those who will actually make the decision, but they have limited applicability in many environmental decisions. Environmental decisions, particularly those to be made by elected decision makers, are often made in a public process, completely open to interest groups, the public, and the media. Sometimes elected decision makers are reluctant to reveal their intentions early in the process. Moreover, the ultimate success of many environmental interventions relies upon the individual decisions of those affected, rather than the intentions or actions of decision makers. Furthermore, large-scale intention surveys can be less costly.

Role-playing is an intriguing approach to teasing out how interested parties might react in response to particular decisions. It offers a more meaningful way than opinion polling to test potential interventions. Public policy and environmental decisions are usually accompanied by conflict among affected groups. Over time, positions can harden, making creative and innovative solutions harder to achieve. Role-playing can help to jostle individuals from much-beloved positions by altering their point of view and asking them to interact with others.

Given the intricate nature of most environmental problems, econometric models are generally more useful than extrapolation. Econometric models enable the practitioner to compare the effects of alternate interventions over time. This is important, especially when making public decisions that will have long-term fiscal and environmental repercussions.

Communicating the Forecast

To participate effectively in any decision-making process, practitioners must be able to justify the use of an analytic tool concisely and convincingly to those who will make the ultimate decision and to those who will live with the result. Truly effective practitioners are also translators who can adapt their discussion to fit the needs of the audience.

Practitioners must be able to discuss the technical details of potential forecasting tools with other practitioners, consultants, and regulators to build consensus among the affected parties on the appropriate tools and to agree upon specific methods and the ultimate use for the results. Then they must be able to concisely explain to decision makers the reason for the forecast, the validity of the method, and the degree of confidence in the result. Throughout the decision-making process, practitioners must also be able to communicate to interest groups, the general public, and the media the reasoning behind the recommendation.

In practice, forecasts are rarely presented alone. Instead, they are a part of an entire package of data, analysis, and recommendations that is presented to decision makers. The practitioner's challenge is to make the package accurate, clear, and meaningful to the audience. Several techniques can be helpful in presenting the results of forecasts and other tools for decision making.

Well-chosen, thoughtfully presented examples and analogies can be some of the most efficient ways to convey concepts, options, and potential outcomes. Using several examples is usually more effective than relying on just one. Using a number of examples can help decision makers to move past the "my situation is globally unique" response.[2]

Scenarios can be a great way to set the scene for significant change. They can help decision makers imagine the future outcomes, both positive and negative, of current decisions. Scenarios can also illustrate the consequences of deciding not to intervene.

Graphics can breathe life into abstract public-policy discussions and can often make the point faster than pages of text. Increasingly, tools like geographic information systems are being used to illustrate the expected outcomes of particular interventions. You should choose graphics as carefully as you choose forecasting methods. Know what the graphic needs to illustrate and how it can best present the message. Designing and creating graphics is another area in which it is useful to know your limitations and when to consult experts. Once the graphic is designed, test it on people who are not part of the study. Find out what it says to them (if anything) and adjust the graphic accordingly until the message is both accurate and clear.

---

[2] Use examples and analogies carefully and do not forget that the devil is in the details. I once watched a citizen planning group reject a potential neighborhood-design strategy because the slides used to illustrate the concept had been taken elsewhere and reflected that other region's predominant architecture, building materials, color choices, and landscaping. It is sometimes difficult to see past surface differences to underlying similarities.

Conclusion

Forecasting can help practitioners and decision makers answer the question "what if . . ." It can bolster a recommendation or provide a basis for argument. Like everything else in planning, it can be argued endlessly to delay or challenge a decision. Forecasting is an important tool for environmental decision making, but it should not be used in isolation.

8
Assessment, Refinement, and Narrowing of Options

MILEY W. MERKHOFER

"Deliberate as often as you please, but when you decide it is once for all."
Publilius Syrus, *Sententiae*, No. 132

Suppose you are a manager who must decide which of your employees to promote to an important, decision-making position. You chose your two top performers as candidates and evaluate their decision-making styles. One, you discover, makes decisions with logic. She collects the relevant information, analyzes the options, and assesses the uncertainties. Her choices nearly always produce good results. Occasionally, however, things have turned out less than perfect because of circumstances impossible to foresee. The other candidate, you find, has had a string of remarkable successes, but bases all of his choices on the flip of a "special" 1964 quarter that was left in his office by a previous employee. What should you do?

Answer: Promote the logical one. Fire the coin-flipper. Keep the quarter.

Terminology

Decision making is the process of making decisions. A decision is an irrevocable allocation of resources. For example, a purchase decision occurs when money changes hands or when a contract is signed. A promotion decision occurs when it is announced to the employee and to the organization. The resources that may be allocated by a decision include money and time (as in a purchase decision) and decision-maker credibility (as in a promotion decision). The allocation is irrevocable in the sense that the choice cannot be altered without at least some additional cost. For example, a buyer who tries to break a purchase contract will invest time, will incur the ill will of the seller, and may have to pay monetary penalties. A manager who demotes a recently promoted employee loses credibility, damages morale, and may adversely impact the employee's chances of succeeding within the organization. The costs of reversing decisions are often high. Some decisions cannot be reversed at any cost. As a result, decision makers must often live with their poor choices.

The goal of decision making is to achieve good decision outcomes. A good decision outcome is one that is desired by decision makers. For example, a purchaser is generally happy when the purchased product performs as well as or better than advertised. Because of uncertainty, there is no guarantee that a good decision will always produce a good outcome. A purchaser can research the market to identify the most reliable brand available. Even so, some chance exists that the purchased unit is a rare lemon. Although good decision outcomes cannot be guaranteed, making good decisions increases the likelihood of achieving good decision outcomes.

Good decisions are produced by a quality decision-making process. Among other attributes, a quality decision-making process:

- Involves the appropriate people
- Identifies good alternatives
- Collects the right amount of information
- Is logically sound
- Uses resources efficiently
- Produces choices that are consistent with the preferences of the responsible decision makers

Bad decisions waste resources and create risks and other costs. For example, a bad environmental decision can increase public-health risks and create legal liabilities for the decision maker's organization. Obviously, making good decisions is important.

Scope

This chapter describes tools for helping decision makers to make good decisions. The tools are aids for assessing, refining, evaluating, and selecting decision options. They rely on underlying assumptions to evaluate and compare alternatives. Some of the tools in this category are intended to weed out unacceptable or less effective alternatives. Others may be used to rank options. Some aim to identify "optimal" alternatives for environmental decisions. The most ambitious purport to define a quality decision-making process.

The types of questions addressed by the tools in this category include:

- What kinds of risks exist and how serious are they?
- How urgent is the need for action?
- Will the risks change, depending on the action that is taken?
- What are the strengths and weaknesses of the available alternatives?

- Should the decision be delayed while additional information is collected?
- Which alternative is best?

The number of available tools for addressing these and similar questions is very large. Space limitations permit only a fraction to be discussed here. Even so, the list of tools to be described is long, as seen in Table 8.1. The relative strengths and limitations of three of the major tools are summarized in Table 8.2.

Tool Users

The tools in this category are intended to aid participants in environmental decisions: government policy makers, corporate decision makers, and participants in collaborative decision-making processes. The tools are also useful for those who, although not directly involved in decisions, wish to persuade others to make defensible choices.

Although decision makers are the primary consumers of the results of these tools, the tools are typically applied by analysts who are specialists in their use. With few exceptions, these tools are complex, and their proper application requires skill and experience. As a consequence, significant differences often exist between the qualities of the best and the typical application practices.

The Environmental Decision-Making Process and Example Tools

Figure 8.1 illustrates a typical environmental decision-making process (Cheshire, 1991; CCPS, 1995). This flowchart is similar to that presented in Chapter 1, but provides details typical of decisions that use the tools in this chapter. A review of this process clarifies the types of decision-aiding tools that may be applied and the typical problem-solving steps that they facilitate.

Tools for Defining the Problem

The first step in a decision-making process that uses tools for assessing, refining, narrowing, and selecting decision options is, typically, problem definition. Defining the problem means clarifying the situation that produces the need or opportunity to make a decision. It includes establishing objectives; clarifying problem scope and significance; identifying stakeholders and their concerns; and outlining applicable political, social, and regulatory issues. The tools described in Chapters 2 through 7 may be useful for this purpose. A section of this chapter ("Tools for Decision Framing") also contains discussion of tools for problem definition.

234   M.W. Merkhofer

TABLE 8.1. Some tools for assessment/refinement/narrowing of options.

Tool	1	2	3	4	5	6	7	8	9	10
Accident Investigation		■	■							
Analytical Hierarchy Process			■		■		■			
Animal Research			■							
Bayesian Probability Methods		■								
Bioassay		■								
Biological Monitoring		■								
Borda Count							■			
Bright Lines						■				
Case-Control Studies		■								
Classical Probability Methods			■							
Confidence Interval		■								
Consequence Model		■								
Contingent Ranking							■			
Contingent Valuation						■				
Controlled Human Exposures		■								
Cost-Benefit Analysis			■		■	■				
Cost-Effectiveness Analysis						■	■			
Crash Simulations		■								
Decision Analysis			■		■		■			
Decision Trees					■					
Discharge Models		■								
Dose-Response Model		■								
Dynamic Models			■							
Econometric Models			■							
Ecosystem Monitoring		■								
Engineering Models			■							
Environmental Impact Assessment	■		■		■					
Epidemiological Studies		■								
Event Trees			■							
Expected Net Present Value							■			
Expected Utility						■				
Expected Value						■				
Expert Systems			■	■						
Exponential Distribution			■							
Exposure Route Models		■								
Exposure Tests		■								
Extrapolation Methods		■								
Fatal Accident Rate		■								
Fate Models		■								
Fault-Trees			■							
Field Tests		■								
Fixed Site Monitors		■								
Gaussian Plume Model		■								
Geographic Information Systems			■	■						
Harvest Models			■							
Hazard Assessment			■							
Hazard Index			■							
Health Surveillance		■								
Hypothesis Testing			■							
Index of Biotic Integrity			■							
Individual Risk		■								
Influence Diagrams			■	■						
Integer Programs						■				
Integrated Assessment			■	■						
Laboratory Exposure Tests		■								

Tool	1	2	3	4	5	6	7	8	9	10
Life Cycle Assessment			■					■		
Lifetime Average Daily Dose				■						
Linear Programs									■	
Maximum Daily Dose				■						
Mental Movies									■	■
Microcosms		■						■		
Moment Estimation Methods								■		
Monte Carlo Simulation								■		
Multiattribute Trade-off Analysis					■				■	
Multiattribute Utility Analysis					■				■	
Multihit Model				■						
Multistage Model				■						
Named Probability Distributions					■					
Networks									■	
Objectives Hierarchies		■								
Opportunity Costs						■				
Performance Measures								■		
Performance Testing		■								
Personal Exposure Monitors		■								
Pollutant Transport And Fate Models		■								
Pollution-Response Models		■								
Population Models		■								
Population Risk		■								
Probabilistic Risk Assessment		■	■	■	■					
Probability Encoding					■					
Prospective or "Cohort" Studies		■								
Queuing Models									■	
Reconstruction Of Surviving Parts		■								
Regression Analysis			■							
Remote Sensing		■								
Risk Assessment		■	■	■						
Risk Communication										■
Risk Comparisons						■				■
Risk Indices		■								
Risk Source Monitoring		■								
Satisficing vs. Optimizing								■		
Screening Tools						■				
Sensitivity Analyses								■		
Single-Value Analysis					■					
Simple Multi-attribute Rating Technique					■		■		■	
Social Choice Theory							■			
Socioeconomic Impact Assessment			■							
Spatial Decision Support Systems			■	■	■					
Strategy Tables									■	
Structured Voting							■			
Tolerance Distribution Model				■						
Utility Cancer Risk				■						
Utility Function						■				
Value Model						■				
Value of Information									■	
"Value of Life"						■				
Variance					■					
Water-Resource Models		■								
Weighted Scoring Methods									■	
Willingness-to-Pay						■				

Key: Primarily used for:

1	2	3	4	5	6	7	8	9	10
Problem definition	Assessing health or environmental risks	Assessing other risks	Determining whether action is needed	Collecting information	Screening alternatives	Identifying alternatives	Evaluating alternatives	Selecting options	Communicating decisions

TABLE 8.2. Characteristics of three major tools for assessing, refining, and narrowing options.

Tool	Strengths	Weaknesses
Probabilistic risk assessment	Provides a systematic, logical process for exploring, understanding, and quantifying risk; when performed well, enables risk managers to identify threats that pose the greatest dangers and to document reasons for conclusions; quantifies uncertainty	Difficult to do well; errors occur when models are too simple, data do not reflect recent changes, or underlying assumptions do not hold; hampered by limited availability of measures for human health and ecological stress; conservative risk estimates may provide misleading conclusions; estimated risk uncertainties can be large; can be expensive and time-consuming; generally requires a multidisciplinary team, including experts in probabilistic methods
Cost-benefit analysis	Systematic process for applying a decision-making logic that appeals to many people; well founded in theory; provides tools for estimating costs and benefits; helps risk managers to direct limited resources in ways that achieve the greatest total benefit; does not require decision makers to make their preferences or beliefs explicit	Concerned only with the net impacts on society, not who pays the costs or enjoys the benefits; relies on market prices that may not reflect people's preferences; ignores the views and preferences of responsible decision makers; provides little opportunity for stakeholders to contribute to the analysis
Decision analysis	Enables decision makers to make decisions consistent with their preferences and beliefs; well founded in theory; capable of accounting for "soft" issues (e.g., uncertainties that cannot be quantified with data); documents the basis for a decision; can help those who disagree to better understand the source of disagreement; provides a means for calculating the value of additional information; produces graphic representations useful for qualitative evaluations and communicating about the decision	Typically requires significant time and involvement from decision makers; makes explicit potentially controversial judgments and viewpoints; cannot represent in a single model the different beliefs and value judgments of multiple individuals; best applied in high-stake, complex decisions when time for deliberation is available

Tools for Estimating Risk and Determining the Need for Action

Action may be needed if the consequences of no action are unacceptable. Action may also be needed if the consequences of taking some action are potentially more favorable than the consequences of doing nothing. Doing

236 M.W. Merkhofer

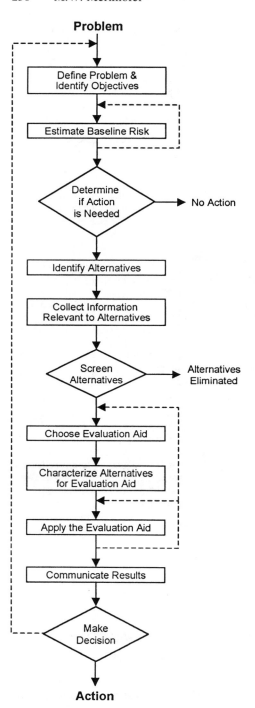

FIGURE 8.1. Typical Environmental Decision-Making Process.

nothing is, in effect, a decision, but it may not be the best decision. Deciding whether to evaluate alternatives to the status quo requires an assessment of whether the potential benefits of taking positive action are sufficiently great to warrant the time and effort necessary to determine exactly what that action might be and to implement it. Relevant considerations include the feasibility and costs of effective interventions, the costs of deciding, and the risks of doing nothing.

The acceptability of the risks of doing nothing depends on the nature and magnitude of those risks. A simple test is to consider whether a worst-case scenario can be tolerated. Envision the sequence of events that might occur if nothing is done. At each step where there is uncertainty, assume the worst outcome, but make the overall scenario a plausible one. What would the consequences be to the parties that would be impacted, to the decision maker, and to the organization? Are there financial impacts, health or environmental impacts, or impacts to the credibility or image to the organizations that would likely be held accountable? Are these consequences acceptable or could they be made acceptable through mitigation after the fact? If worst-case consequences are not acceptable, then a more careful consideration of risks may be needed to decide whether to manage those risks and, if so, what action to take.

Health and environmental risks are significant concerns for many environmental decisions. Regulations may stipulate the legal acceptability of such risks. In such cases, risk acceptability is often determined by comparing existing or projected conditions with the limits allowed by applicable standards. For example, health and environmental regulations may specify maximum allowable emission rates for toxic chemicals; contaminant concentrations in soil, air, or water; or dose exposures for people.

Risk assessment is a tool for describing and quantifying risk. Some health and environmental regulations require the use of risk assessment to ensure compliance. Risk assessment may also be used by decision makers to help determine whether risks are acceptable or whether actions to reduce risk are needed.

Risk assessment provides measures of risk called risk indices. Different risk indices are used for different types of risk. For example, worker accident risk is often expressed as a fatal accident rate (FAR), defined as the number of deaths in every 10 million hours of exposure. Cancer risks are often expressed as an individual risk, defined as the probability of a specified individual dying prematurely from cancer as a result of exposure to the risk agent. Individual risk is often calculated by multiplying a lifetime average daily dose times an estimate of the added probability of cancer per unit of dose. The latter factor is called the unit cancer risk. For noncarcinogenic chemical risk assessment, a frequently computed risk estimate is the hazard index. The hazard index is the ratio of the maximum daily dose received by an individual divided by an estimated acceptable daily dose. More sophisticated forms of risk assessment, sometimes referred to as probabilistic risk assessment, provide probability distributions describing the whole range of possible consequences of risk. Probabilistic risk assessment represents a "megatool" that is discussed more fully later. See Sidebar 8.1 for a detailed example.

Sidebar 8.1
An Example of Probabilistic Risk Assessment: Risks of Exposures to Air Pollution under Alternative Emissions Standards

Probabilistic risk assessment was used to help the State of Florida set standards for emissions of sulfur oxides, an air pollutant produced from burning oil- and coal-based fuels (Merkhofer and Korsan, 1978). The options under consideration ranged from reliance on existing federal emissions standards to new state standards that would be much more restrictive.

The first step was to construct a model for estimating the health consequences under the alternative standards. The model was composed of submodels for the emissions of air pollutants (the risk source); the atmospheric conversion, dispersion, and transport of air pollutants to sensitive populations (exposure processes); and the health consequences of exposures (effects processes).

The submodel for the risk source represented the 14 largest sources of sulfur oxides (fossil-fueled electric-power plants) on a map of the state. Each source was assumed to emit a plume of sulfur dioxide (SO_2) at a rate determined by current and projected electricity-sales growth. New standards were assumed to reduce these emissions, depending on the requirements of the standard.

The exposure submodel simulated the dispersion of the plumes based on historical wind patterns. It also represented the conversion of SO_2 to various oxidation products (e.g., SO_4), some of which are considered more harmful to human health. The exposures of people to the resulting elevated levels of SO_2 and SO_4 were estimated by dividing the state into 72 cells. The number of people in each cell (including the numbers of sensitive individuals, such as children, people suffering from respiratory problems, and the elderly) were then estimated based on census data and population-growth projections. The populations and exposure levels for each cell were estimated individually.

Finally, the health-effects submodel converted the exposures in each cell to numbers of specific health effects, such as increased incidence of lower-respiratory disease in children and fatalities to the

8. Assessment, Refinement, and Narrowing of Options 239

> elderly. The dose-response functions used to convert exposures to health consequences were based on the results of epidemiological, clinical, and toxicological studies. Finally, the numbers of health effects in each cell were summed to obtain statewide totals.
>
> To account for uncertainties, probability distributions were used. Uncertainties related to random variations, such as variations in daily concentrations of contaminants in air, were represented as frequency distributions. Uncertainties related to lack of information were quantified by means of expert judgment. For example, a probability distribution based on judgment was developed to represent experts' uncertainty regarding the incremental elevation to sulfate concentrations caused by a specified SO_2 emission level. Probability and frequency distributions were input to the combined model and used to produce probability distributions over the numbers of health effects related to SO_2 emissions. To simplify the presentation of results to policy makers, the study provided "best estimates" and "95 percent confidence intervals" for the reductions in each type of adverse health effect under each alternative standard. For example, moving to a more stringent emission standard was estimated to result in between about 600 and 2,500 fewer cases of aggravated heart and lung disease to Florida residents. Policy makers used these estimates together with political, economic, and other considerations to choose a state emissions standard.

Risk assessment is most often used to quantify "baseline risks," risks as they currently exist or as they are projected to exist if no actions are taken to control risks. Risk assessment can also be used to quantify the risks that would exist under various risk-reducing alternatives. As indicated by the dotted lines in Figure 8.1, risk assessment is iterative. If a simple, conservative risk assessment indicates baseline risks may be unacceptable, a more detailed and realistic risk assessment is conducted to validate or revise the initial estimates.

Thresholds of risk acceptability, called bright lines, provide useful tools for aiding decisions of whether actions should be taken to reduce risk (CRAM, 1996). For example, some regulators have proposed a threshold of 10^{-5} for excess cancer risk; if a risk assessment predicts more than one incremental case of cancer to a population of 100,000, regulators may judge that risk to be unacceptable. As might be expected, the specification of thresholds of risk acceptability can be controversial.

Risk assessment is also applied in situations where the primary concern is the impact on the natural environment (Suter, 1993). Hazard assessment is a tool used to support regulatory decisions regarding the acceptability of discharges of chemicals to the environment (Cairns et al., 1978). Hazard assessment is conducted as a tiered series of tests and assessments. After each test, the expected environmental concentration to result from the release is compared with the estimated toxic threshold for the chemical. If the estimates are clearly different, a decision about the acceptability of the release is made. If the two estimates are close, including a consideration of the level of uncertainty, the decision is deferred, and more tests are conducted to obtain additional data.

Tools for Identifying Alternatives and Collecting Information

Baseline risk estimates may lead directly to a decision. For example, if a risk assessment of a hazardous-waste site indicates significant dangers, emergency actions, such as evacuating local populations, might be taken without considering other alternatives. Conversely, an estimate of low danger might elicit a decision of no need for further action. Risk assessment often points the way toward risk-reducing alternatives. For instance, a risk assessment of an industrial facility might identify events that could lead to the release of hazardous substances. Cost-effective means for preventing these events might be obvious and implemented without further analysis.

Frequently, though, an assessment of baseline risk is followed by an effort to identify and compare alternatives for reducing risk. Organizational objectives, limits to organizational authority, resource availability, and other considerations can affect the range of alternatives to be examined. Subject to these constraints, it is generally wise to make the list of alternatives broad. A government agency, for example, may wish to consider regulatory and nonregulatory risk-reducing alternatives, such as permits, enforcement actions, pollution prevention, recycling, market incentives, voluntary reductions, and education.

Group participation and structured brainstorming tools are available for identifying alternatives. With strategy tables (Kusnic and Owen, 1992), different mechanisms or types of actions that might be taken are listed as columns in a matrix. Figure 8.2 shows a strategy table for addressing an aging tank farm used for the interim storage of hazardous waste (Bitz et al., 1993). Decisions include what types of waste retrieval and leak-control technologies to develop, what class or types of tanks to address first, and what types of information to collect about the tanks. Alternatives for each decision are listed in the columns. Alternative strategies are developed by selecting compatible combinations of actions from the different columns. For example, one strategy (see Figure 8.2) is to develop hydraulic and mechanical retrieval technologies with leak-detection capabilities; choose a tank containing sludge for a demonstration project and collect tank-leakage data; conduct the demonstration project with a hydraulic retrieval technology; store retrieved wastes in other existing tanks; and finally, remove the leaking tanks and remediate the contaminated soils. The strategy table facilitates the identification and construction of alternatives for complex

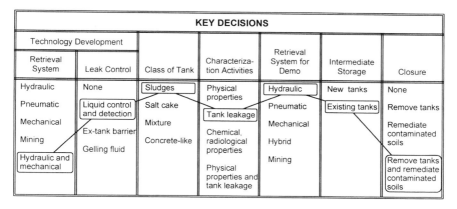

FIGURE 8.2. Sample Strategy Table for Tank Waste Retrieval Decisions.

decisions by encouraging a systematic, comprehensive consideration of options.

Once decision alternatives have been identified, information about those alternatives must be collected. Relevant questions include:

- What are the costs of each alternative?
- What consequences might each alternative produce?
- What laws and regulations apply?
- What groups might be affected?

Chapters 2 through 7 describe tools for addressing these questions. Decision makers often become overwhelmed by the number of facts and opinions raised at this stage. No single alternative looks superior on all dimensions. For this reason, tools are needed to organize information, narrow options, and support a choice that reasonably trades off the accomplishment of competing objectives.

Tools for Screening Alternatives

Screening tools are used to eliminate options from further consideration (Walker, 1986). Screening increases efficiency by limiting the number of options that must be subjected to detailed analysis. With screening, alternatives that fail to meet minimal requirements or levels of performance with respect to applicable criteria are excluded. The key to screening is to identify the performance criteria for which acceptability requirements truly exist. A threshold of acceptability exists if failure to achieve that threshold cannot be compensated for by excellent performance in other dimensions.

Screening criteria are usually applied sequentially. To illustrate, Figure 8.3 shows screens typically used to select sites for energy facilities. Legal

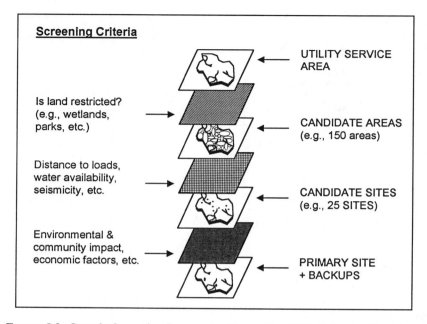

FIGURE 8.3. Sample Screening Process for Power Plant Site-Selection Decisions.

restrictions are considered first; they are applied to identify candidate areas within which the facility could be located. Then, cost and other engineering considerations are used to narrow the areas to a set of practical options. Lastly, environmental and other factors are applied to identify a short list of potentially acceptable sites.

Tools for Evaluating and Comparing Alternatives

The next step after screening is to select the tool to be used to evaluate and compare the remaining options. The choice of tool for evaluating options affects:

- Resource requirements for the analyses, including participants and their skills
- Considerations to be addressed within the analysis
- Time and effort required to perform the analysis
- Nature and form of results

Among the simplest of tools for evaluating and comparing alternatives is structured voting, wherein parties to a decision express their individual preferences (Dummet, 1984). With one approach, each participant expresses his or her preferences by voting for $N/3$ of the options, where N

is the total number of options under consideration. The options are then ranked according to the number of votes each receives.

Although voting is attractive for its simplicity and equity, legal requirements, institutional norms, or prudence may motivate the use of tools that explicitly consider the consequences of alternative actions. For example, the National Environmental Policy Act (NEPA) requires a project conducted with federal funds to provide an environmental impact assessment (EIA). An EIA is a legal document that identifies and evaluates the environmental consequences associated with the proposed project and alternative mitigation measures. EIAs use checklists, matrices of activities and components of the environment, and other devices for identifying, organizing, and displaying the numerous possible effects of complex projects (Beanlands and Duinker, 1983; Westman, 1985). EIAs can force the full disclosure of the potential adverse impacts of proposed projects, but they are often poor decision aids. For example, an EIA of a proposed freeway extension might not include an assessment of the degree to which the project would meet the travel needs of the community. Decision makers must trade off the unintended adverse environmental impacts of a project against the ability of the project to achieve its intended goals. EIAs are generally not organized in ways that make it easy for decision makers to balance the performance of projects using multiple, competing criteria.

Weighted scoring methods (Krawiec, 1984) are tools for applying multiple criteria. They assign numerical scores that rate each alternative on each decision criterion. The scores represent technical judgments about how well the alternatives perform against the criteria. Weights or other value parameters are assigned to indicate the preferences of the decision maker regarding the importance of the criteria. For example, a company might choose a health-insurance plan by rating alternatives on such criteria as costs to the employer, costs to employees, comprehensiveness of coverage, flexibility for the selection of health-care providers, etc. To facilitate ratings, 1-to-5 scales might be defined for each factor, with a 1 being very poor and a 5 being very good. A total score for each alternative is determined by weighting and adding the ratings.

The Simple Multi-Attribute Rating Technique (SMART) (Edwards and Barron, 1994) is one popular weighted scoring method. The approach requires identifying the person whose preference weights should be used, the context and purpose of the decision, and the available alternatives. Then, relevant criteria, or value dimensions are identified. The value dimensions are ranked, and the dimension of least importance is assigned an importance weight of 10. The next-to-least-important dimension is assigned an importance weight representing the ratio of its relative importance to that of the least-important dimension. Weights are assigned to the other dimensions in the same way, preserving importance ratios. The weights are normalized to sum to one by dividing each importance weight by the sum of all the weights. Each alternative is then rated on each dimension using a 0-to-100 scale. The ratings are weighted and summed, and the alternative with the highest aggregate rating is recommended. The approach works well for simple problems, but can produce errors when applied to more-complicated situations. For example, errors can occur if the value dimensions are not independent and when value is not proportional to rating (e.g., if a rating of 50 is not half as good as a rating of 100). (See the section of this chapter on "Tools for Deterministic Analysis" for more discussion of these issues.)

The analytic hierarchy process (AHP) (Saaty, 1980) is another weighted scoring method. It uses a more sophisticated technique for obtaining weights. The first step is to structure the decision problem into a hierarchy, similar to that shown in Figure 8.4 (for a company deciding on a health plan). At the top of the hierarchy is the goal of the decision, in this case, maximizing the overall satisfaction with the choice. Branches identify those value dimensions that contribute to attaining the goal, minimizing costs, and maximizing benefits. Branches from cost distinguish costs to the employer from costs to employees. Branches from benefits represent comprehensiveness of coverage and flexibility to choose health-care providers. The bottom level of the hierarchy identifies the available choices, in this case, alternative health-care plans.

Once the hierarchy is structured, AHP infers weights for the factors in the hierarchy from a series of comparisons between pairs of possibilities. For example, "How much more important to employees is comprehensiveness of coverage compared to flexibility to choose providers?" The response might be a score of three, meaning that comprehensive coverage is judged three times as important as provider flexibility. Such scores must be assigned for each factor at each level of the hierarchy. Inevitably, the assigned scores contain inconsistencies. For example, an inconsistency exists if the scores say that alternative A is twice as good as alternative B,

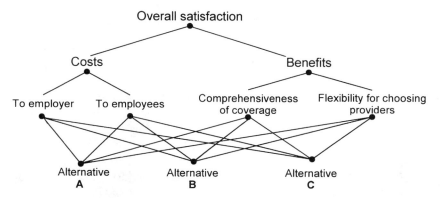

FIGURE 8.4. Sample AHP Decision Hierarchy for Selecting a Company Health Plan.

alternative B is three times as good as alternative C, but alternative A is five (rather than six) times as good as alternative C. To resolve these inconsistencies and to obtain weights, AHP arranges the comparison scores into matrices. The matrices are manipulated mathematically to compute a result known as eigenvectors. The eigenvectors define the weights and provide a measure of the relative consistency of the assigned scores. Proponents of AHP like the structure it provides, the ease of obtaining judgments in the form of comparisons between two items, the mathematical elegance associated with eigenvectors, and the computational speed of programs available to automate the process. However, others criticize AHP for its lack of a sound theoretical foundation and have raised questions about its validity (Dyer, 1990).

In contrast to simple scoring methods, cost benefit analysis and decision analysis are megatools for structuring decision problems and evaluating and comparing alternatives. Decision analysis and cost-benefit analysis are well founded in theory. See Sidebars 8.2 and 8.3 for examples of cost-benefit analysis and decision analysis, respectively.

Social-choice theory, another megatool, is concerned with finding decision rules or procedures (including voting) by which preferences specified by individuals may be directly incorporated into the decision process (Kelly, 1987). The Borda count is an example of a tool from social-choice theory. Suppose there are N alternatives. Each participant ranks the alternatives. Each alternative is then awarded N points for each person who ranked it first, N-1 points for each person who ranked it second, etc., down to one point for each person who ranked it last. The alternative with the most points wins. The advantage of the Borda count compared to standard voting is that the final choice accounts for each person's relative preferences among all the alternatives. A disadvantage of the Borda count is that, because each alternative is judged based on its ranking relative to other alternatives, the winner may be sensitive to "irrelevant" alternatives. Removing a nonfavorite alternative from consideration can affect which alternative receives the most points.

Sidebar 8.2
An Example of Cost-Benefit Analysis: Reducing the Sulfur Content of Gasoline

The Los Angeles (LA) City Council had to decide whether to pass legislation to reduce the sulfur content of gasoline sold in the area (Merkhofer 1987). The proposal would require producers to distribute fuels with sulfur contents no higher than 100 parts per million (ppm). If passed, the legislation would improve the city's air quality. However, it would also produce higher gasoline prices for drivers.

To estimate costs, refineries were surveyed to determine the investment needed to provide low-sulfur (100 ppm) gasoline. For each major refinery, incremental annualized costs were estimated, accounting for necessary capital investment in desulfurization equipment, sulfur-recovery facilities, associated off-site costs, interest during construction, and working capital. Manufacturing costs included return on direct capital investment, energy costs, and labor costs. Total estimated incremental costs averaged roughly two cents per gallon of low-sulfur gasoline produced. These costs were assumed to be passed along as a price increase to LA drivers.

Possible benefits of improved air quality include improved public health and visibility, reduced sulfur-caused damage to materials, and reduced impacts to the natural environment (for example, through reduced acid rain). Because data regarding material damage and environmental impacts were limited, these benefits were ignored in the analysis.

To estimate the reductions in air pollution and impacts to public health, a map of the LA basin was divided into cells. Data on roadway traffic and sulfur emissions from automobiles was used to estimate SO_2 and SO_4 emissions within each cell; plume models were used to simulate the transport and chemical conversion of emissions; population data were used to estimate exposures; and dose-response models were used to convert exposures to estimated reductions in the numbers of specific health effects. Impacts on visibility were estimated with an empirical relationship between concentrations of airborne particulates, including SO_4, and visibility, measured in miles.

To permit a direct comparison of costs with benefits, reductions in health effects with respiratory ailments were converted to equivalent monetary terms with the use of data from a survey mailed to residents in the LA area. The questionnaire contained questions such as, "To avoid severe shortness of breath and chest pains one day per year, the most I would pay is $0, $0.50, $1, $2, $10, $20, $50, $120, $250, $1,000 per year (circle one)." The survey demonstrated that there was great diversity in the willingness of citizens to pay for the benefits of improved air quality. The median values obtained from respondents were used to convert reductions in health effects to representative dollar values. To value visibility improvements, the results of a willingness-to-pay study conducted in New Mexico were used. Finally, total costs were compared with total benefits. The benefits of low-sulfur gasoline were estimated to be roughly half of the total costs. Based on this result and other considerations outside the scope of the cost-benefit analysis, the proposed legislation was not adopted.

Sidebar 8.3
An Example of Decision Analysis: Selecting a Site for a Hazardous-Waste Facility

Decision analysis was applied to help site a facility for disposing of hazardous waste (Merkhofer, Conway, and Anderson, 1997). The waste facility was to be located on Kirkland Air Force Base, just south of Albuquerque, New Mexico. The analysis was conducted as a group decision-making process in two one-day meetings involving representatives from Sandia National Laboratory (the waste generator), regulators, and the local community.

To provide options, an initial list of 156 locations was screened down to five feasible sites by the application of federal siting criteria. These five options were described to participants at the first meeting. A facilitator then led the group through the process of identifying siting objectives. After an hour of discussion, it was agreed that the selected site needs to: (1) protect public and worker health and safety; (2) minimize adverse impacts to the natural environment; (3) meet the necessary technical and regulatory requirements for storing and disposing of waste; and (4) ensure effective and efficient use of resources, including land, money, and time.

Next, influence diagrams were developed for each objective. Participants first agreed on the site characteristics and other factors that influence how well a site achieves each objective. They then agreed on the factor or factors in the diagrams that are the most useful site discriminators. For example, 15 factors were identified as influencing the level of public-health risk. These factors included various geohydrologic characteristics (such as earthquake potential and depth to groundwater), the security of the site against intruders, and the distance of the site from local communities. Only distance was viewed as a useful discriminator, however, because the other factors were judged not to differ significantly from site to site. Finally, one-to-five rating scales were developed for each of the factors identified as a useful discriminator. For example, the scale for distance measured the number of miles between the site and the nearest public community.

At the second meeting, technical specialists from Sandia National Laboratory used the scales to score the sites and explained their reasoning to the group. After a question-and-answer period, the other participants scored the sites, using the same scales but applying their own reasoning.

Techniques from multi-attribute utility analysis were used to determine that an additive equation, in which the performance measures were weighted and summed, was the appropriate form for

> the utility function. Weights were assessed from participants by the use of a technique called swing weighting, wherein participants indicate the value of changing the score on each scale from its worst to best value. Neither weights nor ratings were averaged across individuals. This practice was adopted to ensure that differences in rankings could be traced to differences in ratings, which reflect technical judgments, or differences in weights, which reflect value judgments.
>
> The resulting rankings for the various participants turned out to be virtually identical. Regardless of how the ratings and weights were combined, the site ranking remained the same. Participants unanimously agreed that the top-ranked site, a remote area once used for testing explosives, was the preferred choice. Participants said that they liked the process. It was logical, focused discussion on the issues that mattered, and provided participants with a meaningful role in the decision-making process.

Once a tool for evaluating options is selected, the alternatives to be analyzed need to be characterized or refined to be appropriate to that tool. The way in which the options are defined and specified can affect the choices to be made. If, on the one hand, the decision aid is a voting method, then alternatives must be defined to allow participants to express their preferences for them. For example, the alternatives may need to be collapsed into two main options, or the multiple options may need to be organized and compared in pairs with runoff elections used to identify an ultimate favorite. If, on the other hand, the tool is a cost-benefit analysis, then the alternatives must be sufficiently well-defined to allow their costs (and benefits) to be estimated. In the case of a decision of whether to clean up a hazardous waste site, for example, the specific technologies for removing, treating, transporting, and disposing of the waste may need to be specified.

The application of a tool for evaluating and comparing alternatives requires providing and preparing the data required by the tool. Each tool tends to have its own data requirements. Major categories of data relate to:

- Systems and processes that produce the costs, benefits, and risks
- People and organizations that bear the costs and receive the benefits
- People and environmental resources that are at risk
- Regulatory, socioeconomic, and other impacts of concern
- Value judgments inherent in the decision

For some decision tools, most of the necessary data may be qualitative (e.g., the type of information exchanged in a group discussion prior to taking a vote). Other tools require highly detailed, quantitative data (e.g., the inputs

8. Assessment, Refinement, and Narrowing of Options 249

needed to run a model for predicting the health risks from groundwater contamination).

Given the necessary inputs, the tool for evaluating options is applied. Applications vary in the time required and the complexity involved, depending on the tool. With a voting method, for example, applying the tool may consist of little more than tabulating votes. Other tools require complex numerical computations. Much specialized software is available for automating some or even all aspects of many tools. Expert systems, for example, are computer programs that combine a knowledge base about the problem with a reasoning mechanism (inference engine) to infer new knowledge and to solve the problem (Mumpower et al., 1987). The reasoning may involve rigorous principles of logic or simple heuristics. This structure enables the expert system to replicate the reasoning behavior of experts and to document that reasoning process. Among other applications, expert systems are used as decision aids in system reliability and safety analysis (Poucet, 1990). For example, expert systems exist that pose questions designed to encourage users to identify ways in which industrial workers might be injured or exposed to hazardous materials. Depending on the answers to initial questions, subsequent questions are refined to address those particular subsystems most critical to worker safety.

Applying Tools for Evaluating Alternatives

Tools for evaluating alternatives can be applied in either a satisficing or optimizing mode. With satisficing, the first course of action found to have a satisfactory evaluation is selected (Simon, 1976). With optimization, all options are considered to ensure that the most favorable one is identified. Numerous tools are available for efficiently comparing options, and specialized software is available for solving many of the mathematical forms that a decision model might take. Examples of model forms with specialized solution algorithms include linear programs, which assume performance is proportional to the levels specified for decision and other input variables; networks, which assume that information or material is exchanged between system elements in predictable ways; integer programs, which constrain variables to take on discrete, whole-numbered values; and queuing models, which simulate the effect of delays in systems that sequentially serve users.

As illustrated by the dotted arrows in Figure 8.1, evaluations of decision options often involve iterative analysis. The first evaluation is generally conducted with either a relatively simple tool or a more sophisticated tool applied at a high level of abstraction or simplification. The advantage of multiple applications is that the results of initial analyses may be used to guide and refocus subsequent efforts.

Communicating Results

After the selected decision tool has been used to evaluate and compare options, the results are presented to those responsible for making and/or executing the decision. If the decision must be explained or defended to interested parties, it may be useful to communicate the results of the evaluation to those parties, as well. Important outputs include much more than just which option ranked highest. Depending on the audience and the stakes involved, a presentation of results will generally describe:

- Alternatives considered
- Logic and key assumptions used in the analysis
- Analytic approach and data sources
- Key results expressed in terms of the decision criteria of importance to decision makers
- Sensitivity of the results to key inputs and assumptions
- Recommended alternatives and future actions.

Decision Making

The final step in the decision process is to make the decision. The tool is an aid to the decision-making process. It does not make the decision. Decision makers must overlay on the results of the analysis factors that are not addressed by the analysis before reaching a final choice. A decision might be to undertake one alternative or a combination of alternatives. In some instances, the analysis may show that there is value to delaying the decision for the purpose of collecting additional information, further refining the alternatives, or conducting additional analyses. Delaying and collecting more information is in itself a decision. The entire evaluation process may need to be repeated if new data, considerations, or views come to light. A sound analysis using decision-aiding tools can facilitate implementing a decision by providing a rationale that can be communicated to others.

Key Assumptions Common to All Tools in This Gategory

The tools in this category are all based on the strategy of divide and conquer. "Analysis," according to Webster, has a Greek etymology and means the loosening or breaking up of a whole into its parts. In the context of decision making, analysis involves decomposing the problem into its individual parts, analyzing each part separately, and then drawing conclusions by synthesizing results in a manner appropriate to the parts and their interrelationships.

Decomposition has several potential benefits. If the assessment of uncertainties is conducted separately from the expression of preferences among potential decision outcomes, the expertise of scientists and other experts may be tapped without embroiling them in controversial value judgments. Participants can then contribute according to their specific skills without having undue or inappropriate influence on the decision.

Decomposition, however, has its critics. Limited research has been conducted on the learnability, applicability, or effectiveness of decomposition (Fischhoff, 1979; Armstrong and MacGregor, 1994). The number of ways of decomposing a complicated problem is infinite, and different approaches may produce different answers. Fundamental decompositions, such as the separation of fact and value, are not easily obtained. All alleged facts reflect the perspective of the investigator. Similarly, facts shape values: "The world we observe tells us what issues are worth caring about. . . . Insofar as that world is revealed to us through the prism of science, the facts it creates shape our world outlook." (Fischhoff et al., 1980).

Characteristic Tools Within the Category

The remainder of this chapter is organized around three megatools, major approaches for structuring, assessing, evaluating, and comparing decision options: probabilistic risk assessment, cost-benefit analysis, and decision analysis.

Given all of the tools for assessing, refining, and narrowing options, why emphasize these three? For one thing, they are widely used. For another, these megatools are more powerful than most other tools in this category. The megatools are more like "tool boxes" that contain many tools plus instructions that describe how the individual tools should be used. Many of the tools within each toolbox have value in their own right; they can be used for purposes other than to conduct an application of one of the megatools. Indeed, the preceding chapters have discussed several tools often used within the megatools. However, when these individual tools are combined and applied according to the principles and processes specified by the megatools, they take on added value as aids for environmental decision making.

Another distinguishing feature of the megatools is that they are not intended merely for use on one step of the decision-making process. Instead, and depending on the megatool, they encompass several, most, or nearly all of the steps shown in Figure 8.1 (see Table 8.1). In particular, their purpose is *not* simply to select among alternatives once information collection is complete. For example, each of the megatools includes a subset of tools for obtaining the information that, according to its underlying theory, is needed for decision making. Each also provides concepts and terminology useful for understanding and communicating about environmental decisions. The megatools are, in effect, recipes for quality decision-making processes.

The opportunities for selecting and combining the tools used within the megatools provide enormous flexibility for the time, resources, and expertise needed for applications. At one extreme, each megatool may be regarded as a logical framework for promoting systematic and orderly thinking. At this level, a user need not conduct any quantitative analysis to benefit from the principles and concepts that the megatool has to offer. At the other extreme, sophisticated mathematical models can be developed to implement each megatool in a highly detailed, quantitative fashion. The appropriate choice of the megatool and how it is implemented depends on many factors, including the function that the tool is intended to serve, characteristics of the decision problem, and the time and resources available for the analysis.

Probabilistic Risk Assessment

The section "The Environmental Decision-Making Process and Example Tools" described the typical role of risk assessment within the general environmental-decision-making process. This section describes in more detail a form of risk assessment intended to describe the uncertainties inherent in risk. This form of risk assessment is most useful for decision making when the outcomes of risk are important and highly uncertain.

Definitions of Risk, Risk Assessment, and Probabilistic Risk Assessment

Most writers define risk as the probability that an adverse event will occur or as the product of the probability and the consequences of an event. This is misleading. Probability and consequence are measures of risk, they do not define it. The decibel is a measure of the intensity of sound, but sound is not defined as decibels. One need not know the number of decibels to know whether something is loud. Similarly, risk is not probability or probability times consequence. Risks existed long before the concept of probability was invented.

A better approach is to define risk not in terms of specific measures, but in terms of the basic situation to which the measures apply. Thus, risk is an uncertain situation involving the possibility of an undesired outcome. Risk assessment may then be defined as a systematic process for describing and measuring risk (i.e., uncertainty). Probability and consequence represent one way to measure risk. Rather than provide just two numbers (probability and consequence), probabilistic risk assessment (PRA) indicates the type and nature of all of the possible risk outcomes, the magnitude of these outcomes, their probabilities, and their timing.

8. Assessment, Refinement, and Narrowing of Options   253

The undesired outcomes that are the focus of risk assessment may be of any type. Specialized tools are available for estimating specific types of risk, including financial risks, technical risks, regulatory risks, health and safety risks, and risks to the natural environment. The emphasis of this section is on tools for assessing health and environmental risks, important considerations for many environmental decisions. Similar tools are available for assessing other types of risks.

Tools for health and environmental risk assessment can be organized according to the risk-producing process they address (Covello and Merkhofer, 1993). A health or environmental risk requires (1) a hazard or source of hazardous agents; (2) an exposure process; and (3) a causal mechanism by which exposures to agents may produce adverse health or environmental consequences. Thus, a health and environmental risk assessment involves tools for (1) release assessment, which assesses the potential for a source to release hazardous agents; (2) exposure assessment, which characterizes the movement and change of agents within the environment and the resulting exposures to people and the things they value; and (3) consequence assessment, which identifies the relationships between exposures and the resulting health and/or environmental effects. Whether or not the danger is to health or the environment, risk assessment also involves tools for (4) risk estimation, which provides quantitative measures of risk. PRA tools fall in these same categories. PRA uses data-collection and analysis tools that are similar to those used in standard risk assessment, but includes additional tools that allow for statistical and probabilistic analysis of uncertainty.

Tools for Release Assessment

Tools for release assessment investigate and describe the potential for technologies, products, processes, or systems to release hazardous agents to the environment. Agents may be chemical (e.g., pesticides), physical (e.g., collapsing structures), biological (e.g., viruses), or energetic (e.g., radioactivity).

Risk-source monitoring collects data about a risk source under normal or ambient conditions (e.g., recording radioactivity levels in the off-gas stack of a nuclear power plant). Sophisticated instruments are not the only source of useful monitoring data. Changes in a source observed by plant workers or citizens may be important, so the reporting of such observations should be encouraged.

Performance testing entails collecting data about the risk source or its elements under controlled conditions. For example, electrical and mechanical elements of safety systems are tested to determine the frequency with which they fail. Performance testing is often conducted under harsh or stressful conditions. Such is the case with pressure-release valves, which are tested at high temperatures to improve understanding of their performance under conditions similar to those existing in a fire.

Accident investigation involves reconstructing an accident based on post-accident information. The approach employs structuring tools (e.g., mental movies), examination methods (e.g., reconstruction of surviving parts), and simulation techniques (e.g., crash simulations). For example, such techniques might be used to reconstruct the events leading to an explosion at a chemical plant. A major objective of accident investigation is determining what caused the system to fail.

Statistical tools are used to convert repeated measurements, such as those obtained from monitoring or performance testing, into predictions of future behavior. Examples of statistical tools used for release assessment include named probability distributions, such as the exponential distribution, for representing the time between component failures; regression methods for forecasting dependent variables based on measurements of independent variables; and hypothesis testing. Because statistical methods require few, if any, assumptions based on engineering or cause-and-effect reasoning, they are often seen as attractive because they "let the data speak." In reality, though, statistical models involve potentially important but generally implicit assumptions—for example, that certain types of events occur in a completely random way. Whether such assumptions hold in practice is generally difficult to discern, as are the errors that might be introduced if they do not.

Modeling methods in release assessment are used to construct a mathematical model of a risk source. The model is then used to simulate or predict the behavior of the source. Examples include fault trees, event trees, and discharge models. Fault trees are often used to quantify the probability that a system failure may result in the release of hazardous agents (Vesely et al., 1981). A fault tree is a specialized model that may be represented as a diagram of binary (yes-no) logic that traces backward in time the different ways that a release or other event of concern could occur. Fault trees are well suited to estimating the probability that complicated electrical or mechanical safety systems might fail.

Event trees are useful for identifying possible consequences and estimating probabilities of undesired events (McCormick, 1981). The event tree starts with a particular undesired initiating event, such as the failure of a pump in an oil pipeline system, and projects all possible system responses to that event, including spillage of oil to the environment. Each branch in the event tree represents a possible state (often simply success or failure) for the subsystems (e.g., safety valves and overflow basins) that would be affected as the failure progresses.

Discharge models are useful for characterizing the quantities of materials released in the event of a containment failure, such as a leaking pipeline segment or a ruptured rail tank car. Key considerations reflected in discharge models include the size of the hole through which releases occur and whether the discharge is in the form of a liquid, a gas, or a liquid/gas combination (CCPS, 1995).

Tools for Exposure Assessment

Exposure assessment is concerned with determining what happens to risk agents once they are released to the environment. Exposure assessment is also concerned with sensitive population subgroups, such as children, and with personal behaviors, such as food preparation and eating habits, that affect exposures.

Human exposure can be monitored by providing personal exposure monitors (PEMs) to individuals within the population at peril. PEMs have been used to gather data about people's exposures to carbon monoxide, radiation, and various air pollutants (Akland et al., 1985). More common, though, is the indirect approach of using fixed-site monitors like air-sampling devices installed around a power plant. Useful data can be obtained from monitoring a range of environmental media. For example, to determine exposure potentials from a hazardous-waste site, soil, surface water, groundwater, and air may be monitored for contamination. In addition, biological monitoring may be used to identify food-chain problems (e.g., by measuring chemical residues in the tissues of food crops, livestock, or local wildlife). Remote-sensing data from aerial photography or satellite imagery may also help delineate contaminated areas (see Chapter 3).

Testing to support exposure assessment may be conducted in the field or laboratory. Field tests may be used if the test poses no threat to the environment. For example, a nontoxic dye might be introduced into a water system to improve understanding of pollutant transport. Laboratory tests require constructing physical models, called microcosms, that recreate some portion of the natural environment. Microcosms provide a testbed for experiments that cannot otherwise be conducted. For example, an aquarium may provide a microcosm for the plants and animals living in a lake. A toxic chemical can be added to the aquarium to observe chemical and biological reactions and the resulting partitioning of the chemical among the microcosm's components.

To help predict exposures, exposure models have been developed for a variety of sources, agents, environmental media, and routes of exposure. Pollutant transport and fate models are used to estimate the concentrations of pollutants in air, soil, water, and groundwater. For example, the Gaussian-plume model is often used to estimate spatial and temporal variations of atmospheric concentrations resulting from smokestack emissions. Exposure-route models translate the output of pollutant transport and fate models into the doses actually received by individuals. Population models describe the human and other populations that are in danger and indicate how they may change with time. For example, a population model might recognize different ethnic groups to account for differences in food-consumption practices that might influence susceptibility to various cancers. Because spatial considerations are usually important for exposure assessment, a geographic information system (GIS), a concept discussed in Chapter 6, is often useful for constructing exposure models.

Tools for Consequence Assessment

Consequence assessment investigates the relationship between an exposure to a risk agent and the resulting adverse health and environmental consequences. The effects of risk agents on human health range from minor and temporary (e.g., a minor infection or rash) to severe and permanent (e.g., irreversible organ damage or death). Effects on the environment include:

- Changes to the populations and to the health statuses of important species
- Alterations to animal and plant community structures and habitats
- Damage or destruction to archaeological, historical, or cultural properties
- Aesthetic impacts, such as unpleasant odors from chemical releases.

Health surveillance is the term used to describe monitoring methods for assessing human-health consequences. Health surveillance is widely used in recording causes of death, documenting cases of particular diseases (e.g., HIV-AIDS), and compiling statistics on injuries from accidents. Ecosystem monitoring is used to assess the status of and changes in the quality of the natural environment. The complexity of ecosystems requires that attention be given to trophic levels, feedbacks, successions, and other features of ecosystems. Examples of ecosystem monitoring include recording losses to commercial fisheries from water pollution, estimating visibility reductions from air pollution, and quantifying the health status of wildlife communities. The index of biotic integrity is one of several tools for converting diverse ecosystem data to a single, overall measure of ecosystem stress (Karr, 1991). Although crude, such measures may be adequate for some decision-making purposes.

Epidemiological studies obtain and compare data on the association between human-health effects and exposures. Prospective or "cohort" studies compare people who have been exposed to a suspected risk agent with a control group that has not been exposed. The comparison is generally made over a relatively long time. Case-control studies compare people with a given disease with a control group of persons who do not have the disease.

Testing that uses controlled human exposures may be used to study mild and reversible health effects. For example, the health effects of sulfur dioxide exposure have been investigated in chamber studies in which human volunteers would breathe varying concentrations of the agent while changes in their lung performance were measured. However, most testing for human-health-consequence assessment is based on animal research, in which other species are substituted for humans. Bioassay is the term used to describe a test conducted under controlled conditions to quantify the effects of a substance on a living organism. Test animals (e.g., rats or mice) are divided into a control group and one or more treatment groups. The treatment groups may be given varying doses of some agent, while animals in the control group are not exposed. Extrapolation methods involve assumptions for translating animal data to humans. For example, if the animals are fed toxic substances, extrapolation may include converting animal doses to levels believed to be equivalent for humans with conversion factors based on relative body weight, surface area, or other toxicological considerations.

Exposure tests involving elements of the environment are conducted to investigate the environmental consequences of risk agents. To understand the unintended impacts of pesticides, laboratory exposure tests have been conducted on honeybees, birds, and fish. Field tests are conducted in situations where accounting for all environmental factors is difficult or impossible in the laboratory. For example, mature trees in forests have been fumigated with ozone to determine the concentrations necessary to cause foliar injury (Stewart et al., 1973).

Modeling methods are used in health- and environmental-consequence assessment to translate exposures to a risk agent to adverse health consequences. The principal type of health- and environmental-consequence model is the dose-response model. A dose-response model is a functional relationship between a dose (i.e., a measure of exposure) and an adverse health or environmental response (i.e., the measure of impact). Examples include tolerance distribution models, which are based on an assumption that each person or organism in the population has an individual threshold tolerance for a risk agent, and various mechanistic models, such as the multihit models, which attempt to represent actual biological processes within specific organisms. The biological processes in mechanistic models are typically represented as a series of events that evolve with time, such as processes that control the uptake or residence time of the toxic substance, mechanisms of cellular or organ damage or dysfunction, and mechanisms of repair. For example, the multistage model is a mechanistic dose-response model often used for cancer-risk assessment. It assumes that a tumor originates as a predisposed cell that goes through a series of stages until it becomes malignant. Exposure to a carcinogen is assumed to influence the rate of progression through the various stages. The timing of the transitions through the stages is expressed by a probability function defined by a rate that is proportional to the dose.

Modeling methods used in environmental-consequence assessment differ according to the elements of the environment they represent and the types of risk agents and environmental effects they consider. They also differ depending on the type of mathematics involved. For example, simple curves derived from statistical data are often used to relate the soiling and damage sustained by materials and equipment as a result of exposure to atmospheric air pollution (Silvers and Hakkarinen, 1987). Dynamic models illustrate changes over time. Harvest models are dynamic models for the impact of harvesting on biological populations (e.g., modeling the gradual depletion of whales through harvesting). Pollution-response models have been developed to study ecosystem response to pollution and other habitat disturbances. For example, water-resource models have been developed to represent the degradation of mountain lakes by acid rain.

Tools for Risk Estimation

Risk estimation involves developing quantitative measures of risk. Because risk is uncertainty about adverse consequences, the logical outputs of risk assessment are curves indicating the range and relative likelihood of possible outcomes, which are specified in terms of type, severity, location, timing, and population affected. With PRA, these risk curves typically are probability distributions or frequency distributions. Probability distributions represent uncertainties caused by lack of information and understanding. Uncertainty in outcomes may be caused by lack of knowledge regarding the effectiveness of an unproven technology selected to address an environmental problem. Frequency distributions represent natural variabilities. The health risks of seafood contamination provide an example of variability in risk: the health risks to individuals from contaminated fish differ because, among other reasons, people consume differing amounts of fish.

Risk summary statistics may be computed from risk curves. One standard summary statistic is expected value, the average of possible outcomes weighted by their probabilities. The expected value describes the average or central tendency of a risk curve. Population risk, a common summary statistic, is the expected number of fatalities attributable to a risk per year. Other summary measures indicate the spread or dispersion in outcomes. Common summary statistics for indicating spread and variability include variance and confidence intervals. Qualitative information, such as who and what is at peril, the types of consequences that may occur, the weight of evidence, and the severity and reversibility of effects, is also an important risk-assessment output.

Risk estimation includes tools for linking models for estimating risk outcomes and for using those models to produce risk curves. In the case of health-risk assessment, risk curves may be obtained by linking models for the risk source, exposure processes, and consequence processes to obtain a composite risk model, as illustrated in Figure 8.5. A risk curve is produced by specifying the uncertainties in the model inputs and then propagating these uncertainties through the model to obtain the uncertainties that they

8. Assessment, Refinement, and Narrowing of Options 259

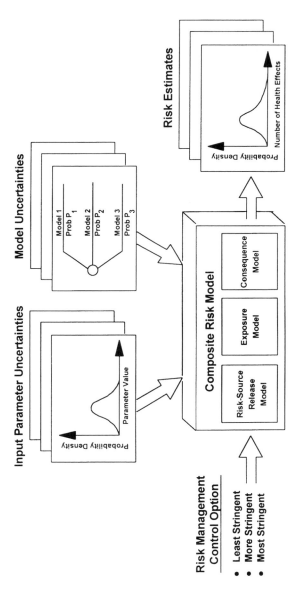

FIGURE 8.5. Using the Composite Risk Model to Obtain Risk Estimates.

imply for model outputs (Morgan, 1981). For example, to estimate the health hazards posed by a nuclear power plant, an integrated model for the releases, exposures, and health consequences of the plant would be used. Uncertainties affecting releases (e.g., the failure rates in a fault tree), exposures (e.g., wind directions and locations of people), and health consequences (e.g., parameters of a dose-response model) are quantified and input to the integrated model. Several methods exist for converting uncertainties about a model's inputs to the corresponding uncertainties that they induce in outputs. One such method is Monte Carlo simulation, which computes the output of a risk model for combinations of inputs that are obtained by randomly sampling values according to the distributions assigned to those input variables. If the component models are themselves uncertain because the underlying physics, chemistry, or biology is poorly understood, this uncertainty can be taken into account by using competing model forms, assigning probabilities to these alternative models, and then combining the results mathematically.

Model coupling not only requires that the inputs and outputs of connected models match, it also requires compatibility of temporal and spatial resolution. For example, the geographic distribution of air pollution is often modeled by dividing a region into cells and estimating atmospheric concentrations within each cell at discrete steps in time. The population models used to estimate exposures must then use similar cell sizes and time steps to ensure model compatibility. Spatial decision support systems (SDSS), discussed in Chapter 6, are designed with these considerations in mind. The models that are connected should also achieve compatibility in their costs and accuracies. For example, it would be inefficient to use an expensive model of a risk source that produces highly accurate, time-varying estimates of release characteristics to which the exposure model is insensitive. If the individual models are designed specifically for the risk assessment, then compatibility can be designed in. Coupling pre-existing models developed for other purposes is often extremely difficult because of model incompatibilities.

Tools for quantifying uncertainty in model inputs differ depending on whether they adopt a classical or Bayesian perspective. Classical methods associate probabilities with events, and statistical methods are used to derive probabilities from the observed empirical frequency with which the events occur. For example, the probabilities of wind directions and speeds at a nuclear power plant are typically frequency distributions ("wind roses") derived from past data. Bayesian methods regard probabilities as subjective and dependent on the information, experience, and theories of the individuals who assign them. With the Bayesian view, probabilities are elicited from experts with probability-encoding methods. Obtaining accurate probability judgments requires understanding the errors and biases that commonly distort beliefs. Probability-encoding methods use a formal process to counter common judgmental biases (Merkhofer, 1987).

8. Assessment, Refinement, and Narrowing of Options 261

Simpler methods for computing risk statistics that do not involve risk curves are also available. If a model's variables are not particularly uncertain, a single-value analysis can be performed in which best estimates are provided as inputs for each variable and the composite risk model is used to produce a best estimate of outcomes. Sensitivity analyses can be conducted in which uncertainties are varied across a range of values to establish bounds on the range of possible consequences. Also, risk statistics can be estimated roughly with moment estimation and other approximation methods without explicitly computing probability distributions.

Strengths and Limitations

At the conceptual level, risk assessment formalizes common sense. In the case of health hazards, understanding and describing risk requires considering the source; its potential to release agents; the movement and changes those agents might undergo in the natural environment; and the likely consequences to people's health. It would be hard to argue that such considerations are not relevant to environmental decision making.

The goal of PRA, to describe and quantify uncertainty, also makes sense. Understanding uncertainty can substantially improve a risk manager's performance. Finkel (1994), for example, uses a sequence of analyses to demonstrate that better representations of uncertainty produce better decisions. In practice, not all risk assessments provide quantifications of uncertainty adequate for decision making. However, risk-assessment professionals are moving toward increased use of probabilistic methods to quantify risk.

Through logic, risk assessment provides a framework that decision makers can use to make more-informed decisions and to "work smarter." It enables risk managers to estimate systematically which environmental threats pose the greatest danger and whether controlling one risk would alleviate (or create) others. Risk assessment is widely used by the federal government, and according to one review, almost half the states use risk assessment to compare and rank environmental threats (Curtis and Michaels, 1996).

PRA not only makes sense conceptually, its mathematics are logically sound and derived from the well-developed disciplines of probability and statistics. In theory, therefore, PRA is not only a useful construct for thinking about risk, it should also be capable of producing accurate quantitative descriptions of risk that account for nearly all relevant considerations.

In practice, though, correctly applying the theory of PRA, and risk assessment in general, is often problematic. Errors in logical soundness occur when models are too simple, when extrapolations from past data are made without regard to recent changes, and whenever underlying assumptions do not hold. For example, problems occur with consequence assessment when dose-response relationships for humans are extrapolated from animal results without regard to fundamental differences between people and animals. Problems also occur when simple linear dose-response functions are used to describe nonlinear, dynamic processes and when correlations and interdependencies are ignored (as discussed in Chapter 3). Epidemiological studies, the linchpin for many risk assessments, have many serious problems that limit their ability to predict disease (Gots, 1992). Consequence assessment for environmental threats is hampered by the limited availability of measures of ecological stress. Although lack of information does not necessarily mean that PRAs will be inaccurate, poor information means that sources of uncertainty are more likely to be overlooked or improperly described. Too frequently, environmental threats are difficult to quantify and, therefore, are left out of the final analysis. Because computing probability distributions is difficult, risk assessment too often reverts to single-value analyses wherein conservative assumptions are used to account for uncertainties. The resulting conservative risk estimates may produce highly misleading conclusions.

Potential users may find even high-quality risk assessments unacceptable. Uncertainties can be so large that results appear useless. For example, a study estimated that the number of bladder cancers in the United States from lifetime consumption of saccharin ranged from .22 to 1,144,000 (NRC, 1978). Also, complete risk assessments can be expensive and time-consuming. For example, a large fault-tree or event-tree model for an industrial facility can cost more than $500,000 and take more than two years to complete. In addition, risk assessment generally requires multidisciplinary research teams, which are difficult to assemble and maintain.

Cost-Benefit Analysis

PRA provides only part of the information that environmental decision makers need. Risk must be balanced against other considerations to reach sound decisions. Cost-benefit analysis (CBA) is a megatool for comparing and selecting alternatives based on their advantages (benefits) and disadvantages (costs). While many decision-aiding tools claim to do this, CBA is distinguished by its foundation in theory and by how it computes costs and benefits. CBA does not view costs and benefits from the specific perspective of responsible decision-making parties. Instead, CBA measures costs and benefits from the perspective of society at large. The best alternative, according to CBA, is the one that leads to the greatest net benefit (benefit minus cost), with benefits and costs aggregated across all individuals in society (Abelson, 1979).

In its simplest form, CBA can be implemented by a decision maker by simply identifying, judging, and comparing the total costs and benefits of alternatives, using best professional judgment. Numerous tools are, however, available for improving the quantification of costs and benefits as part of a CBA. The subsections below describe tools for (1) estimating costs, (2) estimating impacts, (3) valuing impacts, and (4) computing net benefit.

Tools for Estimating Costs

The economic costs of an environmental decision obviously include the expenses borne by the organization that implements the decision (e.g., material and labor costs, ongoing operating and maintenance costs, and interest charges and other transaction costs). According to CBA, however, accounting for the indirect costs incurred by others is also necessary. For example, suppose a government agency implements an environmental regulation. The costs of the regulation include the increased costs incurred by the private sector in the production of goods and services. Providing cost estimates for CBA is often difficult because it requires "second guessing" how impacted parties will react to decisions (Gramlich, 1990). The costs to industry of new clean-air rules, for example, have often been overestimated because of an underestimation of industry's ability to adapt to changing requirements.

Before costs (or benefits) can be estimated, several scoping questions must be answered. What portion of society should be included? If a government decision imposes costs on individuals outside the United States, deciding whether to include these costs may be significant. How far into the future should costs be cumulated? The time frame should be long enough to incorporate all future impacts of sufficient magnitude to be of concern to decision makers. Life-cycle assessment computes the total costs throughout the life of the investment, including any residual deconstruction, decontamination, and disposal costs that might be incurred at the end of its useful service life. The life-cycle perspective is needed to ensure that proposed projects bear full responsibility for all future costs that might be reasonably expected. It also provides a consistent basis for cost comparisons.

The costs (and benefits) of alternatives should reflect incremental impacts compared to a baseline, do-nothing, or status quo option. Furthermore, costs must reflect opportunity costs, the true worth of the resources expended in view of other opportunities to use those resources. For example, if an organization already owns the property for a project, the accounting costs for land may be zero. However, the value of the land to society may be large. For such a project, the opportunity cost of the land might be estimated as the value of that land on the open market.

CBA often uses modeling tools, such as econometric and engineering models, to predict costs. Suppose the government decides to reduce emissions from electric-power plants by taxing sulfur emissions. CBA might use an econometric model (a model based on economic and empirical data) to assess increased industry costs by estimating industry's shift to low-sulfur energy sources (e.g., by comparing supply, demand, and price with interfuel competition). Alternatively, an engineering model might assume some typical production technology, determine the control technology likely to be used, and estimate the cost of that control technology. This process might be repeated for cases representing various types of plants, with the resulting costs aggregated and scaled in accordance with the numbers of plants of each type.

Cost-effectiveness analysis (CEA) is a modified form of CBA that focuses on finding the least-cost alternative for achieving a specified goal. A CEA could be used, for example, to find the least-cost approach for lowering the current ambient ozone standard to .1 part per million, assuming each alternative considered could be assured of achieving the goal. Unlike CBA, CEA avoids the need to estimate noneconomic impacts and to value benefits (Riggs and West, 1986).

Tools for Estimating Impacts

According to CBA, implementing an alternative can produce impacts that are desirable (benefits) or undesirable (disbenefits). These impacts can be primary, secondary, and tertiary effects. Primary effects are the intended and obvious impacts, including the products and services directly provided under the alternative. Secondary effects are those that are not the immediate purpose of the investment. A secondary effect of the ongoing cleanup of hazardous-waste sites at U.S. Department of Energy (DOE) weapon facilities is that it saves jobs that might otherwise be lost during the decline in weapon research and production. Tertiary effects are stimulated even more indirectly; in this case, the economies of local communities will benefit because of fewer layoffs at DOE facilities.

Modeling, data collection, and statistical tools may be used to estimate effects in CBA. For example, risk-assessment models are used to estimate the risk-reducing potential of alternatives, an intended effect of many environmental actions. Risk reduction is the estimated change in some measure of risk (e.g., the expected number of fatalities averted) if the alternative is implemented. Other methods are used to estimate other kinds of effects. Socioeconomic impact assessment is a collection of tools for estimating the social and economic impacts of actions on the community (see Chapter 3). Integrated assessment strives to present the full range of relevant impacts, taking into account cause-and-effect linkages relevant to understanding "end-to-end" integration, as well as contributions and interactions across activities and consequences (Dowlatabadi and Morgan, 1993). Integrated assessment, obviously, requires bringing together a broader set of expertise, methods, styles of study, and degrees of confidence than would typically characterize a study of the same issues within the bounds of a single research discipline.

Tools for Valuing Impacts

CBA argues that effects should be valued according to their worth to individuals. For this reason, the concept of market price is significant for

CBA. In theory, a free market implicitly aggregates individual preferences by balancing aggregate demand with aggregate supply. Each individual adjusts his or her purchases until the value of the last item purchased is just worth what it costs. The equilibrium prices that result then indicate the marginal benefit realized from each individual's consumption of each good. For this reason, CBA assigns dollar values to impacts through a direct or indirect reference to the concept of a market for that impact.

Valuing the products, services, or other effects of environmental actions is relatively easy for CBA when those effects are traded in a free market. For example, the value of property made available for public or private use through the cleanup of hazardous-waste sites might be estimated from the cost of similar-quality real estate in the area. For effects for which no market exists, indirect procedures are used. For example, the value of reducing the number of fatalities to workers might be developed from a value of life estimated from wage differentials for workers in riskier-than-normal occupations. The value of fire protection might be estimated from the prices paid for fire insurance. If there is no direct or indirect market for an effect, CBA uses surveys or interviews with people to estimate their willingness to pay to obtain or avoid the effect, or the difficult-to-value effect is simply omitted from the analysis. Contingent valuation and contingent ranking methods are tools for estimating people's willingness to pay. Contingent valuation postulates a contingent market for the impact measure whose valuation is sought. A change in the level of the outcome measure is specified (e.g., cleaner air), and the subject (a survey respondent) is asked to indicate how much that change is worth. Chapter 2 provides additional discussion of methods for valuing impacts.

Tools for Computing Net Benefit

CBA aggregates costs and benefits occurring at different points in time with the concept of present value. Net present value (NPV) is calculated from the equation:

$$\text{NPV} = B_0 - C_0 + (B_1 - C_1)/(1 + r) + \ldots + (B_1 - C_1)/(1 + r)^T,$$

where B_t and C_t are the benefits and costs in year t, r is the discount rate, and T is the time horizon considered for the evaluation. CBA approaches generally rely on market arguments for selecting a discount rate. For example, resources used for an environmental project in the public sector will, in theory at least, force the displacement of private investments. According to CBA, such investments would be economically efficient only if the rate of return per dollar outlay toward public goals exceeds the opportunities forgone per dollar in the private sector. This logic suggests that benefits and costs of public programs be discounted by the opportunity costs of shifting productive resources out of the private sector. According to one line of reasoning, this argument implies that the discount rate should be set at the before-tax rate of return on private investments. In any case, determining discount rates based on opportunity costs is difficult. As a result, most CBA approaches emphasize the use of sensitivity analyses designed to explore whether varying the discount rate significantly affects the evaluation of options.

CBA usually adopts the classical view of probability (explained above in the section on "Tools for Risk Estimation"). Uncertainties are taken into account by associating probabilities with alternative net present values and then computing expected values (weighting net present values by probabilities and adding). In this case, the decision criterion is expected net present value. According to CBA, any action with a positive expected net present value is desirable, but the one that maximizes this quantity is most desirable.

Strengths and Limitations

Like PRA, CBA is logically sound in the sense that it is founded on a consistent, coherent theory. Properly used, CBA can help focus resources on addressing the most important environmental problems in the most cost-effective manner. Recent initiatives at the state and national levels have argued for statutory mandates for including CBA as a basis for future environmental regulations. States have taken a leading role in promoting increased use of CBA and risk assessment (Curtis and Michaels, 1996). Not all aspects of CBA theory, however, are necessarily attractive. For example, because CBA is concerned only with the net impacts on society, it is unconcerned with exactly who pays the costs or enjoys the benefits. Thus, projects that benefit the rich at the expense of the poor may do well under CBA. Problems with logical soundness also arise to the extent that requirements of the theory do not hold; for example, to the extent that market prices do not reflect people's preferences, the logical appeal of the approach is diminished.

Because CBA ignores the views and preferences of responsible decision makers, it does not normally employ tools that depend explicitly on expert or subjective judgment. Thus, CBA is less effective for problems where little or no data are available for quantifying important uncertainties or where important outcomes exist that do not have immediate, tangible, economic implications. Benefits and costs are often nonquantifiable with CBA not because they are intrinsically nonquantifiable, but because of the absence of relevant and reliable data on the market values of potential impacts. Similarly, uncertainties caused by lack of data are not intrinsically unquantifiable, but CBA may conceal such uncertainties because they cannot be quantified by classical probability methods. CBA also provides little opportunity for stakeholders to contribute to the analysis except, perhaps, in defining the problem (e.g., identifying alternatives). At the same time, however, CBA avoids the necessity of decision makers' providing subjective value judgments. It therefore appeals to some because it appears to be a more value-free guide to decision making. In reality, though, CBA embodies strong value judgments.

Decision Analysis

Like CBA, decision analysis (DA) is a theoretically sound megatool for comparing decision options. Decision theory derives from a set of axioms for what it means to be rational (von Neumann and Morgenstern, 1947). Most individuals find these axioms easy to accept. One such axiom states that all choices are comparable; offer a decision maker any pair of outcomes or lotteries (i.e., outcomes that occur with different probabilities), and the decision maker must be able to identify one or the other as preferable (or state that he or she is indifferent between them). Another axiom states that, if a decision has three possible outcomes, A, B, and C, and if the decision maker prefers A to B and B to C, then the decision maker must prefer A to C.

The axioms lead to a proof of two results central to DA. The first is that a decision maker's preferences can be encoded in terms of a function called a utility function. The utility function represents a scaling of the values the decision maker assigns to decision outcomes. The second result is that a decision maker's preferences for alternatives in decisions involving uncertainty may be measured by calculating expected utility (the sum of the utilities of possible outcomes weighted by their probabilities). Decision analysts adopt the Bayesian, or subjective, definition of probability (see the above section on "Tools for Risk Estimation" and Howard, 1968). As long as the decision maker accepts the axioms of decision theory, the alternative having the highest expected utility must be the one that is preferred.

In practice, DA is concerned with:

- Identifying the alternatives to a decision (what you can do)
- Estimating outcomes and assigning probabilities (what you know or believe will occur)
- Establishing the utility function (what you want)

At the simplest level, DA may be applied by encouraging a decision maker to identify multiple decision alternatives, to specify the range of likely consequences of each alternative, and to apply consistent value judgments when deciding how desirable or undesirable each possibility is. In more rigorous applications, the decision analyst works closely with decision makers and subject-matter experts to encode their judgments into probabilities, utility functions, and other models.

Tools for DA are described according to phases of what decision analysts call the DA cycle (see Figure 8.6).

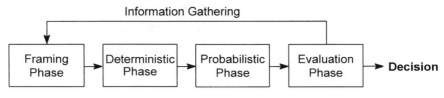

FIGURE 8.6. The Decision Analysis Cycle.

Tools for Decision Framing

Framing tools define the decision problem in a way that allows the alternatives to be evaluated with DA. For example, objective hierarchies (Keeney, 1992) facilitate the definition of performance measures. Performance measures are the consequences of an alternative that determine how good or bad things turn out. They serve as criteria for evaluating options. The decision analyst develops a list of objectives by questioning decision makers and stakeholders and then organizes the objectives into a hierarchy according to their generality. Figure 8.7 provides an example regarding the proposed construction of a highway. General objectives (such as quality of life) are defined in terms of more-specific, lower-level objectives (such as satisfaction of the travel needs of specific groups). Performance measures are defined for each of the lowest-level objectives in the objective hierarchy. Here, the number of commuter trips and the average time commuters spend in transit might serve as performance measures indicating the degree to which the objective "maximize satisfaction of the travel needs of commuters" is achieved. Building an objectives hierarchy helps ensure that no holes (missing objectives) occur in the analysis and helps eliminate situations where double-counting might result (because holes and redundancies are more easily identified from the graphic structure).

Influence diagrams, another DA framing tool, are graphic representations of the relationships among decisions, uncertainties, and performance measures (Oliver and Smith, 1988). Figure 8.8 shows an example regarding an underground pipeline at a chemical processing plant. The diagram was constructed as a group exercise with members of a team tasked to recommend actions to reduce the risks posed by the pipeline (CCPS, 1995).

As illustrated, an influence diagram consists of nodes and arrows. Rectangular nodes represent decisions. In Figure 8.8, there is one decision with three alternatives: (1) do nothing; (2) replace the current pipeline; or (3) insert automated shutoff valves that would limit releases in the event of a pipeline failure. Oval nodes represent uncertainties. Five uncertainties are identified: (1) the number of years until a failure occurs; (2) whether the failure is a slow leak or a rupture; (3) the amount of chemical released as the

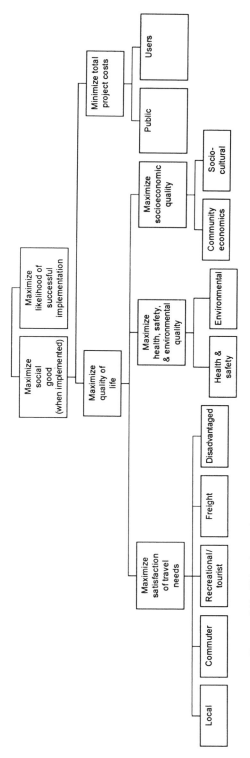

FIGURE 8.7. Objectives Hierarchy Constructed for a Decision Analysis of a Proposed Highway Construction.

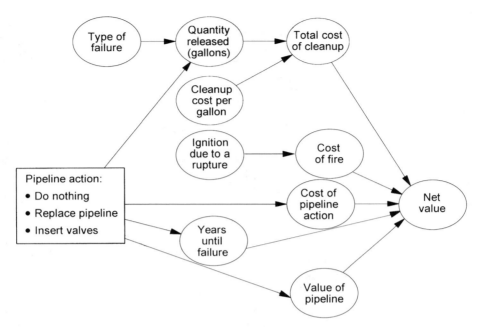

FIGURE 8.8. Influence Diagram for Pipeline Decision. (From Center for Chemical Process Safety. Copyright 1995 by the American Institute of Chemical Engineers and reproduced by permission of Center for Chemical Process Safety of AIChE.)

result of a failure; (4) the risk of ignition if there is a rupture; and (5) the cost to clean up a chemical release. Oval nodes (or rectangular nodes with rounded corners) represent performance measures, decision consequences relevant to the achievement of objectives. The performance measures include three potential costs: (1) improving or replacing the current pipeline; (2) cleaning up a spill; and (3) cleaning up after a fire. In addition, a performance measure is specified to account for any changes to the value of the pipeline. If the current pipeline is replaced, the new pipeline would have a useful service life beyond that of the current pipeline. Therefore, the additional performance measure is (4) the value of the remaining life of the pipeline.

In influence diagrams, an arrow from a node to an oval means that the latter uncertainty is influenced by the outcome of the former uncertainty or the alternative that is chosen. An arrow from a node to a rectangle (not shown in the example) means that the latter decision is made after the outcome of the former uncertainty or decision choice becomes known. Building influence diagrams helps participants understand the problem and organize information prior to developing a quantitative model. Upon completing the diagram shown in Figure 8.8, the team presented the diagram to the plant manager as a means for explaining and gaining approval for the quantitative analysis that was to follow.

Tools for Deterministic Analysis

The purpose of the deterministic phase is to translate the graphic representations of the decision model (e.g., flow charts, influence diagrams, and objective hierarchies) into a mathematical model. As illustrated by Figure 8.9, the deterministic model typically consists of a consequence model and a value model.

The model that combines the consequence and value models is referred to as "deterministic" because at this stage it does not account for the probabilities of key uncertainties. The consequence model predicts the consequences of choosing various alternatives as a function of the outcomes to relevant uncertainties. A consequence model for a decision to clean up a site used to impound chemicals at an oil refinery might estimate the reductions in the numbers of various health effects (e.g., cancers) occurring to plant workers and the local population; the number of fatalities and injuries occurring to cleanup workers; and the total costs of cleanup for each cleanup approach that might be used. These estimates would change, depending on the outcomes of key uncertainties, for instance, estimates of cost and impacts to workers might change depending on the outcome of uncertainty regarding the depth of contamination in soil.

The value model evaluates the consequences according to the preferences of the decision maker (e.g., the decision maker may wish to value or weight potential cancer fatalities to the public more heavily than potential accident fatalities to workers). In addition to accounting for the differing value tradeoffs that decision makers may make, the value model can also account for time preference and risk aversion. Time preference is typically taken into account by discounting impacts that occur in the future and using a present-value concept similar to that used in CBA. However, unlike CBA, the discount rate used in traditional DA is not chosen based on market arguments. Instead, it is chosen to reflect decision-maker preferences for

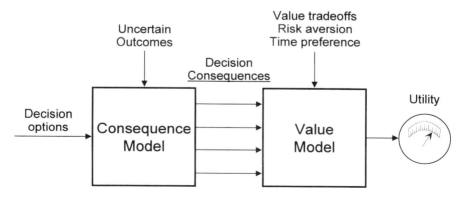

FIGURE 8.9. Form of Deterministic Model Used in DA.

postponing undesirable consequences. Risk aversion refers to the reluctance most people have to take gambles involving the possibility of significant losses, even though the expected value of the gamble seems attractive. Like time preference, risk aversion can be measured and represented by a parameter in a value model.

The various modeling tools discussed in the previous sections are useful for developing consequence models. Multiattribute utility analysis (MUA) (Keeney and Raiffa, 1976) is a set of tools for constructing a value model. The resulting value model is an equation for combining the performance measures to produce an overall measure of value (i.e., utility). An additive equation has the form:

$$U = w_1 S_1(x_1) + w_2 S_2(x_2) + \ldots + w_n S_n(x_n)$$

where U is utility, which is typically expressed in unitless terms (e.g., a 0-to-100 scale) or in equivalent dollars. The x_i are the performance measures, the S_i are scaling functions that indicate the achievement of specific levels of performance on given objectives, and the w_i are weights that reflect the relative importance of the objectives. Weights and scaling functions are assessed from decision makers or stakeholders with formal elicitation techniques. The additive form is typically used in multi-attribute tradeoff analysis, a form of MUA sometimes used to support collaborative decision-making processes involving multiple stakeholders with diverse interests. The additive form of the utility function is only appropriate if the value of doing well on any performance measure does not depend on the level of performance achieved on any other measure (which is necessary for the weights to be constants). Tests are available to determine if the additive or some other form is appropriate.

Tools for Probabilistic Analysis

In a sensitivity analysis, each uncertain input to the deterministic model is varied across its range of uncertainty. Input variables that have little or no effect on the final value are left at their nominal or most likely outcomes throughout the probabilistic phase of the analysis. Probability distributions are developed for those uncertainties that (1) have a large impact on the resulting utility for one or more alternatives, or (2) cause the preferred alternative (the alternative producing the highest utility) to change as the variable is moved from high to low settings. Probability distributions are derived from data or assessed from experts and then represented graphically in a decision tree. Figure 8.10 shows a decision tree corresponding to the influence diagram in Figure 8.8. As illustrated, a decision tree con0sists of decision nodes and chance nodes. The branches emanating from deciion nodes indicate the alternatives that are available. The branches emanating from chance nodes indicate the possible outcomes to the key uncertainties.

8. Assessment, Refinement, and Narrowing of Options   273

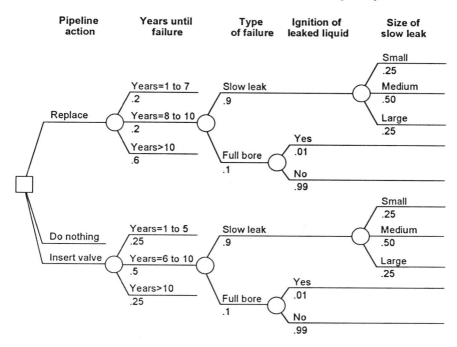

FIGURE 8.10. Decision Tree for Pipeline Decision.

Tools for the Evaluation Phase

To conduct an analysis with a decision tree, the analyst uses the deterministic model (i.e., the model developed in the deterministic phase) to specify decision outcomes corresponding to the sequence of decisions and events represented by each path through the decision tree. The nodes in the tree correspond to the model's sensitive input variables, so each path through the tree requires a separate run of the model. The utility function is used to assign utilities to each outcome, and these are displayed at the tree's endpoints. A risk curve describing the range and likelihood of the utilities of the possible outcomes to each alternative is calculated by computing the probabilities of each path through the tree (by multiplying the probabilities assigned to each branch along the path). Because the usual risk curve produced for a decision tree is a cumulative probability distribution, the probabilities of utilities less than each possible utility are summed or cumulated. Thus, the cumulative probability distribution is a nondecreasing curve that shows the probability of obtaining an outcome with utility less than or equal to any specified value. A decision tree may also be "solved" to identify the decision strategy that maximizes expected utility. This solution is accomplished by the "roll-back" procedure. The decision strategy that maximizes expected utility is found by starting with the utilities assigned to the endpoints of the tree and successively computing expected utility at each chance node and selecting the highest expected utility at each decision node.

Strengths and Limitations

Like the other megatools, DA can be regarded as a formalization of "common sense." It provides a rigorous and logically sound alternative to CBA that may be more (or less) appropriate than CBA, depending on the circumstances. More so than other megatools, DA is capable of accounting for "soft" issues, including the preferences of the decision maker and uncertainties that cannot be quantified from historical data. Conversely, compared to CBA, applications of DA typically require more time and involvement from decision makers. DA also produces explicit statements regarding potentially contentious judgments, such as the value of life and the likelihood of public deaths. Such statements can create controversy or, possibly, even legal liabilities for decision makers. Furthermore, applications of DA to problems involving multiple decision makers can be problematic. Individuals may have different beliefs and value judgments that cannot be simultaneously represented in a single model. On the other hand, DA can help those who disagree to better understand the source of the disagreement.

The various graphic representations produced by DA, including objective hierarchies, influence diagrams, and decision trees, are valuable aids for conducting qualitative evaluations and for communicating about the decision. Also, DA provides some additional capabilities not available with other tools, including a means for calculating the value of information. Value-of-information analysis places a dollar value on what it is worth to the decision maker to resolve uncertainties prior to committing to a decision (Demski, 1972).

Hazardous-waste-site cleanup is an example of a context where value of information is useful. The long-term health and environmental impacts caused by a hazardous-waste site are often highly uncertain. These uncertainties complicate the decision of what technologies to use to clean up the site. Therefore, site remediation typically includes studies intended to reduce uncertainties and to better characterize the site. How much money should be spent reducing site uncertainties prior to beginning cleanup? Value-of-information analysis can help answer such questions. According to DA, additional information should be collected only if the value of that information is greater than the costs of obtaining it.

Many of the criticisms directed at DA and related tools are easily refuted. The most common criticisms are that DA produces modeling bias (by emphasizing those considerations that are most easily modeled); produces results that have been manipulated by the analyst; is incompatible with institutional and political realities; and conflicts with basic ethical principles. Although such faults do occur, they do not represent flaws in the tools so much as problems with the way the tools are sometimes applied. Other common criticisms, such as reliance on judgmental inputs; inability to deal with unknown unknowns; difficulty of measuring costs and benefits; and promotion of anthropocentric values, are inherent shortcomings of virtually any decision-making process. The question is not whether DA and other tools are perfect, but whether the proper use of such tools can provide a significant improvement over unaided decision-making processes.

Obviously, DA and the other megatools offer little value to decision makers faced with routine decisions that must be made expeditiously. Such tools are probably not going to be useful to emergency-room physicians, firefighter rescue teams, or hazardous-material response teams. However, the tools can be enormously valuable for high-stakes, complex decisions when time for deliberation is available. When properly applied, DA and the other tools that are discussed in this chapter can help environmental decision makers choose scientifically sound, cost-effective, integrated actions that reduce or prevent risks while taking into account social, cultural, ethical, political, and legal considerations. The megatools are obviously complex, but they work when simpler tools do not.

Communication of Tool Use and Results

Decision-aiding tools are of little value if their results are not effectively communicated or are not persuasive to the parties to the decision. Communication is, therefore, critical to the successful application of tools for the assessment, refinement, and narrowing of options. The most effective means for communicating tool use and results is to involve interested parties in applications of decision-aiding tools. More generally, remember that effective communication is a two-way street: it means both listening and speaking. Communicators should learn about the concerns, values, knowledge, and experience of their audiences. Stakeholders, for example, might suggest new alternatives or have knowledge of risk sources and mechanisms of exposure that risk assessors do not have. The degree to which information provided by stakeholders is incorporated into decisions can influence trust, a key to the acceptance of decisions. Communicators must be prepared to explain and answer questions about any specific, relevant aspect of the analysis, not just to rely on "canned" presentations.

Risk communication is a particularly important and challenging aspect of communication. Decision situations are often emotionally charged. For example, citizens generally experience a sense of outrage on learning that a local industry has put them at peril through the release of pollutants, and outrage inevitably magnifies the perception of danger. Risk communication should include clear messages about the nature, severity, and likelihood of dangers and should demonstrate genuine concern for stakeholders and their reactions to risk problems (NRC, 1989). Given their importance, risk communications with the public should involve well-tested methods; an untested communication should no more be released than an untested product (Morgan et al., 1992).

Complexity is a major barrier to understanding for most of the tools in this category. Environmental issues are generally complex, so the tools and models for effectively addressing these problems are often necessarily complicated. Congress is considering various proposals to increase the transparency of the logic underlying government decisions. For example, consideration is being given to requiring agencies to compare a risk to be regulated with other dangers that might be experienced by the public. Risk comparisons can help people understand threats and place them in perspective. More generally, transparency means revealing and characterizing the assumptions, uncertainties, default factors, and methods used in the analyses that support decisions. In public debates, the decision rule needs to be clearly articulated. CBA or DA may be acceptable for those that believe efficiency or utility is an appropriate basis, but not to those who feel the choice should be rights-based (e.g., zero imposed risk) or procedure-based (e.g., best available control technology).

Making Tools Used and Useful

Numerous opportunities exist for applying the tools in this category. For example, Congress could use the tools to help develop risk-management legislation and to provide oversight of existing agency programs. The EPA could use the tools to support development of an integrated air-toxics strategy for urban areas or to support its Common Sense Initiative to integrate permitting requirements for manufacturing facilities. States and regional airshed or watershed authorities could use the tools to address current environmental problems. A city or county could use the tools to address community-level concerns, such as the cleanup of hazardous-waste sites. Business and industry could use the tools to select cost-effective strategies for meeting regulatory requirements and otherwise controlling risks to workers and the public.

A variety of strategies exists for improving the application of the tools in this category. Organizations can send staff members to courses and training seminars to develop in-house capabilities. Decision makers can hire management consultants who specialize in the application of the tools. Pilot projects involving a team of internal staff and consultants is often an efficient way to gain experience and understanding. During applications, sponsors and users can minimize technical problems by insisting that crucial assumptions be clearly identified and subjected to peer review. Time and costs can be controlled by selecting methods that meet schedule and economic constraints. Simple risk assessments based on expert judgment and unvalidated assumptions can be conducted in a matter of days or even hours. Such methods may not produce precise risk estimates, but they may be capable of inexpensively producing useful risk bounds.

Multidisciplinary problems can be dealt with as in other efforts requiring multidisciplinary teams. Good leadership, clear lines of authority and responsibility, and adequate mechanisms for communication and review can overcome the difficulties of harnessing the talents of multiple experts. Cross training can improve integration among participating specialists. For example, training in CBA or DA may be useful for risk assessors. Many risk assessors are unfamiliar with the information about risks that is needed by other decision tools. As a result, the questions asked in and the results of risk assessment often do not match the needs of CBA or DA.

Innovations

The tools in this category are being advanced through work in numerous disciplines, including applied mathematics, business management, computer science, economics, finance theory, operations research, and psychology. Much of the relevant research may be characterized by the term "decision science."

Because decisions that address poorly specified problems are especially difficult, improved structuring and framing tools are of high value. Research is underway to develop improved methods of representing problems visually to facilitate communication and analysis. Techniques for aiding the development and screening of alternatives are key areas of research. Extensions to address specific technical difficulties for risk decision making are also under way, including better techniques for assessing extremely-low-probability and high-consequence events, evaluating short-term investments with risks and rewards that occur over very long time frames, and eliciting values from stakeholders and the public. Research to develop improved tools by integrating concepts and techniques from distinct fields is also proving successful. For example, options theory (theory for valuing financial options) is being combined with DA to produce methods for more accurately estimating the value of risky projects and for identifying optimal dynamic and adaptive strategies for managing those projects (Smith and Nau, 1995). The approach integrates CBA and DA by using market data to quantify risks related to the market (e.g., commodity prices) and the subjective judgment of decision makers to quantify risks that are unique to the decision maker's organization.

Behavioral or descriptive decision-making research is needed to better understand how people process information and reach decisions. Developments in this area support the design of practical tools that better accommodate the decision-making processes actually used by individuals and groups. Research on tools for facilitating real-time decision making and negotiation is an active area of development. For example, work is ongoing to improve and streamline the modeling, data-gathering, and analysis tasks required for existing megatools.

Within health and environmental risk assessment and risk management, the focus needs to shift from the traditional chemical-by-chemical, medium-by-medium, threat-by-threat strategy to an integrated approach that recognizes the collective impact of multiple risk agents. People and the environment encounter multiple hazards, and effects are not always additive. Thus, information and risk assessment methods are being developed that address chemical mixtures and combined chemical-microbial-radiation exposures.

Computer technology has benefitted tools in this category tremendously, and further developments are likely. Key areas of research include:

- Improvements in the usability of software to help guide someone with little or no training in the decision-aiding tools through the analysis
- More powerful solution algorithms and software to enable larger and more complex problems to be solved and to better address real-time decision making
- Software that is customized to specific-risk decision problems
- Applications of artificial intelligence (AI), expert systems, learning systems, and decision-support systems to environmental decision making

Although the tools in this category have been successfully applied numerous times, organizations are understandably unsure and reluctant about extending applications to new areas. Empirical research to document the use and acceptance of various approaches to specific problems, agencies, and industries would be helpful. Also useful would be:

- Experimental applications that demonstrate the use of specific tools on representative environmental decisions and that identify ways to improve the applicability of the methods
- Research on organizational contexts and decision situations to identify which tools are most likely to succeed or have been most successful
- Legal research on the hazards of formal decision making to determine, for example, whether revealing the explicit basis for a safety-related decision increases or decreases an organization's legal vulnerability
- Improved training programs that expose decision makers to formal processes for decision making and that enable organizations to implement more-sophisticated decision aids with in-house staff.

Acknowledgments. The author thanks the reviewers who provided helpful comments on this chapter, including Howard Kunreuther and Charles Van Sickle.

Key Resources

In *Value-Focused Thinking* (Cambridge, MA: Harvard University Press, 1992), Ralph L. Keeney describes a systematic process for defining decision problems, creating alternatives, and articulating objectives.

The Center for Chemical Process Safety (CCPS) in its *Tools for Making Acute Risk Decisions* (New York: American Institute of Chemical Engineers, 1995) compares various decision-aiding methods, including voting, weighted scoring methods, cost-benefit analysis, and decision analysis.

In *Uncertainty* (Cambridge, England: Cambridge University Press, 1990), M. Granger Morgan and Max Henrion provide a guide to dealing with uncertainty in quantitative risk and policy analysis.

Miley W. Merkhofer's *Decision Science and Social Risk Management* (Norwell, MA: D. Reidel, 1987) contains a detailed discussion of strengths and limitations of cost-benefit analysis, decision analysis, and other decision-aiding approaches.

In *Acceptable Risk* (Cambridge, England: Cambridge University Press, 1981), Baruch Fischhoff, Sarah Lichtenstein, Paul Slovic, Stephen L. Derby, and Ralph L. Keeney provide an analysis of approaches to making acceptable-risk decisions.

Relevant websites include those of the Decision Analysis Society of INFORMS (http://www.fuqua.duke.edu/faculty/daweb), the Decision Science Institute (http://www.gsu.edu/~dsiadm), and the Society for Risk Analysis (http://www.sra.org).

References

Abelson, P. 1979. *Cost-Benefit Analysis and Environmental Problems*, Westmead, England: Saxon House.

Akland, G.G. et al. 1985. Measuring human exposures to carbon monoxide in Washington, D.C. and Denver, CO., during the winter of 1982–83. *Environmental Science Technology* 27:369.

Armstrong, J.S. and MacGregor, D. 1994. Judgmental decomposition: When does it work? *International Journal of Forecasting* 10:495–506.

Beanlands, G.E. and Duinker, P.N. 1983. *An Ecological Framework for Environmental Impact Assessment in Canada*. Halifax, Nova Scotia: Institute for Resources and Environmental Studies, Dalhousie University.

Bitz, D.A., Berry, D.L., and Jardine, L.J. 1993. *Decision and Systems Analysis for Underground Storage Tank Waste Remediation System, SAND94-0065*. Albuquerque, NM: Sandia National Laboratories.

Cairns, J., Jr., Dickson, K.L., and Maki, A.W. (Eds.). 1978. *Estimating the Hazard of Chemical Substances to Aquatic Life*, STP 657. Philadelphia: American Society for Testing and Materials.

Center for Chemical of Process Safety (CCPS). 1995. *Tools for Making Acute Risk Decisions*. New York: American Institute of Chemical Engineers.

Cheshire, R.A. 1991. Introduction to Environmental decision making. In: R.A. Cheshire and S. Carlisle (Eds.), *Environmental Decision Making: A Multidisciplinary Perspective*. New York: Van Nostrand Reinhold. Pp. 1–13.

Churchman, C. 1961. *Prediction and Optimal Decision: Philosophical Issues of a Science of Values*. Englewood Cliffs, NJ: Prentice-Hall.

Commission on Risk Assessment and Risk Management (CRAM). 1996. *Risk Assessment and Risk Management in Regulatory Decision-Making.* (Draft.) Washington, D.C.: Commission on Risk Assessment and Risk Management.

Covello, V.T. and Merkhofer, M.W. 1993. *Risk Assessment Methods: Approaches for Assessing Health and Environmental Risks.* New York: Plenum.

Curtis, T. and Michaels, T. 1996. Risk assessment and cost-benefit analysis: The states' perspective. *Inside EPA's Risk Policy Report* 3(8):29–30 (August 23).

Demski, J. 1972. *Information Analysis.* Menlo Park, CA: Addison-Wesley.

Dowlatabadi, H. and Morgan, M.G. 1993. Integrated assessment of climate change. *Science* 259:1813.

Dummet, M. 1984. *Voting Procedures.* Oxford: Clarendon Press.

Dyer, J.S. 1990. Remarks on the analytic hierarchy process, *Management Science* 36(3):249–258.

Edwards, W. and Barron, F.H. 1994. SMARTS and SMARTER: Improved simple methods for multi-attribute utility measurement. *Organizational Behavior and Human Decision Processes* 60:306–325.

Finkel, A.M. 1994. Stepping out of your own shadow: A didactic example of how facing uncertainty can improve decision-making. *Risk Analysis* 14(5):751–761.

Fischhoff, B. 1979. Informed consent in societal risk-benefit decision. *Technological Forecasting and Social Change* 13:347–357.

Fischhoff, B., Slovic, P., and Lichtenstein, S. 1980. Knowing what you want: Measuring labile values. In: T.S. Wallsten (Ed.). *Cognitive Processes in Choice and Decision Behavior.* Hillsdale, NJ: Erlbaum. Pp. 117–142.

Gots, R. 1992. *Toxic Risks: Science, Regulation, and Perception.* Boca Raton, FL: Lewis Publishers.

Gramlich, E.M. 1990. *A Guide to Benefit-Cost Analysis*, 2nd ed. Englewood Cliffs, NJ: Prentice Hall.

Howard, R.A. 1968. The foundations of decision analysis. *IEEE Transactions on Systems Science and Cybernetics* SSC-4(3):211–219.

Karr, J.R. 1991. Biological integrity: A long-neglected aspect of water resource management. *Ecological Applications* 1:66–84.

Keeney, R.L. 1992. *Value Focused Thinking.* Cambridge, MA: Harvard University Press.

Keeney, R.L. and Raiffa, H. 1976. *Decisions with Multiple Objectives: Preferences and Value Tradeoffs.* New York: John Wiley and Sons.

Kelly, J.S. 1987. *Social Choice Theory.* New York: Springer-Verlag.

Kepner, C.H. and Tregoe, B.B. 1981. *The New Rational Manager.* London: John Martin Publishing.

Krawiec, F. 1984. Evaluating and selecting research projects by scoring. *Research Management* 27(2):21–25.

Kusnic, M.W. and Owen, D. 1992. The unifying process: Value beyond traditional decision analysis and multiple decision-maker environments. *Interfaces* 22(6):150–166.

McCormick, N.J. 1981. *Reliability and Risk Analysis.* New York: Academic.

Merkhofer, M.W., Conway, R., and Anderson, R.G. 1997. Multi-attribute utility analysis as a framework for public participation in siting a hazardous waste management facility. *Environmental Management*, 21(6):831–839.

Merkhofer, M.W. 1987. Quantifying judgmental uncertainty: Methodology, experiences, and insights. *IEEE Transactions on Systems, Man, and Cybernetics* SMC-17(5):741–752.
Merkhofer, M.W. and Keeney, R.L. 1987. A multiattribute utility analysis of alternative sites for the disposal of nuclear waste. *Risk Analysis* 7(2):173–194.
Merkhofer, M.W. and Korsan, R.J. 1978. *Florida Utility Pollution Control Options and Economic Analysis—Volume 2: Cost-Benefit Analysis of Alternative Florida Sulfur Oxide Emissions Control Policies*. SRI final report, project 5,080. Menlo Park, CA: SRI International.
Morgan, M.G. 1981. Probing the question of technology-induced risk. *IEEE Spectrum* 18(11):58–64.
Morgan, N.G. et al. 1992. Communicating risk to the public. *Environmental Science and Technology* 26:2048–2056.
Mumpower, J.L. et al. 1987. *Expert Judgment and Expert Systems*. New York: Springer-Verlag.
National Research Council (NRC). 1978. *Saccharin: Technical Assessment of Risks and Benefits*. Washington, D.C.: National Academy Press.
National Research Council (NRC). 1989. *Improving Risk Communication*. Washington, D.C.: National Academy Press.
Oliver, R.M. and Smith, J.Q. 1988. *Influence Diagrams, Belief Nets, and Decision Analysis*. New York: John Wiley & Sons.
Poucet, A. 1990. STARS: Knowledge based tools for safety and reliability analysis. In: G.E. Apostolakis and P. Kafka (Eds.). *The Role and Use of Personal Computers in Probabilistic Safety Assessment and Decision Making*. New York: Elsevier. Pp. 379–397.
Riggs, J.L. and West, T.M. 1986. *Essentials of Engineering Economics*. 2nd ed. New York: McGraw-Hill.
Saaty, T.L. 1980. *The Analytic Hierarchy Process*. New York: McGraw-Hill.
Silvers, A. and Hakkarinen, C. 1987. Materials damage from air pollutants. *EPRI Journal* 12(6):58–59.
Simon, H.A. 1976. *Administrative Behavior: A Study of Decision-Making Process in Administrative Organization*. 3rd ed. New York: The Free Press.
Smith, J.E. and Nau, R. 1995. Valuing Risky Projects: Option Pricing Theory and Decision Analysis. *Management Science*, 41(5):795–815.
Stewart, D., Treshow, M., and Harner, F.M. 1973. Pathological anatomy of coniferous needle necrosis. *Canadian Journal of Botany* 51(5):983.
Suter, G.W. (Ed.). 1993. *Ecological Risk Assessment*. Chelsea, MI: Lewis Publishers.
U.S. Environmental Protection Agency (EPA). 1990. Hazard ranking system, Final rule. *Federal Register* 55:51532–51667.
Vesely, W.E., Goldberg, E.F., Roberts, N.H., and Haasl, D.F. 1981. *Fault Tree Handbook*, NUREG/CR-0492. Washington, D.C.: U.S. Nuclear Regulatory Commission.
von Neumann, J. and Morgenstern, O. 1947. *Theories of Games and Economic Behavior*. Princeton, NJ: Princeton University Press.
Walker, W.E. 1986. Use of screening in policy analysis. *Management Science* 32(4):389–402.
Westman, W.E. 1985. *Ecology, Impact Assessment, and Environmental Planning*. New York: John Wiley and Sons.
White, D.J. 1975. *Decision Methodology*. New York: John Wiley and Sons.

Decision-Maker Response

Lynn C. Maxwell

The tools described in this chapter fit into three categories: risk assessment, cost-benefit analysis, and decision analysis. To this list there is a fourth that has been found to be quite useful in situations involving stakeholders with diverse and strongly held opinions: multi-attribute trade-off analysis.[1] Briefly, multi-attribute trade-off analysis moves from the identification of issues and concerns about a particular decision to the development of preferred strategies. The approach includes the following steps:

- Identifying public issues and relevant concerns
- Translating public issues and concerns into quantitative evaluation criteria, decision options, and uncertainties
- Crafting options into strategies
- Identifying possible future conditions (uncertainties)
- Constructing scenarios (combinations of strategies and future conditions)
- Using trade-off analysis to find the best strategies for the future

There are generally several participants in the decision-making process, including the analyst who uses the tools, gathers data, and identifies results; a manager or team leader who is responsible for oversight of the analysis process; the decision maker; and the stakeholders who have a definite interest in the outcome and can influence the decision maker.

The choice of the decision-making tool generally falls to the analyst and team leader. The success of the decision outcome can hinge on the appropriate choice of the tool. A simple tool may not provide the necessary information, and a complex tool may be too cumbersome and may produce results that are not well understood. Such a choice can lead to the wrong decision or to no decision at all (which may be the wrong decision).

The decision characteristics influencing the choice of tools are presented in Table R8.1. The tools are shown along the top of the table and include risk assessment, simple cost-benefit or net-present-value analysis, decision analysis, and multi-attribute trade-off analysis. Moving from left to right, these tools are generally ranked from simple to complex with the possible exception of risk assessment.

The analyst should ask about several decision characteristics or questions before choosing a decision tool. First, what is the impact of the results on

---

[1] Andrews, C.J. 1991. Spurring inventiveness by analyzing tradeoff: A public look at New England's electricity alternatives. *Environmental Impact Assessment Review* 12:185–210.

the criteria? Will the impact be large or small? Will the judgment be based on a single, easily understood criterion or on multiple, hard-to-quantify criteria? Will the alternatives be simple or will they involve several options? How important is uncertainty, and what are the potential risks? Does the decision need to be made immediately, or is time available for information gathering and deliberation? What is the importance of the decision to stakeholders; is it minimal, or do the stakeholders hold either diverse or strong opinions?

As shown in Table R8.1, a simple cost-benefit analysis, such as net present value, may be the appropriate tool if the impact on the results is small; there is one criterion (e.g., dollars); the decision alternatives are simple; the risk is small; the planning horizon is close; and stakeholders have only a casual concern. More complex tools, such as decision analysis or multi-attribute trade-off analysis, can be used when the impact on results is large, decision alternatives are complex, risks are significant, or time is available for deliberation.

Multi-attribute trade-off analysis may be preferred over the decision analysis tool when there are multiple criteria that also may be difficult to quantify or there are stakeholders with diverse or strong opinions. This approach is useful when stakeholders participate in the process.

The presentation of results and the types of results presented can make or break a decision. One of the largest risks is that the results presented are too

TABLE R8.1. Choice of tools.

Decision characteristics	Risk assessment	Net present value	Decision analysis	Multi-attribute trade-off analysis
		Result impact (e.g., $)		
Small	X	X		
Large	X		X	X
		Criteria		
Single	X	X	X	
Multiple			?	X
		Decision alternatives		
Single	X	X		
Complex			X	X
		Importance of uncertainty		
None		X		
Risky	X		X	X
		Immediacy of decisions		
Now	X	X		
Later		X	X	X
		Importance of stakeholders		
Somewhat		X	X	
Strong opinions	X			X
Diverse opinions				X

TABLE R8.2. Need or desire for understanding tools and results.

Who	Detailed			General			
	Tools	Analytical results	Interpreted results	Alternative criteria	Detailed alternatives	What ifs	Nonquantitative information
Analyst	Must	Must	Must	Must	Must	Must	Somewhat
Manager/ team leader	Somewhat	Must	Must	Must	Must	Must	Must
Decision maker	Little	Little	Must	Must	Somewhat	Somewhat	Must
Stakeholder	Somewhat	Somewhat	Must	Must	Somewhat	Somewhat	Must

complex for the decision maker to understand in the time the decision maker has available. This situation generally leads to decision gridlock or to a poor decision. The participants that need to understand the tools and the type of results are shown in Table R8.2. The participants are shown along the left of the table, and differing results are indicated at the top. The detailed results include information about the tools used in the analysis and analytical results (model output or simplified model output). Other information is targeted to the decision maker. Most importantly, this information includes the results interpreted in the decision maker's language. Additional information may include details about the criteria, one or more of the alternatives considered, what ifs, and nonquantifiable information that relates to the decision. In Table R8.2, the degree to which each participant needs or desires to know the various types of information is rated as a must, somewhat, or little. These categories are obviously subjective.

Generally, the detailed information must be known by the analyst and the team leader. The decision makers and stakeholders are largely interested in the interpreted results, the key criteria used to make the recommendations, and nonquantifiable information that supports the recommendation. The decision maker may also want to know what alternatives were considered and the results of one or more what ifs. One exception is that stakeholders may want to know about the tools and analytical results, particularly if there is a lack of trust between the stakeholders and decision makers.

Thus, the key to the successful use of decision tools is the appropriate presentation to the decision maker.

9
Post-Decision Assessment

GILBERT BERGQUIST and CONSTANCE BERGQUIST

Federal, state, and local governments spend billions of dollars annually in attempts to implement public policies, and private businesses and citizens spend billions more complying with the programs and regulations spawned by those policies. Almost as staggering as the amount of money that is spent is how little effort is made to find out what and how much is being achieved. This chapter will focus on a series of tools that should be useful to environmental decision makers in determining the character and level of impact that their policies, programs, and activities are having on the environment.

The Context

Post-decision analysis is a neglected part of the environmental decision-making process. Five reasons account at least partially for this neglect.

First, the most obvious and most egregious environmental problems that have been the focus of environmental policy for the past 27 years did not demand rigorous post-decision analysis. The problems were clear, and the results of the response were fairly obvious. Granted, a significant investment was made at the federal level in environmental monitoring systems to track the progress of corrective actions. But the long-term results are spotty; some areas (notably air) have relatively good data, and others (water quality, for example) have relatively weak documentation.

Second, post-decision analysis, particularly in large organizations, requires management sophistication and political will to be effective. Many managers do not have the training or the practical experience required. Other managers are apprehensive about the potential negative impacts that might arise from exposing their programs and activities to the assessment and evaluation of post-decision analysis.

Third, the organizational discontinuity of the American political system contributes powerfully to the de-emphasis of planning and post-decision analysis. Elected executive officials and their appointed administrators generally have four to eight years to carry out their programs. With no certainty that their policies or practices will continue past their tenure, they often focus on achieving limited, short-term agendas. Similarly, legislative elected officials have short terms and a focus on annual or biennial budgets restraining their vision and limiting their interest in the long-term management process.

Fourth, the management context in which environmental decision makers work mitigates against using long-term planning and management tools. Today's management context is marked by extraordinary complexity. Legal, social, cultural, economic, and political side issues and concerns effectively prevent important decisions from being simple. Further, the rapid pace of change in the decision environment and the pure volume of issues to dealt with forces many decision makers into a constant state of "crisis management." Taking the time and expending the effort to conduct more contemplative, longer-term planning and management activities are greater commitments than many managers feel they can make.

Fifth, with environmental budgets growing slowly or declining, some managers have little interest in engaging in planning and management activities that seem to require growth and change. With limited resources, they feel constrained to do the best they can in managing existing programs and activities.

After a quarter of a century of environmental protection, we find ourselves laboring in a policy environment where the issues are rapidly changing, the current programs and activities are losing their relevance, and the information and management processes needed to redirect our efforts are limited. In the past, expanding budgets allowed the absorption of new issues without the need to evaluate previous program results. Today, the circumstances are substantially different.

Federal, state, and local environmental organizations are clearly in the midst of some very difficult times. Continuing increases in the demand for services and in the cost of delivering those services in the face of low-growth, static, or declining appropriations have been responsible, at least in part, for the perception of lower levels of performance and for reduced confidence on the part of the citizenry. The management changes that governments must make are not modifications of their business-as-usual practices. Rather, fundamental management changes are needed in how governments approach and implement their programs and activities.

Five years of growing improvement in environmental-management techniques and processes at the federal and state levels were snapped into focus by *Setting Priorities, Getting Results: A New Direction for EPA* (National Academy of Public Administration, 1995, 1997). That report was prepared by the National Academy of Public Administration for the U.S. Environmental Protection Agency (EPA) to examine the development and implementation of environmental policy. The report is an excellent analysis of environmental-policy development and the state of environmental-management practice at EPA and in the United States in general. A number of its many recommendations have particular relevance to postdecision-making processes and techniques:

- Establishing a single organizational unit reporting directly to the Administrator that centralizes all of the relevant management capabilities of EPA
- Using risk and relative risk as the means of setting environmental priorities in the agency
- Developing national environmental goals with measurable benchmarks to set national direction and to serve as the foundation of agency strategic planning
- Institutionalizing the Governmental Performance and Results Act requirements to tie future budgets to environmental results
- Requiring performance reporting to be a part of EPA's internal budget process
- Re-establishing a formal accountability system
- Enhancing program evaluation within the agency
- Creating an environmental-statistics program capable of providing high-quality environmental information, data, and indicators
- Developing program measures
- Restructuring the relationship with states to provide flexibility to states that achieve measurable program and environmental results.

The importance of these recommendations must not be underestimated. They are profound and can lead the way to a fundamental recasting of the way environmental policy is developed, implemented, and evaluated. The increased emphasis on ranking issues and activities with risk-based tools, on evaluation through measurement, and on programmatic and environmental accountability all contribute to an important redirection in environmental management. While these proposals are presently focused on EPA, they will influence the standards and direction for environmental management that will affect states, tribes, regional and local governments, and the private sector. The remainder of this chapter will focus on a series of tools that are being used to bring about this small revolution in environmental management.

Post-Decision-Assessment Tools

Overview

Post-decision-assessment tools in their simplest form attempt to answer the question, "How are we doing?" Every day, federal, state, tribal, substate and local governments, the private sector, and nonprofit environmental organizations make hundreds of thousands of decisions directly and indirectly affecting the environment. The scope, scale, character, and significance of those decisions vary widely. In their totality, they represent the implementation of environmental policy in the country. When grouped with other decisions made by an organization, they reflect the policy of that agency and the results of its programs. As individual decisions, they represent an attempt to solve some specific environmental problem.

At all of these levels, they reflect attempts to change environmental conditions. And it is the measurement of that change that is the essence of post-decision assessment. Without such assessment, we cannot know what results we are achieving, and we remain uncertain about the merits of our collective and individual decisions. Without such certainty, decision makers cannot make confident decisions as to whether to continue, change, or terminate the path selected by their decisions.

Framework

The organization of this discussion has three elements: the organizational level, the type of organization, and the tools.

Organizational Level

The term "decision" is broadly applied in this book. To facilitate discussion of the tools, let us consider three types of decisions, based on the organizational level at which the decision is being made.

- Mission-Based Decisions: While the specific impact of each environmental decision may be important to assess, policy makers, organization leadership and staff, and the public will likely want to assess the collective or cumulative effect of all the environmental decisions of the organization toward meeting its overall environmental mission. Post-decision assessment at this level focuses on whether the organization is achieving the broad results it was designed to achieve.
- Program-Based Decisions: From the point of view of decision making, an environmental program may be seen as a structure and/or process designed to produce certain types of environmental decisions. The collective or cumulative impact of those decisions, measured in terms of environmental results, reflects the worth of that program.
- Individual Decisions: Many individual decisions are of sufficient importance to public and private decision makers or the public to require individual assessment to guide further action and to show results.

Type of Organization

Chapter 1 identifies eight types of organizations that make environmental decisions:

- Federal government
- Regional government
- State government
- Local government
- Large businesses and business associations
- Small businesses
- State, regional, and national citizen groups
- Local citizen groups

To reduce redundancy, all of the governmental categories will be collapsed into a single classification—governmental—because, with some variation in the sophistication of their application, the tools applicable for governmental organizations are appropriate for all but the smallest of local governments. Thus, the types of organizations considered here are:

- Governmental
- Large businesses and business associations
- Small businesses
- State, regional, and national citizen groups
- Local citizen groups

Governmental organizations will most often be concerned with the assessment of their own decisions. Private organizations and citizen groups are interested in assessing their own activities, but may have an interest in assessing the decisions and the results of decisions of other organizations, particularly governmental organizations.

The Tools

This discussion will focus of four postdecision-assessment tools: goals and goal systems, indicators, budgeting systems, and evaluation.

Goal Systems

Overview

Goals are generally thought of in terms of their direction-setting capabilities. They are established to give us targets, to tell us where we are going, and to indicate what we want to achieve. They have the potential to be one of the most powerful management tools an organization can employ. In addition to this prospective role, goals can also be an effective post-decision-assessment tool by providing a standard against which progress can be periodically measured. In this chapter, goals will be viewed from the perspective of their role as a post-decision-assessment tool.

To reach their potential, goals should not only be used to set a direction and a target, they should also be used interactively to measure and document progress toward meeting the goal. For many agencies and their decision makers, this measurement is a scary proposition. Publicly stating what you want to achieve by a particular time in measurable terms and then regularly reporting on your progress is fraught with political dangers and the prospect of embarrassment. As a consequence, agencies have avoided developing public, visible, measurable, aggressive environmental goals, choosing to develop dramatic and, often, poetic qualitative goals that bind them to little. An example comes from the 1978 Florida State Comprehensive Plan in which the overall goal for water management was to "manage water and related resources to achieve maximum economic and environmental welfare for all the state's citizens on a long-term basis" (Askew, 1978, p. 174). Such goal statements may be acceptable if they are supported by quantified objectives. An example of a supporting objective for this goal would be to "manage surface waters to allow for reasonable beneficial uses while maintaining and, where necessary, reestablishing natural water resource and biological relationships to provide for a balance of urban, agricultural, and natural systems, recognizing that natural productivity is optimized under unaltered conditions" (Askew, 1978, p. 178). While both statements might be of value as expressions of policy, neither provides any foundation for measuring progress toward achievement. Further, a review of the remainder of the document reveals no other mechanism for measurement. Even where effective goals are written, they are often underplayed and are seldom reported on after their development.

Goals should reflect achievements of critical importance to the environment, achievements that measure the overall success of the organization. In addition, they should be:

- Directly measurable and supported by one of the indicators
- Quantitatively stated
- Capable of being graphically displayed
- Time limited
- Capable, ideally, of displaying a target condition that can be sustained

While conservatism and fear on the part of decision makers is understandable in many contexts, indicator-supported goal systems have provided the foundation for some of the most successful environmental-management systems in the world. In identifying effective models for goal development and use, two projects stand out: The Netherlands's *National Environmental Policy Plan (NEPP)* and The Chesapeake Bay Program. As different as these projects are, they share a level of success generally not matched by other environmental-planning processes, a success that results from a planning logic founded in the effective use of environmental indicators.

National Environmental Policy Plan (The Netherlands)

The *NEPP* is a national environmental-planning process. Through the 1970s and the early 1980s, it became readily clear to the Dutch that they were living in one of the most polluted and environmentally damaged countries on the planet and were in danger of polluting The Netherlands to an unlivable state within a few decades. High population densities; high automobile densities; a highly industrial economy with a strong chemical component; intensive agriculture highly dependent on fertilizers; structural water management; and their geographic position in the midst of other European industrial giants had placed the nation's environmental future in jeopardy.

With the precarious nature of the country's environmental future, the initiation of the NEPP began with a rather profound goal, to "reverse environmental degradation and achieve sustainable development within one generation" (The Netherlands Ministry of Housing, Spatial Planning, and the Environment, 1993, p. 1). The strategy of the NEPP included adoption of quantified (measurable) targets and time frames, the integration of environment into decision making by all sectors of society, clear identification of responsibility for actions, creativity in the design and use of policy instruments, a commitment to long-term reshaping of social and economic structures, and recognition of The Netherlands' dependence on international cooperation (The Netherlands Ministry of Housing, Spatial Planning, and the Environment 1993, p. 2). The Dutch identified a series of preeminent environmental issues or, in their terms, themes:

- Climate change (global warming and ozone depletion)
- Acidification (acid deposition)
- Eutrophication (excessive nutrient buildup in surface water)
- Diffusion (uncontrolled spread of chemicals)
- Waste disposal
- Disturbance (noise and odor)
- Dehydration (damage from lowered groundwater levels)
- Squandering (inefficient use of natural resources, energy, and raw materials)

For each of these themes they developed broad composite indicators capable of measuring performance. Specific national goals for each indicator were set to achieve environmental sustainability within 20 years. Because the goals were based on quantitative measures (indicators), they were displayable in graphic form. Figure 9.1 shows a general representation of a NEPP goal. Inherent in each goal is the trend with regard to each theme, the present status, interim objectives on the path to achieving the goal, and the point at which the goal of sustainability is achieved.

As the plan progresses, an additional line can be added to the graph: actual environmental performance. The performance line can be compared

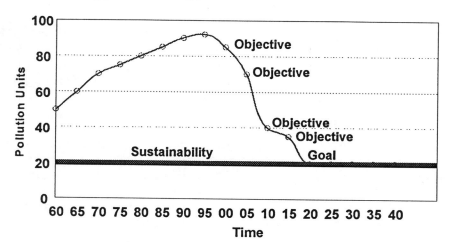

FIGURE 9.1. A generic environmental goal and its objectives, expressed in terms of units of pollution.

to the objectives, and a ready assessment can be made regarding the *environmental* success in reaching the goal. This assessment then serves as the foundation for revising the plan and reassessing action commitments.

The Dutch then broke down each of these broad goals into a series of more specific measures that were components of the larger composite indicator. They identified economic sectors (target groups) that were associated with each of the themes. Target groups include agriculture; traffic and transport; industry and refineries; gas and electric supply; construction; and consumers and retail trade. Industries within each of the sectors took responsibility for their own part in achieving the national goals, and specific targets to measure their performance were developed. Action plans based on a collaborative process involving governmental agencies and the economic sectors were also developed and implemented.

The point to be noted here is the absolutely pivotal role of indicator-based goals in the process. As used by the Dutch, indicators:

- Focus planning attention on the achievement of environmental results
- Provide the mechanism for measuring success in carrying out the planning process
- Provide a visible, appealing, and scientifically sound focal point for all stakeholders and the public
- Provide the basis for negotiation in bringing the private sector to the table and ensuring its participation
- Create the mechanism by which all of the participants in the process can focus on results as the means of reassessing and revising the plan.

Chesapeake Bay Program

The Chesapeake Bay Program is a federal state local program, generally coordinated by the EPA, that has become a model resource-based planning system. It is of interest here because of its near-total dependence on environmental indicators as the source of its dynamism (Chesapeake Bay Program, 1995).

Although the logics that define the NEPP and the Chesapeake Bay Program are very similar, the structure and formality of the processes are strikingly different. The Chesapeake program was formally initiated in 1983 as a means of dealing with the environmental degradation of the bay. Working cooperatively with the relevant federal program and the state and local governments whose watersheds affect the bay, the program initiated a long-term restoration of the resource.

The central focus of the project was the development of an extensive monitoring system that yielded a rich array of environmental indicators capable of documenting the health of the system. Though with far less formality than the Dutch, the program used indicators in much the same way to achieve demonstrable environmental results. Indicators were given three important roles in the process: (1) as key measures of success; (2) as facts to support goal-setting and program management; and (3) as targets and endpoints for the restoration effort. With some of the indicators used as a measurement tool, quantitative goals were set for key concerns. Measurement of the achievement of the goals was reported each year, and those results became the basis for the readjustment of the whole process.

Although considerably different in many ways, the NEPP and the Chesapeake Bay Program are virtually identical in their basic logic—the logic that makes their respective processes successful. Both emphasize the achievement of environmental results. Both use indicators as a means of measuring success and setting measurable goals. Both use the analysis of the goals and the indicators as the means of ensuring that the planning process is adaptive.

National Goals for America

The EPA has for the past three years been attempting to develop a broad set of environmental goals for the United States. Although still in draft condition, *Environmental Goals for America with Milestones for 2005* (USEPA, 1996) has evoked much discussion, and although somewhat controversial, represents a positive step for results-oriented environmental management. The document identifies 12 major issues and develops several benchmarks (measurable goals) for each. The goals, in brief, are:

1. Every American city and community will be free of air pollutants at levels that cause significant risk of cancer, respiratory disease, or other health problems.

2. All of America's rivers, lakes, and coastal waters will support healthy communities of fish, plants, and other aquatic life and uses like fishing, swimming, and drinking water supply.
3. America's ecosystems will be safeguarded to promote the health and diversity of natural and human communities and to sustain America's environmental, social, and economic potential.
4. Every American public water system will provide water that is consistently safe to drink.
5. The foods Americans consume will continue to be safe for all people to eat.
6. All Americans will live, learn, and work in safe and healthy environments.
7. All Americans will live in communities free of toxic impacts.
8. Accidental releases of substances that endanger our communities and the natural environment will be reduced to as near zero as possible.
9. Wastes will be stored, treated, and disposed of in ways that prevent harm to people and other living things.
10. Currently contaminated places will be restored to uses desired by surrounding communities.
11. Significant risks to human health and to ecosystems from global environmental problems will be eliminated at the transboundary level.
12. Americans will be empowered to make informed environmental decisions and to participate in setting local and national priorities.

The goal statements themselves are qualitative, but the benchmarks are, with some exceptions, reasonably effective, quantitative reflections of the dimensions of the goal. While some of the benchmarks are program measures instead of environmental measures, and some benchmarks lack both the program and environmental data that give them real measurability, the system is plainly a step in the right direction. How EPA will employ these goals and the associated benchmarks is not yet clear, but the potential of these goals is powerful. It is hoped that the successful adoption and use of these goals will serve as a model for other environmental agencies to follow.

Goals and Post-Decision Assessment

The three examples of goal systems are all large-scale, two national systems and one multistate regional system. However, the use of goal systems is appropriate and fully feasible for virtually any organization. While the development of a comprehensive, fully integrated, national goals system, such as the NEPP, is extraordinarily complicated and difficult, less ambitious versions are sufficiently simple, straightforward, and relatively nontechnical to be within the range of any organization. At the end of this chapter, a summary process for developing an integrated goal-based

TABLE 9.1. Strengths and weaknesses of goal systems.

Strengths	Weaknesses
Focus attention and measures achievement on organization priorities	Expose organization to criticism if goals are not met
Excellent for projecting results to policy makers and the public	Setting targets can be scientifically inexact
Conceptually simple and easy to understand	Associated process can become extensive and expensive
Can be used effectively by virtually any organization	Require ongoing attention and maintenance to be effective
Presents achievements in explicit, measurable terms	

planning system is provided that could be implemented by virtually any organization.

Theoretically, goal systems could be used to support any type of decision, but most commonly they will support mission-based and program decisions. Their attributes are summarized in Table 9.1.

Indicators and Indicator Systems

Overview and Concepts

Indicators are one of the most powerful management tools available and really are the foundation for all the other tools discussed in this chapter. A working definition of an indicator is, "A parameter, or value derived from a parameter; which points to; provides information about; or describes the state of a phenomenon, environment, or area with a significance extending beyond that directly associated with a parameter value" (Group on the State of the Environment, 1993, p. 6).

The key point about an indicator is that it is a measure that has significance that is broader than the measure itself; that is, the measure represents a much wider issue, condition, phenomenon, or circumstance than what is being directly measured. The original intent was to begin this discussion with monitoring systems instead of indicators. Much, if not most, of the measurement capacity of environmental organizations is rooted in monitoring systems—structured, regular data-collection systems designed to provide measurements of environmental values or program achievements. Ultimately, however, monitoring systems produce only data and this data must be reduced to a few directed, summary measures (indicators) before it can be of use to decision makers.

The relationship between data and its reduction into more refined forms is summarized in Figure 9.2 (Braat, 1991).

The increasing reduction of data represented in this pyramid shows the progression from raw data to analyzed data (used by scientists and program

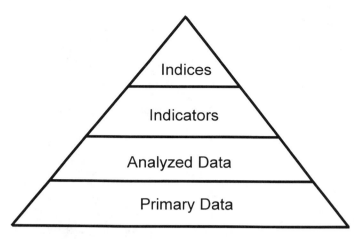

FIGURE 9.2. The refining of data.

managers for specific program, resource-protection, and research purposes) to indicators (used by decision makers to make strategic and operational decisions and to summarize environmental conditions to other decision makers and the general public) to indexes (mathematical combinations of individual indicators). As data become more and more condensed, their utility in conveying environmental information to different stakeholders in environmental conditions broadens and becomes more powerful, as shown in Figure 9.3 (Hammond, 1995).

Environmental organizations need to measure performance at five levels:

- **Administrative.** Are their financial-management systems working well? Are the employees motivated, well trained, and performing well? Is the organization meeting its legal obligations? All of these questions are important to organizations of any type and can be supported by indicators, but they do not measure any type of mission-based result.
- **Program activity and efficiency.** Are their programs running smoothly? Have the completed activities been cost-effective? Program-efficiency measures document how well the program is operating mechanically and what and how many outputs it is producing, but they do not show if the program is achieving the results that its mission specifies. These are the infamous "beans" of measurement systems.
- **Program performance.** Program-performance measures are focused on measuring program results or outcomes. Each program has a mission, a purpose that it is trying to accomplish. Program-performance measures document how well the program is doing in fulfilling its mission, in achieving the outcomes and results it is intended to achieve. In some cases a program-performance indicator can be expressed as a direct or indirect

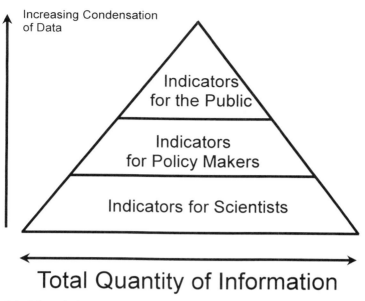

FIGURE 9.3. The relationship between appropriate quantity of information and the audience for that information.

environmental result; but in most cases, results are expressed as program outcomes.

Program-performance indicators are direct or indirect measures of the achievement of the intended purpose of a program, expressed as either an environmental result or a program activity. They might include:
- Number of square feet of floor space illuminated by Greenlights lighting upgrades
- Number of homes and schools tested for radon
- Number of industrial categories meeting NESHAPS (National Emission Standards for Hazardous Air Polluting Substances).

• **Policy.** Policy indicators are a hybrid of environmental and program-performance indicators. Most environmental agencies, in addition to measuring program and environmental results, also attempt to implement policies across all of their programs and activities to achieve some purpose that may not be purely environmental. Two examples are environmental justice and pollution prevention. Both policies have, in addition to their environmental and programmatic components, broader social, economic, and cultural concerns. Agencies are increasingly drawn to including policy indicators in their systems.

Policy indicators are direct or indirect measures of either environmental variables or program activities that can be used to assess the status and trends in the accomplishment of an environmental achievement set in a broader social, economic, and cultural context. Policy issues might include pollution prevention and environmental justice; policy indicators might include:

- Number of tons of chemical pollutants recycled by participants in a pollution-prevention program
- Average pollutant releases, grouped by racial composition of the exposed population.
- **Environmental.** The most important type of indicator for our purposes is the environmental indicator because these measures tell us what we ultimately want to know: What is happening to the environment? Is it getting better or worse? What changes are occurring in the environment? Are we achieving our mission?

Environmental indicators are direct or indirect measures of environmental quality that can be used to assess the status and trends in the environment's ability to support human and ecological health. They might include:

- Emissions of sulfur dioxide
- Exceeded limits of the ambient standard for ozone
- Toxic releases into the air of heavy metals and metal compounds

All five of these categories are important parts of an organization's management system. However, administrative and program-activity indicators do not play a role in post-decision assessment and will not be addressed. Rather, attention will be focused on program-performance, policy, and environmental indicators, the three types of indicators that deal with the achievement of *results*.

What are the characteristics of a good indicator? The Florida Center for Public Management uses the selection criteria found below for the indicators they develop. Indicator criteria are divided into essential criteria (requirements an indicator *must* meet) and preferable criteria (specifications an indicator *should* meet). Essential criteria include:

- **Measurability.** The indicator measures a feature of the environment that can be quantified simply with standard methods with a known degree of performance and precision.
- **Data Quality.** The data supporting the indicators are adequately supported by sound collection methods, data-management systems, and quality-assurance procedures to ensure that the indicator is accurately represented. The data should be clearly defined, verifiable, scientifically acceptable, and easy to reproduce.
- **Importance.** The indicator must measure some aspect of environmental quality that reflects an issue of major importance in demonstrating the current and future conditions of the environment.
- **Relevance.** The indicator should be relevant to a desired, significant policy goal, issue, legal mandate, or agency mission (e.g., emissions of air pollutants) that provides information of obvious value that can be easily explained to the public and decision makers.

- **Representativeness.** Changes in the indicator are highly correlated to trends in the other parameters or systems that they are selected to represent.
- **Appropriate scale.** The indicator responds to changes on an appropriate geographic (e.g., national, state, or regional) and/or temporal (e.g., yearly) scale.
- **Trends.** The data for the indicator should have been collected for a sufficient period of time to allow some analysis of trends, or they should provide a baseline for future trends. The indicator should show reliability over time, bringing to light a representative trend, preferably annual.
- **Decision support.** The indicator should provide information to a level appropriate for making policy decisions. Highly specific and special parameters, which may be useful to technical staff, will not be of much significance to policy staff or managerial decision makers.

Preferable criteria include:

- **Results.** The indicator should measure a direct environmental result (e.g., an impact on human health or ecological conditions). Indicators expressing changes in ambient conditions or changes in measures reflecting discharges or releases are acceptable but not preferred. Process measures (e.g., permits, compliance and enforcement activities, etc.) are not acceptable.
- **Understandability.** The indicator should be simple and clear, and it should be sufficiently nontechnical to be comprehensible to the general public with brief explanation. The indicator should lend itself to effective and appealing display and presentation.
- **Sensitivity.** The indicator can distinguish meaningful differences in environmental conditions with an acceptable degree of resolution. Small changes in the indicator show measurable results.
- **Integration of effects or exposures.** The indicator integrates effects or exposures over time and space and responds to the cumulative impacts of multiple stressors. It is broadly applicable to many stressors and sites.
- **Data comparability.** The data supporting an indicator can be compared to existing and past measures of conditions to develop trends and define variation.
- **Cost-effectiveness or availability.** The information to compile an indicator is available or can be obtained with reasonable cost and effort and provides maximum information per unit effort.
- **Anticipativeness.** The indicator is capable of providing an early warning of environmental change; that is, it has predictive strength.

Uses

The development of an indicator system capable of providing measures of environmental, policy, and program results is one of the most powerful general-purpose tools that any agency can possibly develop. Among their many uses are as a:

- Mission-level tool to provide a broad evaluation of environmental agencies' performance in protecting and managing the environment
- Foundation of measurements for structuring environmental goals
- Basis for measuring environmental achievement and progress
- Basis for making strategic budget decisions
- Means of evaluating the performance of individual programs and activities
- System to monitor the health of individual ecosystems within the context of broader environmental conditions
- Measure of environmental, programmatic, and personal accountability
- Means of constituency building
- Structure around which to develop environmental-education programs
- Tool for public relations and information dissemination

In the past five years, the growth in the use of indicators at both the federal and state levels has been explosive. In 1990, only a handful of states were using indicators in any direct sense, and only two, Florida and North Carolina, had made any explicit attempt to systematically develop and document a comprehensive environmental-indicator system; EPA was only beginning to develop explicit indicator systems.

That situation has radically changed. Almost 30 states have now developed or are finishing initial work on environmental indicators or closely related environmental-reporting documents, and virtually all states report they expect to undertake indicator-development projects in the near future. At the federal level, a number of interagency and intra-agency organizations are at work developing indicator systems and, perhaps more importantly, beginning the process of redesigning federal environmental-monitoring systems. EPA is developing a set of national environmental goals that are measured by indicator-supported benchmarks. The cornerstone of a revised federal-state relationship at EPA is a program of Performance Partnership Grants that are founded on the achievement of environmental results, again, measured explicitly by environmental and program indicators. Just beginning is a movement at the local level, and even at the community level, to use indicators. The potential for the institution and application of indicator systems by various types of environmental organizations is demonstrated in Table 9.2.

Activity is occurring across a range of organizations. A good summary of governmental indicator activity can be found in *The Directory of Environmental Indicator Practitioners* (Florida Center for Public Management, 1997a), prepared by the Florida Center for Public Management. This guide identifies and summarizes hundreds of indicators at the local, regional, state, national, and international levels.

TABLE 9.2. Indicator uses by type of organization and decision.

	Governmental	Large business and business associations	Small businesses	State, regional, and national citizen groups	Local citizen groups
Mission-based decisions	Indicator systems; benchmark systems; state-of-the-environment reports	Environmental indicator systems; specific environmental indicators in corporate reports; corporate state-of-the-environment reports	Focused indicator systems; individual indicators	Indicator systems; benchmark systems; state-of-the-issue reports	Community-based indicator systems
Program decisions	Program-based indicators	Program-based indicators	Program-based indicators	Program-based indicators	Program-based indicators
Individual decisions	Monitoring-based indicators	Monitoring-based indicators	Monitoring-based indicators	Monitoring-based indicators	Monitoring-based indicators

At the international level, the activities of the Dutch have already been identified. A number of European nations have begun the process of developing indicators, but the most active nation is Canada, with sophisticated indicator development being carried out at both the national and the regional level.

At the federal level in the United States, the Office of Water in EPA has identified a group of core indicators to represent water issues (Office of Water, 1997), and the Office of Air and Radiation, through the National Air and Radiation Indicators Project (NARIP), is concluding an intensive stakeholder process to develop a set of 80 to 90 national air and radiation indicators for local, state, regional, and national use (Florida Center for Public Management, 1997b).

Numerous states are involved in indicator development. Perhaps the best-developed systems have been produced by Florida (Florida Department of Environmental Protection, 1994), North Carolina (State Center for Health and Environmental Statistics, 1995), and Illinois (Critical Trends Assessment Project, 1994). Some states have developed indicator systems that are broader than environmental issues and that attempt to provide the best summary measures for a wide range of policy issues. Generally known as benchmark systems, they are the highest-level policy use of indicators. States with benchmark systems include Oregon (Oregon Progress Board, 1994), Florida (Florida Commission on Government Accountability to the People, 1996), and Minnesota (Office of Strategic and Long-Range Planning, 1991). A number of states are using limited sets of indicators to support state-of-the-environment reports. Perhaps the most complete and extensive state-of-the-environment report is produced by Kentucky (Kentucky Environmental Quality Commission, 1992). Many other states have produced shorter, highly focused, colorful, graphically interesting documents primarily designed for public consumption. Vermont (Agency of Natural Resources, 1996), Washington (Ragsdale, 1995), and Montana (Environmental Quality Council, 1996) have produced good examples of this type of document.

Indicator systems lend themselves to supporting specific ecosystems. The National Estuarine Programs, in particular, have used indicators effectively. Two examples of state funded ecosystem projects are indicator systems for the Hillsborough River and Bay (Florida Center for Public Management, 1996a) and the Apalachicola River and Bay (Florida Center for Public Management, 1995) ecosystems in Florida.

An example of an industry-based indicator system is a system designed to support a strategic plan for the Florida Institute for Phosphate Research, a university-supported information-support research program on problems and issues concerning the phosphate industry in Florida (Florida Center for Public Management, 1996b).

Considerable incipient indicator development is occurring at the local level. One good example of a completed local indicator system is found in Sitka, Alaska (City and Borough of Sitka, 1996).

TABLE 9.3. Strengths and weaknesses of indicators.

Strengths	Weaknesses
Capable of supporting a variety of management tools and uses	Can require significant technical expertise to develop and maintain
Have high flexibility, allowing close tailoring to an organization's specific needs	Require significant time, effort, and (often) financial resources to develop
Provide measurable evidence of progress	Require commitment to support periodic maintenance
Are within the technical capabilities of most organizations	Indicator-quality data are often not available
Focus attention on results, not process	Data are often not available for important new and emerging environmental issues

The use of indicators is governed by their attributes, which are summarized in Table 9.3.

Budget-Accountability Systems

For most agency managers, the budget is the most important planning process. Historically, the budget has either driven the planning process or, worse, been considered separate from broader agency planning. This perception reflects the lack of trust in planning processes or, at least, ensures that planning processes do not work well. In reality, the budget process is merely that part of a much larger and broader planning process where decisions are made about what resources are needed to carry out programs and activities for the next year or two. Until budgets are viewed as extensions of larger planning efforts, success in making significant improvements in environmental management is likely to be marginal. Several techniques are available to allow budget processes to employ postdecision-assessment tools to improve the quality of budget decisions.

Zero-Based Budgeting

The size and complexity of public programs conspires against measuring their effectiveness. Given the large size and character of governmental programs and the limited resources available to legislative and executive bodies, funding of public programs has been dominated by incremental budgeting. Under incremental-budgeting schemes, programs are not rigorously reviewed on a regular basis to determine the appropriateness of their resource allocations or the success of their activities. Unless there is some extraordinary consideration, budget increases or decreases are made on the margin for continued operation. New programs are occasionally created and become part of the base of the budget. Over time, this process can produce serious misallocations of resources as circumstances change. Unfortunately, agency managers have little incentive to correct these misallocations.

One potential tool that legislative or agency decision makers could employ is zero-based budgeting. Instead of looking only at marginal increases or decreases to the corpus of the budget, each year every position and budget allocation in every program needs to be justified in terms of measures of program efficiency and activity, program performance, and environmental results. While the process may be too onerous to complete every year for large agencies, its use once every four years could be effective in periodically rectifying resource misallocations. The key to making zero-based budgeting work is effective measurement systems at the program and at the environmental levels. If program managers and agency managers know that they will need to justify the continuance and expansion of their programs and activities based on the achievement and demonstration of results, their attention to performance measurement could be greatly increased.

Goal-Based Budgeting Systems

Legislative bodies are beginning to employ result-based measurement directly into their management of the budget process. The most prominent use of the process is the federal Governmental Performance and Results Act (GPRA). The act schedules all federal agencies to modify their budget processes over the next several years to include: strategic plans; annual performance plans for FY 1999 that establish accountability for results (i.e., targeted performance levels and/or indicators) for FY-1999 budget requests; and measurement systems to collect data on performance. The core of the process is the development of measurable goals and other measures of environmental and programmatic results. In theory, environmental information will be used to justify both the selection of environmental programs and their continuance. This is a new process that is still under development, and how it will play out is still conjecture. It has, however, the capability to be revolutionary, forcing consideration of the achievement of mission-based results to the forefront of budget decisions.

A similar program is presently being implemented in Florida, and doubtlessly, varieties of this process are being developed in other states.

Uses of Budget-Accountability Systems

The use of budget-accountability systems is governed by their general attributes, which are summarized in Table 9.4.

Program Evaluation

Another effective post-decision-making tool is program evaluation. One common definition of evaluation is: the process of delineating, obtaining, and providing useful information for judging decision alternatives (PDK

TABLE 9.4. Strengths and weaknesses of budgeting systems.

Strengths	Weaknesses
Systematically review performance of the entire organization and all of its activities and programs	Can be complex to organize and carry out
Have the capability to reallocate resources to areas of need or greater priority	Can be extremely time consuming and expensive
Facilitate integration of the budget with broader planning processes	Can be highly disruptive of staff routine
Reduce waste associated with cumulative resource misallocations produced by incremental budgeting	Can expose organizational and financial weaknesses to unsympathetic, outside decision makers
	Only useful to larger, usually governmental, organizations

National Study Committee on Evaluation, 1971). After goals and indicators for environmental systems are specified, decisions must be made whether to continue funding programs as they exist, to modify the programs, to expand them, or to discontinue them. Program evaluation can be used to make these decisions.

Few examples exist among environmental programs of formal, systematic, consistently implemented program evaluations. Factors mitigating against evaluations are administrative concerns about the use of the results (they may be used to defund programs), lack of advanced planning and systematic processes, lack of trained staff to conduct program evaluations, and previous negative experiences with program evaluations.

Perhaps the most common form of program evaluation is prompted by "sunset laws." Sunset laws are legislative requirements built into a program's statutory authorization that terminate the program as of a specific date. The continuance of the program is contingent upon legislative reauthorization after some sort of evaluation. The evaluative processes used, however, are spotty. While the sunset date sets an evaluation date for the agency, direction or resources for establishing a formal evaluative process are seldom given. As a result, as the sunset date grows near, the agency starts pulling together the best case it can with available information and data. Alternatively, some sort of executive or legislative inspector general or auditor general may conduct an evaluation. However, such evaluations are often viewed as being hostile in intent and evoke a defensive and/or political response from the organization. Properly used sunset laws could be an effective foundation for program evaluation, but two conditions need to be met. First, the process needs to be depoliticized through the use of politically neutral, professional evaluators. Second, agencies need to be required, and provided the resources, to design and implement an evaluation process capable of providing the information necessary to adequately evaluate the program.

Two broad types of program evaluations are formative evaluations, for ongoing improvement as part of an internal feedback loop, and summative evaluations, for go/no-go decisions as part of an external decision loop (Scriven, 1967).

Some form of formative evaluation may be employed in some organizations, but systematic, periodic evaluation programs are scarce. Often, whatever evaluation is carried out is performed by auditors, and the atmosphere of the evaluation is hostile. Managers of environmental units can benefit from formative evaluations of their programs to make needed changes and improvements in the processes and structures of the programs. But until those conducting the audits can establish helpful or value-neutral evaluations of programs, program managers will likely remain suspicious and resistant to program evaluation.

Even less commonly used are summative evaluations. Agency directors and legislators could use summative evaluations to determine funding priorities. Typically, environmental-program evaluations have not addressed the difficult issues of terminating programs that are not demonstrating appropriate changes in the quality of the environment. The only common form of summative evaluation may be the legislative sunset, which frequently takes the form of a hostile evaluation from an outside entity. Program managers must have the freedom to evaluate the overall success of their programs in an atmosphere free of punitive actions from their superiors. Otherwise, summative evaluations are unlikely to be used to weed out or correct deficient or ineffective programs.

Program evaluations integrate information collected through indicator systems with additional data collected through other sources to document the effectiveness and outcomes of environmental programs. The basic steps in a program evaluation are:

1. Describe the purpose of the evaluation.
2. Determine the specific objectives of the evaluation.
3. Develop the evaluation design or model to be used.
4. Write or document the plan for the evaluation.
5. Identify or select data-collection instruments or measures.
6. Collect data.
7. Analyze and summarize data.
8. Develop the evaluation report.
9. Translate the report conclusions and recommendations into action steps for change.

Many evaluations can be conducted in-house with existing expertise. Other steps, such as data collection and analysis, may require assistance from evaluators with specialized training. Environmental offices can increase their expertise in program evaluation through training by evaluators. An example of some training materials that an organization could use to train its own managers and staff to perform evaluations is a set of

TABLE 9.5. Strengths and weaknesses of systematic program evaluations.

Strengths	Weaknesses
Support all environmental decisions: strategic, program, and individual	Can be methodologically complex, requiring specialized staff or consulting assistance
Provide clear, factual documentation	Can be expensive and time consuming
Remove political overtones when conducted by independent professional evaluators	Can create political liabilities by exposing organizational problems

materials prepared for the Florida Department of Education (Florida Department of Education, 1995).

Some commonly used data-collection methods are surveys, community polls, individual interviews, focus-group interviews, other group interviews, and document reviews. These methods are discussed at length in other chapters of this book.

Much time and effort can be saved by well-designed instruments and planning in advance for data collection and analysis. Too often, agencies will disseminate surveys or conduct interviews without considering in advance how the data will be summarized and analyzed.

Conducting systematic program evaluations can provide environmental-program managers with timely information about the effectiveness of their programs, recommendations for improvements, and evidence for use with agency heads and legislators concerned about the justification of the program. As such, program evaluation provides the feedback loop in a strategic planning system. Information from the evaluations inform decision makers about the effectiveness of their planning and corrective actions needed to keep the agency moving toward the long-range vision. The attributes of evaluations are summarized in Table 9.5.

A Performance-Planning Model Supporting Post-Decision Analysis

Post-decision assessment and management techniques can be effectively applied on individual systems, but their impact can be enhanced and mutually reinforcing if they are explicitly designed to be an integral part of an organization's management system. The Florida Center for Public Management (FCPM) of Florida State University has developed a simple indicator-driven model environmental-management planning process that is founded in the logic inherent in the NEPP and Chesapeake Bay processes (Florida Center for Public Management, 1996c). The intent of this model process is to provide organizations with a performance-planning structure that provides consistent, comparable products and results across all ecosystem planning efforts while accommodating wide variations in political circumstances, planning sophistication, public involvement, previous planning activity, and media attention. Because the model deals with planning performance and results, and not with organization and process management, the model should work equally well in integrating the work that has occurred in extensively studied and well-documented systems, as well as serving as the core development process for systems that have had relatively little previous planning and research. The principal features of the model include:

- A *strategic* orientation to set long-term guidance and direction
- An *operational* component to set the action agenda for achieving results
- An *issue-based* orientation that identifies a limited number of strategic *environmental* concerns around which all of the planning revolves
- An *indicator-supported* system capable of quantitatively measuring trends and progress in dealing with the issue
- A *goal-driven* approach that focuses the activities associated with each of the issues on the achievement of specific, quantitative and graphically displayable environmental results
- *Accountability* systems that require participants to identify their specific roles and actions and to be held responsible for the fulfillment of those duties
- An *iterative process* that ensures that the plan is reviewed annually and that strategic- and operational-planning provisions are modified on the basis of the environmental success of the project as measured against the project's goals

The products of the model closely resemble relatively standard strategic-planning formats. But the success of the model will reflect how well it is carried out, not its form. In general, the model has three basic components: (1) a strategic component that identifies the critical issues, sets goals for measuring the achievement of success, and sets the long-term direction necessary to guide the actions of the principal participants; (2) an operational component that specifies the individual and collective actions that each of the participants must take and holds them accountable for their performance; and (3) a review component that continually assesses the process and directs changes in the strategic and operational components, as required. In implementing the model across these three components, the strategic component should:

- Identify eight to 12 strategic issues that reflect the priority environmental concerns that must be dealt with in the next five to 20 years if the values of the ecosystem are to be maintained or improved
- Identify indicators that accurately measure critical trends associated with each issue and that will provide measures of success
- Develop a goal and objectives for each issue that are quantitative and indicator-based and that (therefore) lend themselves to graphic display

- Develop a series of strategies for each issue that reflect the achievements that must be accomplished to meet the goal

The operational component should:

- Negotiate among the participants the specific actions that must take place within a two-year period to make appropriate progress in meeting the strategic goal for each issue
- Develop an accountability system for tracking and annually reporting on the achievement of the specific, time-based commitments made by each participant

And the review component should:

- Review annual progress in meeting each of the goals
- Make changes in the strategic and/or operational plans if progress is not satisfactory
- Conduct an annual update of the operational plan to add a new second year to the plan.

Conclusions and Summary

As promising as developments for result-based management tools are at the present, many serious impediments exist to their successful implementation. The three major ones are described here: the difficulty of establishing monitoring systems, political influences, and the availability of role models.

Monitoring-System Development

All the good intentions in the world to use indicators in profound and effective ways can be limited or thwarted if the quality of supporting information is poor or unavailable. Old issues are often not being measured well, and new and emerging issues are commonly not measured at all. Serious attention needs to be paid to examining our monitoring and indicator data-collection systems. Fundamental decisions need to be made concerning how much and what kind of data we will collect on what issues, and a commitment (moral, physical, and financial) has to be made to collect the needed data.

Political Impediments

The use of environmental information faces structural and attitudinal barriers. The lack of faith in planning processes and the fear of information being used against them are powerful disincentives to the use of results-based measurement by decision makers. Distrust of executive agencies by legislative bodies creates an environment of hostility that is not conducive to the most effective uses of post-decision-assessment tools. Results-based tools can be used to support a positive decision process, but they also harbor the possibility of being used for punitive purposes, which can destroy or limit their use and effectiveness.

Effective Examples

Environmental agencies and organizations in the United States need role models in setting up their management processes. It is difficult to point to examples of organizations that have established long-term, comprehensive systems that have achieved notable results. While it is unpopular to recommend that the federal government serve as the model for anything in our current political context, the EPA, in its developing role as a cooperative leader, may be the best hope. The EPA's development of national goals; its centralization of environmental-management authorities (including the budget) into a new Office of Planning, Budgeting, and Accountability; and the restructuring of its role with the states to emphasize environmental achievements may presage the development of an effective, complete environmental-management system capable of setting the standards for all environmental organizations.

Summary of Post-Decision-Making Tools

Four tools have been cited and described for their ability to review or follow up on environmental decisions after those decisions have been finalized and, in some cases, acted upon. Each of these tools is applicable in a specific circumstance and has a particular purpose. And each has particular strengths and weaknesses that govern when and under what circumstances the tool can be effectively employed.

Key Resources

In *Environmental Policy Performance Indicators, A Study on the Development of Indicators for Environmental Policy in The Netherlands* (The Hague: Ministry of Housing, Physical Planning and the Environment, 1993), Albert Adriaanse presents an excellent discussion of the Dutch system and The Netherlands' use of indicators for policy development.

In *An Ecosystem Planning Model* (Tallahassee: Florida Center for Public Management, Florida State University, January, 1996), Gilbert Bergquist briefly discusses a simple but adaptable model for structuring a planning process to use measurable goals and indicators as evaluative tools.

Setting Priorities, Getting Results: A New Direction for EPA (Washington, D.C.: National Academy of Public Administration, 1995) and *Resolving the Paradox of Environmental Protection: An Agenda for Congress, EPA, and the States* (Washington, D.C.: National Academy of Public Administration, 1997) are thorough and well-considered analyses of environmental policy and decision making at the EPA and elsewhere in government that include excellent recommendations for improving those processes.

Two fundamental and excellent texts on evaluative processes in education are Ralph W. Tyler, Robert M. Gagne, and Michael Scriven (Eds.), *Perspectives of Curriculum Evaluation*, AERA Monograph Series on Curriculum Evaluation, No. 1. Chicago: Rand McNally. (especially Scriven's chapter on the methodology of evaluation) and PDK National Study Committee on Evaluation, *Educational Evaluation and Decision Making*. Itasca, IL. F.E. Peacock Publishers. 1971.

References

Agency of Natural Resources. 1996. *Environment 1996: An Assessment of the Quality of Vermont's Environment*. Waterbury, VT: The Vermont Agency of Natural Resources.

Askew, R.O'D. 1978. *The Florida State Comprehensive Plan*. Tallahassee, FL: Office of the Governor.

Braat, L. 1991. The predictive meaning of sustainability indicators. In: Onno Kuik and Harmen Verbruggen, *In Search of Indicators of Sustainable Development*. Boston: Kluwer Academic Publishers. pp. 57–70.

Chesapeake Bay Program. 1995. *The State of the Chesapeake Bay, 1995*. Washington, D.C.: USGPO.

City and Borough of Sitka. 1996. *Sitka Coastal Indicators Project*. Sitka, AK: City and Borough of Sitka.

Critical Trends Assessment Project. 1994. *The Changing Illinois Environment: Critical Trends*. Springfield, IL: Illinois Department of Energy and Natural Resources and The Nature of Illinois Foundation.

Environmental Quality Council. 1996. *Our Montana Environment: Where Do We Stand?* Helena, MT: Environmental Quality Council.

Florida Center for Public Management. 1995. *Apalachicola River and Bay Ecosystem Management Plan: Environmental Indicator System*. Tallahassee, FL: Florida Department of Environmental Protection and Florida State University. Also available on the World Wide Web at http://www.fsu.edu/~cpm/APALACH/index.html, visited Nov. 11, 1997.

Florida Center for Public Management. 1996a. *Hillsborough River and Bay Ecosystem Management Plan: Environmental Indicator System*. Tallahassee, FL: Florida Department of Environmental Protection and Florida State University. Also available on the World Wide Web at http://www.fsu.edu/~cpm/HILLS/index.html, visited Nov. 11, 1997.

Florida Center for Public Management. 1996b. *Strategic Indicators of Success: Florida Institute of Phosphate Research*. Tallahassee, FL: Florida Institute of Phosphate Research and Florida State University. Also available on the World Wide Web at http://www.fsu.edu/~cpm/FIPR/index.html, visited Nov. 11, 1997.

Florida Center for Public Management. 1996c. *An Ecosystem Planning Model*. Tallahassee, FL: Florida State University.

Florida Center for Public Management. 1997a. *Environmental Indicator Technical Assistance Series, Vol. 4, Directory of Environmental Indicator Practitioners*. Tallahassee, FL: Florida State University.

Florida Center for Public Management. 1997b. *Final Review Draft, National Air and Radiation Indicator Manual*. Tallahassee, FL: Florida State University.

Florida Commission on Government Accountability to the People. 1996. *Florida Benchmarks Report*. Tallahassee, FL: Executive Office of the Governor.

Florida Department of Education. 1995. *A Resource Manual for the Development and Evaluation of Special Programs for Exceptional Students*. Tallahassee, FL: Florida Department of Education.

Florida Department of Environmental Protection. 1994. *Strategic Assessment of Florida's Environment (SAFE)*. Tallahassee, FL: Florida Department of Environmental Protection. Also available on the World Wide Web at http://www.fsu.edu/~cpm/safe/safe.html, visited Nov. 11, 1997.

Group on the State of the Environment. 1993. *OECD Core Set of Indicators for Environmental Performance Reviews*, OECD/GD (93) 179. Paris: Organization for Economic Cooperation and Development. October 15, 1993, p. 6.

Hammond, A.L. 1995. *Environmental Indicators: A Systematic Approach to Measuring and Reporting on Environmental Policy Performance in the Context of Sustainable Development*. Washington, D.C.: World Resources Institute.

Kentucky Environmental Quality Commission (KEQC). 1992. *State of Kentucky's Environment: A Report of Progress and Problems*. Frankfort, KY: Kentucky Environmental Quality Commission.

National Academy of Public Administration. 1995. *Setting Priorities, Getting Results: A New Direction for EPA*. Washington, D.C.: National Academy of Public Administration.

National Academy of Public Administration. 1997. *Resolving the Paradox of Environmental Protection: An Agenda for Congress, EPA, and the States*. Washington, D.C.: National Academy of Public Administration.

Netherlands Ministry of Housing, Spatial Planning, and the Environment. 1993. *National Planning for Sustainable Development: The Netherlands Experience*. The Hague: The Netherlands Ministry of Housing, Spatial Planning, and the Environment.

Office of Strategic and Long-Range Planning. 1991. *Minnesota Milestones*. St. Paul, MN: Minnesota Planning.

Office of Water. 1997. *Environmental Indicators for Water Quality in the United States*, EPA-841-R-96-002. Washington, D.C.: U.S. Environmental Protection Agency.

Oregon Progress Board. 1994. *Oregon Benchmarks: Standards for Measuring Statewide Progress and Institutional Performance*. Salem, OR: Oregon Progress Board.

PDK National Study Committee on Evaluation. 1971. *Educational Evaluation and Decision Making*. Bloomington, IN: Phi Delta Kappa.

Ragsdale, D.P. 1995. *Washington's Environmental Health 1995*. Olympia, WA: Washington State Department of Ecology. Also available on the World Wide Web at http://www.wa.gov/ecology/weh95.html, visited Nov. 11, 1997.

Scriven, M.S. 1967. The methodology of evaluation. In: R.W. Tyler, R.M. Gagne, and M. Scriven, *Perspectives of Curriculum Evaluation*, AERA Monograph Series on Curriculum Evaluation, No. 1. Chicago: Rand McNally. pp. 39–83.

State Center for Health and Environmental Statistics. 1995. *North Carolina Environmental Indicators*. Raleigh, NC: Department of Environment, Health, and Natural Resources.

U.S. Environmental Protection Agency (USEPA). 1996. *Environmental Goals for America with Milestones for 2005*. Washington, D.C.: U.S. Environmental Protection Agency.

Decision-Maker Response

KATHARINE JACOBS

This chapter provides a comprehensive overview of the state of the art of post-decision assessment, including a description of the techniques, an evaluation of their application in the environmental context, and some success stories. Extensive information regarding management issues, the effect of changing environmental priorities, the use of environmental indices, and selection criteria for indices is included. Other topics discussed include goal systems, budget-based systems, and program evaluation.

The authors have a good understanding of the issues associated with post-decision assessment, and delve into the motivations of environmental managers in the context of politics and budgetary limitations. They also appear to have had substantial opportunities to observe decision making in the "real world." This section is useful in that it should make practitioners aware of the pitfalls that have been encountered by others. However, the authors appear to presume that the issues faced by decision makers are similar at the state, local, and national levels. Although these issues may be similar in nature, one suspects that some tools used in post-decision assessment may be more useful at one level than another. For example, identifying environmental indicators or benchmarks at the local ecosystem or watershed level is much easier than at the national level.

It is true, as the authors indicate, that inadequate attention is paid to assessing the success of environmental programs and that those programs are frequently focused inappropriately. However, the regulatory processes required by some federal programs are frequently more cumbersome and less easily redirected than those at the state and local levels.

In the discussion of the changing character of environmental issues, the role of the "expert" vs. the role of the "public" in decision making has changed dramatically. Historically, many technical issues were decided by people who had substantial technical training and who were generally deferred to because of their expertise. Parallel to the rise in public skepticism regarding government in general is a rise in distrust of "experts." Such experts were disenfranchised, in part, because of their inability to incorporate the public's concerns into their technical world view.

The rise in citizens' referendums and initiatives comes from a distrust of politicians and experts, as well as the empowerment of the public that has come through the opportunity for public input in environmental impact statements and a host of other programs. Because the public does not have access to the same technical information that the "experts" do, its input is affected strongly by perception of environmental quality. This perception may, in fact, be more relevant in the long run than the opinions of the experts, but it makes development of measurable goals and objectives more challenging.

Along with the general empowerment of the public comes a new group of participants in the decision-making process: the "professional" public, people who attend public meetings regularly, either representing themselves or a group. They may or may not be directly or substantially affected by a particular decision, but they comment regularly and vociferously. This phenomenon has substantially changed the nature of public meetings and the decision-making process.

In the environmental arena, the positions of individuals, businesses, and political organizations have become somewhat entrenched over time, limiting creative problem solving and good design of regulatory programs and feedback mechanisms. A perception that various interest groups will fight particular approaches reinforces the status quo. The incorporation of regulatory impact assessment, risk-based priority setting, and market considerations at both the state and federal levels will probably provide substantial opportunities for redistribution of resources in environmental programs, which in turn should provide opportunities to design regulations that incorporate the tools suggested in this chapter. However, the continual shrinkage of budgets for environmental programs may result in lack of resources and time for appropriate design of data-collection systems, information analysis, and program review. The average environmental program manager is overwhelmed by an ever-expanding workload with a diminishing budget. The "crisis-management" atmosphere that results may not allow adequate time to focus on the program-assessment phase.

The demand for accountability may actually diminish the amount of time that a manager has to implement effective programs. The "bean-counting" mentality is a problem not only because it measures actions and not results, but also because it becomes an end in itself and takes up valuable time. This diversion of resources could happen inadvertently if too much focus is placed on the evaluation tools, as well.

The authors focused on improvements that were required in managers to develop effective programs. An aspect not developed by the authors is the lack of political will to support good environmental decisions that cost money or that are unpopular with particularly vocal interest groups. In the absence of good political leadership, the managers may not be the ones responsible for program failures.

Another component of the changing world of environmental decision making is that the level of complexity of many decisions has increased. We now have access to data on minute quantities of contaminants, but the health and environmental implications of such information are not adequately understood. The anxiety level of the public and its political representatives has changed the focus and narrowed the alternatives available in many arenas, increasing the costs of the alternatives while the budgets of the affected agencies are, at the same time, decreasing. Complexity also leads to disagreement between experts, as in the global-warming controversy. If experts do not agree, it is difficult to make defensible decisions.

One of the unfortunate effects of the lack of foresight in program design is that substantial amounts of data have been produced that are unavailable for use. In Arizona, the Department of Environmental Quality manages water-quality programs, and the Department of Water Resources manages water quantity. They have no comprehensive, linked database, which means that the consultants involved in groundwater recharge, remediation, and water supply have to do independent research to develop a project. Although reams of information exist regarding the quality of water in the vicinity of Superfund and state-initiated remediations, no comprehensive data repository and no comprehensive data-management policy exist. Therefore, it is very difficult to establish indices or even trends in water quality over time. Database quality may actually decrease over time because of incompatibility of computer systems, lack of quality control, and changes in detection techniques.

This lack of correspondence between the resources to be managed and the jurisdiction of the agencies that are charged with managing them is everywhere. Fragmented management may be one of the primary causes of ineffective programs. Some optimism is justified, however, as we move toward ecosystem- and watershed-based planning. Some excellent examples of overcoming institutional fragmentation are in the recent federal-state collaborative planning processes, such as the CAL-FED Bay Delta Program and cooperative habitat-protection programs in response to the Endangered Species Act.

The authors suggest that complex environmental organizations need to measure performance at five levels: administrative, program activity and efficiency, program performance, policy, and environmental. The most important are the last three, which measure results. However, actually measuring success at the policy-objective level is difficult. Policy indicators that measure progress toward goals, such as "environmental justice," are likely to be subjective.

The section of the chapter that describes the essential criteria for a good indicator is very useful. Clearly, consideration of these factors is imperative for an appropriate assessment of program effectiveness. It was interesting, given the highly technical nature of most environmental issues, that "understandability" was among the preferable criteria and that "the indicator should lend itself to effective and appealing display and presentation." This choice clearly acknowledges the role of the public in environmental decision making, but also brings up the question of who the real audience is for such decision-making tools. The tools should be appropriate for use by the person or group that actually makes the decision; if it is a political decision, then the tools do need to be accessible to the lay person.

Program evaluation is not uncommon at the state and local levels, though it may not be implemented under ideal circumstances. Arizona has a "sunset law" that requires ending regulatory agency functions after a seven-year cycle if they are found not to be useful. The enforceable management plans that are developed to manage the groundwater in the "Active Management Areas" of the state were designed to incorporate an iterative goal-based assessment so that groundwater overdraft can be eliminated incrementally in these areas. In other words, although the tools may be labeled by some other name, participation in post-decision assessment may be changing in form without a major change in substance. Still, it would be useful for the tools described in this chapter to become used as regular components of program design.

Possibly a root cause of the lack of adequate attention to achievement of environmental objectives is the disconnection between the actions of private citizens and the impacts of their actions. Very few Americans are aware of the implications of the decisions that they make on a daily basis, from the decision to drive their cars to work to deciding how to eradicate unwanted insects from their property. If decisions made at this level are not adequately understood, it is not surprising that the decisions that are made collectively are inadequately analyzed.

10
Next Steps for Tools to Aid Environmental Decision Making

MARY R. ENGLISH and VIRGINIA H. DALE

As the prior chapters suggest, information-gathering and analytic tools can be used by a variety of participants in environmental decisions, including the heads of the decision-making organizations, their staffs, and other interested groups and individuals. As has been illuminated in this book, a variety of tools can be used to improve the ways in which information for environmental decisions is obtained, organized, and analyzed. And the use of these tools increases the likelihood that, despite sometimes large uncertainties and unknowns, wise decisions will be made.

Tools can be used to collect information from existing sources; they also can be used to identify and obtain essential new information. Tools can be used to elucidate the limitations, as well as the strengths of information input to the decision process; they also can reveal assumptions guiding the process that might otherwise go unexamined. Tools can be used to make large quantities of diverse information intelligible; they also can be used to make estimates and predictions when important information is lacking. And tools can be used to set priorities in timing, resource allocation, and data needs; they also can be used to evaluate options and retrospectively assess decisions. But effective and appropriate tools are not enough; they also require good human judgment and a high level of interaction among participants in the decision at hand.

The main theme of this book has been a comprehensive assessment of information-gathering and analytic tools now available to aid environmental decisions. In this chapter, we consider *future* priorities in tool development. Some of the tools needed for tomorrow's environmental decisions may be wholly new; others may be adapted from existing tools. In either case, despite a recent explosion of tools caused by the widespread availability of powerful personal computers, multifaceted software packages, and the Internet, the environmental decision maker's tool kit is not complete, nor do all participants in environmental decision making have access to tools. At this juncture, it is appropriate to step back and examine factors that should be taken into consideration as tools are developed and refined over the coming decade.

This chapter first discusses factors that constrain the use of tools today and then turns to trends in the way environmental decisions are made. From these discussions, criteria are derived for the information-gathering and analytic tools that are likely to be needed most during the coming decade. Finally, we draw upon the discussions of the authors in this book to briefly assess where directions in tool development are headed today and whether they are likely to meet the needs of tomorrow.

Factors Constraining the Use of Tools

The factors that constrain the use of information-gathering and analytic tools can be grouped into four categories: information, time and resources, accessibility, and communication and trust-building. Together, these factors affect which tools are likely to be most useful as aids to environmental decision making at the subnational level.

Information: Too Little, Too Much, or the Wrong Kind

When information is limited, participants in an environmental decision must choose between asking for more information and forging ahead without additional information (perhaps by extrapolating from existing data or analogizing to estimate missing information). The choice helps to determine the tool to be used.

Sometimes a paucity of information provides a convenient reason for deferring a decision while more information is gathered. The United States government thus far has not confronted the issue of whether greenhouse gases, especially carbon emissions, should be limited, arguing that more scientific evidence is needed to confirm their link to global climate. In some cases, however, collecting necessary information may be difficult or impossible. While the socioeconomic information discussed by Freudenburg in Chapter 4 is vital to good environmental decision making, it may not be possible to get an adequate baseline of the socioeconomic conditions that prevailed before an environmental decision was contemplated because the mere prospect of the decision may have altered those conditions (Gramling and Freudenburg, 1992).

Paradoxically, too much information can be as bad as too little. Large amounts of information are being collected by many different individuals and organizations, using increasingly precise measurements, and are becoming available via the Internet, CD-ROM, and other computer-based media as well as in print. Often, however, this information simply overwhelms rather than informs. The data may be incommensurate with each other or not fully relevant to the situation at hand; they may be unverifiable and of suspect quality; they may not be organized in a manner that is useful for the situation; and gaps or imperfections in databases may not be evident if the database is extremely large. As Lyndon noted in Chapter 5, as a plethora of information becomes available, the task of integrating everything related to a particular issue becomes virtually impossible. Similarly, as Armstrong indicated in Chapter 7, too much information may simply muddy the waters and may actually be misleading. Furthermore, information of poor quality can detract from good decision making.

Information *scale*, both spatial and temporal, is an especially important problem for environmental decision making. Often, data are aggregated into spatial units that prevent their use for subnational decisions. As Freudenburg noted in Chapter 4, the accuracy of economic and demographic projections will often be poorest at the smaller scales that are often the most relevant for environmental decisions. Air-quality information needed by a neighborhood may only be expressed by county, and economic information needed for a watershed that straddles several states may only be available on a state-by-state basis. Or data providing a portrait of *today* may be available, but a characterization of changes over time may not be possible because the data were not gathered at the appropriate scale until recently, or because data have been gathered too infrequently to provide a sufficient number of data points.

Clearly, some information-gathering and analytic tools will be needed to collect, organize, and distill the massive quantities of information that are increasingly available. Now more than ever, however, tools that do not rely on large or extremely precise amounts of data are also needed as high-uncertainty issues are tackled, as more factors are deemed relevant, and as place-based environmental decisions are made at smaller scales.

Time and Resource Limitations

Limited time, untrained people, and small budgets all constrain which tools can be used. In some cases, these hurdles can be lowered; but in others, they are likely to remain part of the context of environmental decision making.

Some decisions must be made within days or weeks; there may not be time to wait while more data are gathered or people become acquainted with a complex new decision-aiding tool. Other decisions may be more deliberative and allow the use of elaborate and unfamiliar tools, but tool use may still be constrained by other factors.

Those who typically use information-gathering and analytic tools for environmental decisions are likely to be paid staff to decision makers within government and business or within environmental advocacy organizations. While the staff within large organizations is likely to be specialized and, possibly, either skilled in the use of decision-aiding tools or equipped with the background to learn these new skills, other organizations (including many small governments and businesses) are more likely to have a staff of only one or two "jacks of all trades" with neither the time nor the background to become masters of complicated decision-aiding tools. In addition, as environmental decision making becomes more open (a trend already in evidence and likely to continue), other participants within and outside the decision-making organization will need access to the decision-aiding tools. Yet they, too, may have neither the time nor the capacity to become tool experts. Tool simplicity and ease of use will remain an overriding consideration for many.

Closely related to the factor of limited training is that of tight budgets, which often means fewer staff, fewer training opportunities for staff, fewer resources to gather primary data, less fancy equipment to process data, and fewer (and possibly less expert) outside consultants. To overcome some of these limitations, federal support is being supplied, especially by such organizations as the National Science Foundation and the U.S. Environmental Protection Agency, to aid in the use of tools for environmental decision making by local governments and nonprofit organizations. But typically, this support is only seed money, and it is not available universally or on a sustained basis. Budgets will remain a big factor in tools for environmental decision making.

Lack of Access

Even if participants in an environmental decision have the time and resources to use a particular decision-aiding tool and the right information to plug into it, they still must have access to the tool. As noted above, the opportunities to acquire various tools have greatly expanded, especially because of the Internet but also because of demonstrations at professional and trade association meetings, workshops, and so forth. Nevertheless, all tools are not widely available. Some may currently be accessible only to relatively few people, either because the tool is experimental or because it is not well-publicized.

Tools that are still in an experimental stage, where the assumptions and approaches are not yet worked out, are typically restricted in use to those people involved in their development. While some tools may evolve into wide usage, as computerized database-management systems have during the past 10 years, others may never become commonly available. Some tools may be discarded as flawed attempts; others, while workable, may never become "commercialized."

Three key issues facing tool developers and tool users are (1) whether tools should become widely available while they are still being developed and tested (a debate akin to that concerning experimental drugs); (2) how tools that have been developed for one client can be adapted for broader use; and (3) how "commercialization" or "technology transfer" can best be accomplished, especially if those developing the tools are researchers in nonprofit institutions with little incentive to try to market the tool or otherwise disseminate it.

10. Next Steps for Tools to Aid Environmental Decision Making 321

The Hurdles of Communication and Trust

Participants in an environmental decision need to understand the output of or results from tools used in reaching the decision. It will be important to understand not only the basic meaning of the output (e.g., the difference between "median" and "average"), but also its limitations (e.g., its degree of precision, spatially and temporally; its sensitivity to particular conditions; and its level of uncertainty). To reach a robust, well-informed understanding of the tool's results, participants should be familiar with key assumptions undergirding the tool, as well as the strengths and weaknesses of information inputs.

For participants to attain this understanding, they will need truthful, clear, and concise explanations from those who "ran" the tool. Often, a decision-making organization needs to be able to communicate not only with people *external* to that organization, but also with those within the organization who ultimately will make the decision.

Closely related to the issue of communication is the issue of trust. To trust means to relinquish an element of control over one's life (Luhmann, 1979). Quite understandably, people are reluctant to relinquish control until they feel they have good grounds for doing so. If the strengths and limitations of tools and their results are clearly communicated, interpersonal and interorganizational trust is more likely to be built over time. Increased trust does not necessarily lead to agreement on a decision, for different values will continue to figure importantly. It does lead, though, to greater common ground about the informational basis for the decision.

Criteria for Tools for Tomorrow

From these constraints, a few criteria for the tools of tomorrow can be inferred. Those crafting tools must be sensitive, however, not only to constraints, but also to changes that are occurring in environmental-decision processes. These changes were alluded to in Chapter 1 and have been noted by contributing authors to this book, such as Gregory, Freudenburg, Lyndon, and Merkhofer; they also have been a subject of inquiry for NCEDR researchers.

In a companion effort to this overview of decision-aiding tools, NCEDR researchers have developed a typology, summarized in Sidebar 10.1, of environmental decision-making modes (Tonn et al., forthcoming; English, 1998). These six decision modes described in the sidebar are not unique to environmental issues; furthermore, they all take place within a larger cultural context (particularly, the individual values and beliefs, as well as the collective norms and knowledge of those most interested in the decision at hand). The ways in which decisions are made also are affected by (and, in turn, may affect) the structures and other activities of the institutions

Sidebar 10.1
Modes of Environmental Decision Making

Analysis-centered. Analysts within the decision-making organization develop a carefully crafted technical or policy recommendation for the ultimate decision maker (typically, the head of the organization). Quantifiable information often is preferred, and elaborate methods for considering components of the situation, and then weighing alternatives, often are employed. While other people internal or external to the decision-making organization may participate in the decision process, they typically do so only by providing input on their goals and values.

Élite corps. Senior members of the decision-making organization reach either agreement or a majority view on the issue at hand. Staff presentations are followed by discussion and negotiation among the senior members; "bottom line" information is sought, including information about the views of special interests. But while these views may figure importantly, outsiders typically do not participate in the decision process.

Emergency action. Emergency managers within the decision-making organization make a rapid decision concerning a crisis situation. Knowledge of the situation is gathered quickly and may be incomplete; instead, predetermined procedures and "seat of the pants" judgments are used. While other people and organizations may participate in emergency preparations or mopupactivities, few others participate in the emergency action itself.

Routine procedures. Administrative or technical staff within the decision-making organization make day-to-day decisions concerning familiar situations following predetermined procedures. The decision typically requires specified, standardized information. Although others within and outside the decision-making organization may have participated in establishing the broad policies that led to the routine procedures, few others participate in implementation of the procedures.

Conflict management. Staff or leaders within the decision-making organization seek to resolve a controversial issue using a decision process that is open and often protracted. The process typically begins with a meeting of people internal and external to the organization who represent various sides of the conflict. The process may be kicked off with an issue scoping, which may itself be a source of conflict. Typically, information is presented by a variety of people, followed by discussion and negotiation. This dialogue may lead to more informa-

10. Next Steps for Tools to Aid Environmental Decision Making   323

> tion being sought, leading to further discussion and negotiation, and so forth.
>
> **Collaborative learning.** Various members internal and, sometimes, external to the decision-making organization work together as equals to address an issue that is widely acknowledged to be neither easily addressed nor well-understood. The process is likely to be long and iterative: As information is obtained, people are encouraged to revisit their original goals and beliefs, and the nature of the issue may be collectively rethought. Decisions are subject to change over time as new collaborative learning occurs.

participating in the decision. And none of the modes exists as a discrete type; instead, they are likely to act in combination. For example, an analysis-centered mode may be in support of an élite-corps mode, which may precipitate a conflict-management mode. Nevertheless, despite the fluid nature of environmental decision making, typologizing its modes can help to clarify where environmental decision making is today and where it is headed.

During the past several decades, environmental decision making has tended to follow the routine-procedures, emergency-action, analysis-centered, and élite-corps modes. Of these, decision processes that center on either analysts or a cadre of senior managers have been preeminent for controversial issues with potential large-scale and long-term consequences. Nevertheless, the relatively new modes of conflict management and collaborative learning are gaining in importance. Since the early 1980s, conflict management has become a vital part of environmental decision making in a pluralistic society that strives to be open and participatory (Fisher and Ury, 1981; Talbot, 1983; Meeks, 1985; Bingham, 1986). And now, in the late 1990s, the concept of collaborative learning (also called adaptive learning or adaptive management, depending upon the emphasis) is receiving increasing attention, especially as a way to deal with highly complex issues where values are diverse and knowledge is limited (Senge, 1990; Heifetz, 1994; Gunderson et al., 1995).

Many of the information-gathering and analytic tools developed to date have been in support of the first four, more conventional modes, especially the analysis-centered mode. This situation leads to two questions that are worth posing, even though answering them would require an additional book: (1) Can currently available information-gathering and analytic tools also be used in the conflict-management and collaborative-learning modes of decision making, or will they need to be modified? (2) Will completely new tools be required as various organizations address the hard environmental questions that are arising in a society that is changing, with environmental issues becoming more pressing and complex?

Despite these questions about the implications of new modes of decision making, a few criteria can be derived about tools that are likely to be needed by subnational organizations as they confront tomorrow's environmental issues within a climate of decision making that is information-intensive and highly analytic, but also more open and deliberative. These criteria vary somewhat depending upon the type of organization. Some may be able to use extremely complex, sophisticated tools; while others, especially relatively small public- or private-sector organizations with limited budgets and staffs, will need either to use much simpler tools or to rely on experts for analysis and interpretation.

For tools to be useful to organizations with severe staff and budget constraints, the tool should meet the following three criteria:

- The tool should not be difficult to use and should require little prior training.
- The equipment required should be widely available and either inexpensive or, if costly, highly versatile.
- The data requirements should not be extremely extensive or specialized; furthermore, the tool should take advantage of standard, well-documented data sources now commonly available through such means as the Internet.

In addition, regardless of the types of organizations for which they are intended (i.e., whether or not the organizations are heavily burdened with resource constraints), tools should meet the following ten criteria:

1. Tools (and tool users) should be explicit about what the tool can and cannot accomplish, the assumptions that are built into the tool, and how terms used in the tool's application are defined.

2. Tools should clearly specify the types of data to be used, including their spatial and temporal scales, along with possible data sources.

3. Qualitative information, expert judgments, and sources of "soft" information such as role-playing should be considered as integral to tools rather than as add-ons.

4. Tools should be able to integrate the perspectives of various disciplines (e.g., economics and ecology) and various interests (e.g., economic growth and environmental protection); their viewpoints should be as encompassing as possible, and feedbacks and linkages across disciplines should be fostered.

5. Tools should be able to incorporate new knowledge and new understanding as they become available.

6. Tools should take advantage of the new capabilities offered by technological advances.

7. Ideally, tools should proceed from input to output fairly rapidly.

10. Next Steps for Tools to Aid Environmental Decision Making   325

8. Both the results of tools and how they work should be clearly communicated via diverse approaches.

9. Tools' results should be accurate and clear, not misleading; factors affecting their validity and reliability (including assumptions, data accuracy and precision, sensitivity to altered conditions, and sources of uncertainty) need to be explicit parts of the results.

10. Tools should be easily explained and disseminated; the dissemination plan should be part of the tool design rather than an afterthought.

Clearly not every tool will be able to rate high on each of the above criteria; some trade-offs may have to be made. Nor should *all* new tools strive to meet these criteria. Instead, some specialized tools may need to sacrifice meeting one or more of the criteria in order to serve particular purposes. These criteria do, however, indicate the direction in which the development of information-gathering and analytic tools should be headed in the coming years.

Furthermore, these criteria may be equally applicable to tools intended to promote creativity, communication, and involvement, which, as National Center for Environmental Decision-Making Research (NCEDR) research has indicated (Wolfe et al., 1997), are vital for participants in environmental decisions. The purposes of information-gathering and analysis, on the one hand, and creativity, communication, and involvement, on the other hand, should not be regarded as mutually exclusive. As a number of contributing authors to this book have suggested, information-gathering and analytic tools can and should be designed to foster a creative, participatory decision process.

New Directions

Chapters 2 through 9 focused chiefly on currently available tools. In each chapter, however, the author(s) also briefly discussed tools that are on the horizon. Taken together, these discussions provide a sketch of what can be expected in tool development.

Several authors (Gregory, Freudenburg, and Merkhofer) spoke about the desirability of *expanding options* rather than narrowing the number of alternatives to be considered, and of the ability of some new tools to do so. As Gregory suggested, information about environmental values formerly was used to help select from a small set of alternatives; but now, with newer tools that can organize massive amounts of data, a broader set of options can and often should be considered. Gregory also commented that *in-depth values elicitation* can reveal areas of agreement as well as disagreement and that differences in values can lead to a better set of alternatives based on trade-offs across objectives.

Gregory, Freudenburg, and Merkhofer also spoke of the positive effects of *integrative work across various disciplines*, such as ecology, economics, engineering, and sociology—integration that enables both the uncovering of new choices and a more holistic approach to decision making. As Freudenburg noted, tools that contribute to the new synthesis of environmental impacts with economic prosperity are one example of this integration, as are tools that build on geographic information systems (GIS).

Osleeb and Kahn reinforced the importance of *geographic information systems* as building blocks for new tools that couple GIS with spatial mathematical models to produce spatial decision-support systems, noting that these advances have been made possible in part by powerful new personal computers and workstations. They also commented that progress is being made in *three-dimensional representations* and the use of *orthodigital photography* to supplement the more common digitized map. These advances are welcome; as Merkhofer noted, improved methods to *visually represent problems* and thereby facilitate both analysis and communication are an important new research area. Together with new ways of presenting information, new ways of obtaining, organizing, and analyzing environmental conditions are becoming available. Dale and O'Neill noted the growing availability of increasingly sophisticated tools to supplement human senses, take measurements, consolidate and manipulate information, and model essential features of the environment.

In a similar vein, but having to do with the legal rather than the physical world of environmental decision making, Lyndon commented on how computerization of environmental law is enabling the *specialization of legal requirements* to deal with different facets of environmental quality, while at the same time, a *more holistic approach to environmental regulations* is being taken. Lyndon noted that new information technologies have enabled a rethinking of environmental law in information terms through voluntary actions (such as ISO 14000 audits) as well as mandated actions, and she went on to suggest that the impetus to produce and share knowledge about environmental impacts is likely to grow. She also commented that new information technologies are facilitating broader participation in legal processes affecting the environment; in addition, the networking capability of the Internet appears to be fostering a new kind of environmental politics.

Tools for auditing are not limited to the legal arena. Armstrong, for example, pointed to *new auditing procedures* that can identify areas where forecasting methods are being applied improperly or not at all. These auditing procedures are akin to the relatively new *results-based management tools* described by Bergquist and Bergquist, tools that use goals and indicators to match promise with progress. Tools like these have built-in flexibility to adapt to new knowledge and improved understanding: Armstrong, for example, commented that expert systems that incorporate the latest forecasting methods are one innovation on the horizon.

10. Next Steps for Tools to Aid Environmental Decision Making   327

As Freudenburg noted, some of the toughest challenges are those where policy actors and scientists must work together on issues that are stubbornly and inherently unanswerable. Merkhofer, similarly, commented that these "poorly specified" issues are especially difficult, and that *tools for structuring and framing issues* are an important area of research. Merkhofer also noted that *tools for real-time negotiation and decision making* is a developing area in decision aids.

Conclusion

Many of the tools now under development appear to support more well-informed, integrated, and inclusive decision-making processes. Insofar as they do, they are likely to aid and foster new trends in environmental decision making, particularly the trend toward wise conflict management and open-minded collaborative learning and away from closed decision processes more exclusively centered on analysts or an élite corps of decision makers.

Nevertheless, the horizon is not cloudless. As Freudenburg noted, many of the most important environmental decisions are those involving no obvious analogues or precedents. This impairs our collective ability to grapple with these issues and to know what to expect. Similarly, Merkhofer pointed out the fundamental issue of how people act, react, and interact: A better understanding is needed of how we individually and collectively process information and reach decisions. These comments suggest that tools can only achieve so much; that wisdom is also necessary.

In addition, tools are not of much help if they are sitting on the shelf. As Armstrong commented, a number of innovative tools are available; the problem is that often they are not used. In a similar vein, Bergquist and Bergquist remarked upon two serious impediments to the use of post-decision assessment tools: (1) often, old issues are not measured well and emerging issues are not measured at all; (2) decision makers' lack of faith in post-decision tools and their fear that the resultant information will used against them are powerful disincentives.

While the intractability of some environmental issues and some people means that optimal, universal use of decision-aiding tools may not be possible, much still can be done. We can improve our chances that tools will be used and useful in crafting good environmental decisions. The hurdles described at the beginning of this chapter (information, time and resources, access, and communication) need not be insuperable. Tools can be developed that minimize these hurdles and meet the challenging criteria enumerated here. Some of these tools are already in the works. There is much room for improvement.

> For more information on tools to aid environmental decisions, visit the website of the National Center for Environmental Decision-making Research: http://www.ncedr.org

References

Bingham, G. 1986. *Resolving Environmental Disputes*. Washington, D.C.: The Conservation Foundation.

English, M.R. 1998. Environmental decision making by organizations: Choosing the right tools. In: K. Sexton, A.A. Marcus, W. Easter, D.E. Abrahamson, and J.L. Goodman (Eds.). *Better Environmental Decisions*. Washington, D.C.: Island Press.

Fisher, R. and Ury, W. 1981. *Getting to Yes*. Boston: Houghton Mifflin.

Gramling, R. and Freudenburg, W.R. 1992. Opportunity-threat, development, and adaptation: Toward a comprehensive framework for social impact assessment. *Social Forces* 57(2):216–234.

Gunderson, L.H., Holling, C.S., and Light, S.S. 1995. *Barriers and Bridges to the Renewal of Ecosystems and Institutions*. New York: Columbia University Press.

Heifetz, R.A. 1994. *Leadership Without Easy Answers*. Cambridge, MA: Harvard University Press.

Luhmann, N. 1979. *Trust and Power*. New York: John Wiley and Sons.

Meeks, G. Jr. 1985. *Managing Environmental and Public Policy Conflicts*. Denver: National Conference of State Legislatures.

Senge, P.M. 1990. *The Fifth Discipline*. New York: Doubleday.

Talbot, A.R. 1983. *Settling Things*. Washington, D.C.: The Conservation Foundation.

Tonn, B.E., English, M., and Travis, C. In preparation. Environmental decision making: A comprehensive view, draft manuscript.

Wolfe, A.K., Schnexnayder, S.M., Fly, M., and Furtsch, C. 1997. *Developing a Users' Needs Survey Focusing on Informational and Analytical Environmental Decision-Aiding Tools*. Knoxville, TN: National Center for Environmental Decision-Making Research.

Biographies of Contributing Authors

J. Scott Armstrong, Professor of Marketing at The Wharton School, University of Pennsylvania, received his Ph.D, from MIT in 1968. He wrote the most frequently cited book on forecasting methods: *Long-Range Forecasting: From Crystal Ball to Computer.* He is one of the founders of the International Institute of Forecasters, the *Journal of Forecasting*, and the *International Journal of Forecasting.* The mass media has covered his work in forecasting, the communication of research, social responsibility in management, the prediction of consumer behavior, and the myth of market share as an objective. Dr. Armstrong is an international lecturer, serves as a visiting professor at numerous universities around the world, and provides consulting services in the public and private sectors. His forecasting work has been published in *Information Systems Research*, *Management Science*, *International Journal of Forestry*, and *Interfaces*, among other journals. Other areas of publication are marketing, scientific methods, strategic planning, education, social responsibility, applied statistics, and organizational behavior. His website is www_marketing.wharton.upenn.edu/faculty/armstrng.html

Constance Bergquist currently serves as Vice President for Education Programs, Correctional Services Corporation. She received a PhD in Evaluation and Measurement from Florida State University. She has more than 20 years of experience consulting in education and human-services programs. Dr. Bergquist has been the Committee Chair for Research and Evaluation for the American Evaluation Association as well as President of the Florida Educational Research Association and of the Southeast Evaluation Association. She has conducted numerous formative and summative program evaluations for state and local education and human-services agencies. She designed and conducted an evaluation of the delinquent youth programs for the Florida Department of Health and Rehabilitative Services. She designed and implemented a research study for the Highlands County Schools (Florida) on dropouts and dropout prevention and successfully conducted evaluations of five Florida education programs, leading to their approval by the National Joint Dissemination Review Panel as exemplary education programs. She designed and conducted five impact studies of efforts stemming from educational-reform legislation for the Illinois State Board of Education. She also was a lead researcher for the Florida Evaluation of the Child Abuse and Neglect Prevention Programs, and has twice coordinated the development of Florida's Child Abuse and Neglect Prevention Biannual State Plan.

Gilbert Bergquist is a Senior Management Consultant and Assistant Professor of Public Management at the Florida Center for Public Management, Florida State University. He is responsible for the development and management of the Center's Environmental Management program. Projects have included the Southeast Regional Environmental Indicators Conference, National Environmental Goals and Indicators Conference; Strategic Assessment of Florida's Environment, Florida Comparison of Environmental Risks, ECOSAFE (a prototypic process for use in planning for designated ecosystems), and Florida Institute of Phosphate Research Indicator Development. In his previous position as Planning and Research Administrator to the Florida Department of Environmental Regulation, Dr. Bergquist managed agency-based strategic and operational planning activities, managed federal grants, and provided oversight to the agency's research activities. He has also served in two administrator positions (of the Housing Assistance and Data Development and Information sections) within the Florida Department of Community Affairs and served on the staff of the Florida Senate Commerce Committee.

Virginia H. Dale is a Senior Scientist in the Environmental Sciences Division at Oak Ridge National Laboratory and has served as a co-team leader for the tools research team of the National Center for Environmental Decision-Making Research. She is also an adjunct faculty member in the Department of Ecology and Evolutionary Biology at the University of Tennessee. She obtained her PhD in mathematical ecology from the University of Washington. Her primary research interests are in environmental decision making, forest succession, land-use change, landscape ecology, and ecological modeling. She has worked on vegetation recovery subsequent to natural and anthropogenic disturbances, tropical deforestation in southeast Asia and the Brazilian Amazon, and integrating socioeconomic and ecological models of land-use change. She has served on the Environmental Protection Agency's Scientific Advisory Board and the Governing Board of the Ecological Society of America. She currently is a member of the National Research Council's Committee on Ecosystems and the USDA Forest Service's Committee of Scientists. Her community-involvement activities have included serving as chair of the steering committee of Citizens for Quality Growth and Chair of the Environmental Quality Advisory Board for the City of Oak Ridge.

Mary R. English is a Research Leader at the Energy, Environment, and Resources Center (EERC) at the University of Tennessee (UT), Knoxville, and has served as a co-leader of NCEDR's tools research team. In addition, she is a member of the UT Waste Management Research and Education Institute, codirector of the EERC Program for Environmental Issues Analysis and Dialogue (Pro-Dialogue), and adjunct to the UT Sociology Department. She currently serves on the National Research Council's Board on Radioactive Waste Management. Her research since the 1970s has focused on environmental policy analysis, particularly in the areas of waste management, land use, and energy. During the past decade, she has concentrated on the social, political, and ethical aspects of environmental management (e.g., analyzing alternative mechanisms for involving people in open decision-making processes, assessing the political and social problems of siting controversial facilities, and investigating institutional arrangements for remediating contaminated sites). She has a PhD in Sociology and an MS in Regional Planning.

William R. Freudenburg is Professor of Rural Sociology and Environmental Studies at the Department of Rural Sociology at the University of Wisconsin. Dr. Freudenburg has held official positions with the American Association for the Advancement of Science, the American Sociological Association, and the National Academy of Sciences, among others. He is the winner of the 1995–1996 Award for Distinguished Contributions to the Sociology of Environment and Technology from the American Sociological Association. Recent publications have focused on perceptions of offshore oil development, media coverage of hazard events, social-impact assessment, and risk analysis and communication.

Robin Gregory is a Senior Researcher at Decision Research in Vancouver, Canada. He works on problems of environmental-policy analysis and applied decision making. In addition to his appointment with Decision Research, he is an adjunct professor at the University of British Columbia and Director of Value Scope Research; previous positions include Special Advisor to the United Nations Institute for Training and Research and Manager of the National Science Foundation's program in Decision, Risk, and Management Sciences (DRMS). Dr. Gregory's specializations include valuation of nonmarket environmental resources, risk management and perceptions of hazardous technologies, and understanding small-group negotiation behaviors. Recent publications focus on the role of multiattribute techniques and behavioral decision research in facilitating broadly acceptable solutions to multidimensional, multistakeholder environmental policy conflicts.

Sami Kahn, Esq., Rutgers University Science Education Specialist, has an MS in Ecology and Evolutionary Biology and a JD from Rutgers University. As a law clinic associate at Rutgers, she authored position papers to EPA, presenting scientific and legal arguments opposing the development of Hackensack Meadowlands Estuary. She is currently developing a national environmental science curriculum under a grant from the U.S. Department of Education. She is a member of the American Bar Association, New Jersey State Bar Association, and the New Jersey Science Teachers Association.

Mary L. Lyndon of the St. John's University School of Law has written extensively on law and the information systems that support environmental and occupational-health regulation and toxic torts. Her publications include an analysis of the information economics of toxicity-data production, a critique of the current law on proprietary information, and an exploration of the role of tort law as a monitor of technical change. Professor Lyndon has a JD degree from Northeastern University and a JSD from Columbia University, where she was the Julius Silver Fellow in Law, Science, and Technology in 1985–1986. Before joining the law faculty at St. John's, she was an assistant attorney general for the State of New York and litigated a wide variety of environmental cases. In that capacity, she was an active advocate for environmental legislation at the state and federal levels. She has also practiced broadcasting and telecommunications law.

Miley W. Merkhofer, Principal, Applied Decision Analysis, Inc., Price Waterhouse, Coopers LLP, has more than 20 years of experience in decision and risk analysis. He formerly held a position as Associate Director and Manager of Research Programs in the Decision Analysis Group of SRI International and has authored two books on quantitative methods for analyzing health and environmental risks. Key projects include a cost-benefit analysis of alternative cleanup standards for the Nevada Test Site, analysis of strategies for removing waste from Hanford underground storage tanks, evaluation of alternative methods for health and environmental decision making, multiattribute evaluation of Superfund hazardous waste sites for the EPA, and development of a method for setting national ambient-air-quality standards. Dr. Merkhofer has published chapters in six books; authored or coauthored more than 30 research reports; and published numerous articles in *Risk Analysis, Technological Forecasting and Social Change, The Environmental Professional,* and *Management Science.*

Robert V. O'Neill is a Corporate Fellow at Oak Ridge National Laboratory. He earned his PhD at the University of Illinois in 1967 and came to Oak Ridge National Laboratory as a Ford Foundation postdoctoral fellow. He has been a pioneer in ecosystem modeling, risk analysis, hierarchy theory, and landscape ecology. He has more than 200 publications, including four books. He has served on advisory committees for the National Academy of Sciences, the National Science Foundation, and interdisciplinary programs at major universities. He has earned several awards, including Distinguished Statistical Ecologist from the International Association for Statistical Ecology and Distinguished Landscape Ecologist from the International Association for Landscape Ecology. His current research involves developing a continental monitoring system based on landscape principles and interdisciplinary work in economic ecology and integrated assessment.

Jeffrey P. Osleeb is Professor and Chair of Geography and Director of the Energy and Environmental Policy Studies Program, Hunter College, City University of New York. He obtained his PhD in Economic Geography with a minor in Economics and Operations Research from the State University of New York at Buffalo. In addition to his position at Hunter College, he is the president of Management Technology and Data Systems, Inc. He has also served as consultant to the World Bank in the development of geographic information systems (GISs) to assess forest resources and biodiversity in Nepal; is a national officer in the University Consortium for Geographical Information Sciences; and has served on a number of academic and professional committees, including the National Academy of Sciences Subcommittee on Marine Transportation and the Environment. Recent work has focused on the development of a GIS for the New York City Department of Environmental Protection. His research in spatial analysis, location decisions, and transportation planning is published in *Urban Systems*, *Geographical Analysis*, *Economic Geography*, and a journal of the New York State Banker's Association, among others.

Wendy Hudson Ramsey earned her Bachelor of Arts degree in Biology from Wells College and the Master of Environmental Management in Resource Ecology from Duke University. She served on the research staff of Oak Ridge National Laboratory's Environmental Sciences Division (ESD) for five years. During that time, she led a preliminary investigation of programs for the derivation and application of cleanup criteria in eight European countries and provinces of Canada; participated in integrated-assessment planning for the National Acid Precipitation Assessment Program; provided analysis of proposed environmental regulations and their potential impacts to DOE; drafted Clean Water Act guidance memoranda for national distribution by DOE's Office of Environmental Guidance; and cochaired a workshop series on integrated assessment of climate change impacts in the southeastern United States. She also served as detailee to the DOE Office of Policy, Assessment, and Guidance and as civilian team lead for water-quality protocols during environmental-compliance assessment at U.S. Air Force bases. Ms. Ramsey is currently working as an independent consultant for individuals and organizations with environmental project communications and research needs.

Claire Van Riper-Geibig has served as a Research Associate with the National Center for Environmental Decision-Making Research (NCEDR). Here, she has been involved with the decision-aiding tools research team assisting with the examination of an overview of existing decision-aiding tools. At the Energy, Environment, and Resources Center at The University of Tennessee, she assisted with the Common Ground Project, a land-use study funded by the Department of Energy (DOE) assessing the public's opinions on future uses for the Oak Ridge Reservation. Prior to her work with NCEDR, she was a Technical Information Specialist at DOE's Office of Scientific and Technical Information (OSTI). Her responsibilities included writing and editing the *InForum Newsletter* for the Office of Science Education and Technical Information (OSETI), coordinating placement of various materials on the OSETI and OSTI web pages, and membership on the planning committee for the annual InForum meeting. She has an MLS in Library and Information Science and a BS in Journalism from Wayne State University.

Index

A

Accident investigation, 254
Acid Rain Control Program, 134, 135
Adaptive management, 25
Aesthetic resources, 7
Agency actions, 135–136
Air quality, 6
Alliance of American Insurers, 18
Alouette River Stakeholder Committee (ARSC), 51–52
Alternative dispute resolution (ADR), 152
American Petroleum Institute, 18
Analogies, 207–208
Analytic hierarchy process (AHP), 244–245
Archival research techniques, 102–103
Areal Locations of Hazardous Atmospheres (ALOHA) model, 168–169
Arizona v. California, 17–18
Attitude surveys, 42

B

Bayesian statistics, 85
Biological data, 67
Biospheric values, 41
British Columbia, 47
Brown's model, 209
Budget-accountability systems, 303–304
  goal-based budgeting, 304
  zero-based budgeting, 303–304
Built environments, 8–9
Businesses
  associations of, 11
  large, 11
  small, 11

C

California's Proposition 65, 154,
CD-ROM, 146
Centers for Disease Control and Prevention, 154
  Central Arizona Project, 17–18
Chemical Manufacturers Association, 18
Chesapeake Bay Agreement, 17
Chesapeake Bay Foundation, 17
Chesapeake Bay Program, 17, 293–295
Citizen groups
  advisory committees, 47–48
  local, 11
  national, 11
  regional, 11
  state, 11
Clean Air Act (CAA), 33, 131, 132, 133, 134, 135, 158
Clean Water Act (CWA), 132, 133
Clean Water Action, 17
Clean South Bay, 16–17
Collaborative learning, 323
Columbia–Snake River system, 66
Community, 141–142
Comprehensive Environmental Response, Compensation, and Liability Act (CERCLA). *See* Superfund
Conflict management, 323
Conjoint analysis, 44, 208

336 Index

Consequence assessment, tools for, 256–258
Contingent-valuation (CV) techniques, 43, 115
Cost-benefit analysis (CBA), 24, 245–246, 262–267
  costs, tools for estimating
    life-cycle assessment, 263
    modeling tools, 263–264
  impacts
    tools for estimating, 264
    tools for valuing, 264–265
  net benefit, tools for computing, 265–266
  strength and limitations of, 266–267
Cost-effectiveness analysis (CEA), 264
Court decisions, 137–138
Critical natural areas, 6
Cross-disciplinary integration, 50, 325–326
Cultural resources, 7

D

Data collection, 253, 306–307
Daubert v. Merrill-Dow Pharmaceuticals, Inc., 159–160
Decision analysis (DA), 21–26, 245, 247–248, 267–275
  decision framing, tools for, 268
    influence diagrams, 268, 270
    objective hierarchies, 268, 269
  deterministic analysis, tools for, 271
    consequence model, 271
    deterministic model, 272
    value model, 271–272
  probabilistic analysis, 272
    decision tree, 272–274
  strengths and limitations of, 274–275
Decision making, 231–232, 288
Decision pathway survey, 44–46
Delphi, 205–207
Discharge models, 254–255
Disqualification heuristic, 111
Dose-response functions, 69–71
  communities, complex, 70
  conditions, changing, 71
  life histories, complex, 70
  mixed stressors, 70
  observation, difficult, 71
  prior conditions, 70
  spatial complexity, 70–71

E

Economic/demographic data, 101–102
Elite corps, 322
Emergency action, 322
Emergency Planning and Community Right to Know Act (EPCRA), 133, 141–142
Endangered Species Act, 33
Energy distribution, 7
Energy production, 7
Environmental data, importance of, 91–93. *See also* Environmental setting, tools to characterize
Environmental decision-making modes, typology of, 322–323
  analyis-centered, 322
  collaborative learning, 323
  conflict management, 323
  elite corps, 322
  emergency action, 322
  routine procedures, 322
Environmental decision-making process, 236, 276–277, 282–284
  innovations in, 277–278
  megatools for, 251–252
    cost benefit analysis, 262–267
    decision analysis, 267–275
    probabilistic risk assessment, 252–262
    multi-attribute trade-off analysis, 282–284
  results, communication of, 275–276
  situations, 14–19
    global, 18–19
    local, 16
    national, 18
    regional, 16–18
  tools for, 233–250, 234, 235, 250–251
    alternatives, evaluation of, 242–249
    alternatives, identification of, 240–241
    alternatives, screening of, 241–242
    decision making, 250
    problem, definition of, 233

results, communication of, 250
risk estimation, 235, 237–340
See also Government
Environmental impact statement (EIS) process, 49, 94, 243
Environmental issues, types of, 4–7
Environmental Policy Institute, 17
Environmental Protection Agency (EPA), 63, 118, 132, 133, 134, 135, 136, 138–139, 141, 147–148, 150, 154, 293–294
Evaluations, 306
Event trees, 254
Expert systems, 212
Exposure assessment, tools for, 255–256
Extrapolation, 209–210
Exxon Valdez oil spill, 36–37
  and contingent-valuation techniques, 43
  and natural variability, 68
  nonmarket damages, quantification of, 115

F

Federal Insecticide, Fungicide, and Rodenticide Act (FIFRA), 132, 133
Federal Power Commission, 66
Fieldwork, 103–107
  nonrandom interviews, 105
  participant observation, 106–107
  surveys, 103–105
Focus group, 47–48, 105
Forecasting, 192, 226–230
  checklist, 220
  framework for, 193–195
    and interventions, 195–196
    use of, 196
  innovations in, 219
    auditing, 219–221
    software, 219
  methods
    characteristics of, 194
    principles of, 197–198
    types of, 199–200, 215–216
  role of, 226–227
    communication of, 228–229
    limitations of, 227
  method, choice of, 227–228
  uncertainty in, 216
  use of, 216–217
    presentation techniques, 217–218
    scenarios, 218–219
Freedom of Information Act (FOIA), 141

G

Gains and losses, valuation of, 36–37
Gaps and blinders, 107–113
  interdisciplinary double-checks, 110–111
  public-involvement techniques, 111–113
  research-sensitivity analyses, 109–110
Geographic information, integration of, 161–162, 163, 190–191
  case study, 170–186
  spatial-information-integrating technologies, 162–170
    geographic information systems (GIS), 116–118, 164–166
    geographic plume analysis (GPA), 168–170, 183–184
    spatial decision-support systems (SDSS), 166–168
  See also Greenpoint/Williamsburg, Brooklyn
Geographic information systems (GIS). *See* Geographic information
Geographic Information Systems/Land Information Systems Conference, 185–186
Geographic plume analysis, 168–170, 183–184
Goal-based budgeting, 304
Government
  federal, 13, 33, 104–105, 132–135, 161
  local, 11–12, 16
  regional, 12–13, 16–18
  state, 12
Grand Canyon Visibility Transport Commission, 158
Green technologies, 7
Greenpoint/Williamsburg, Brooklyn, case study, 170–186
  background, 170–174

Citizens Advisory Committee, 170, 176
geographic plume analysis, 183–184
GIS, development of pilot, 176–180
metadata development, 174–176
SDDS, use of, 180–183
Group elicitations, 47–48, 50

H
Hazardous Waste Treatment Council, 18
Hedonic methods, 39–40
Heuristic models, 80–81
Historic resources, 7
Homocentric values, 41
Human environment, 94–95

I
Indicators and indicator systems, 295–303, 301
  characteristics of
    data, quality of, 298
    decision support, 299
    importance of, 298
    measurability, 298
    relevance of, 298
    representativeness, 299
    scale, appropriateness of, 299
    trends, 299
  criteria for
    anticipativeness, 299
    cost effectiveness, 299
    data, comparability of, 299
    effects, integration of, 299
    results, 299
    sensitivity, 299
    understandability, 299
  performance, measurement levels of, 297
    administrative, 297
    environmental, 298
    policy, 297–298
    program activity and efficiency, 297
    program performance, 297
  strengths and weaknesses, 303
  uses of, 303–303
Individual decisions, 9, 288

Industrial Sources Complex model (ISC3), 169–170
Influence diagrams, 268, 270
Information. *See also* setting, tools to characterize environmental
Infrastructure, 6
Intention surveys, 203–204
Interest groups, role of, 314
International Organization for Standardization (ISO), 140, 151
Internet, 146–147
Interviews, 46, 105

J
Judgmental bootstrapping, 208

K
Key-informant interviews, 105
Keystone effects, 68

L
Landsat Thematic Mapper, 65
Law, 130–131, 143–144, 157
  innovations in, 150
    economics, 150–151
    information strategies, 151–152
    participation, 152–153, 157
    trends, 152–155, 157
  research, sources for, 142
    CD-ROM format, 146
    computerized services, 145
    internet, 146–147
    printed sources, 142, 143–145
  survey of, 131–132
    agency actions, 135–136
    court decisions, 137–138
    information production and access, 139–142
    procedural law, 138–139
    statutes, 132–135
League of Women Voters, 17
Leslie matrix model, 81
Life histories, complex, 70
Local Governments for Superfund Reform, 18

M
Management tools, result-based, 326
Maps, 74